PLANT GENETIC ENGINEERING

Vol. 5

Improvement of Vegetables

Editors

Rana P. Singh **Pawan K. Jaiwal**

Department of Biosciences
M. D. University,
Rohtak-124001 India

2003

SCI TECH PUBLISHING LLC, U.S.A.

ISBN : 1-930813-14-7
SERIES ISBN : 1-930813-17-1

First Published 2003

Published by
SCI TECH PUBLISHING LLC.
9207, Country Creek Drive
Houston, Texas - 77036 (U.S.A.)
Tele. : 713-541-94-- Fax : 713-541-9401
E-mail : sci@hotmail.com

Printed by
RAINBOW PROCESSORS & PRINTERS
New Delhi-110059 India

PREFACE

The consumption of vegetables in human diet is increasing. Vegetables are rich and comparatively cheaper source of minerals, vitamins and antioxidants. The production of vegetables, in general, is inadequate to meet the growing demands of the ever increasing population. Though conventional breeding has played a significant role in developing high yielding hybrid cultivars of different vegetable crops, a quantum jump in the improvement of productivity and nutritional quality of these crops have been realized by adopting the recent methods and techniques of plant molecular biology and biotechnology.

Large number of molecular markers have been developed, in recent years, to identify the gene of interest in vegetable crops (Chapter 1). During the last decade several transgenic vegetable crops have been developed for various agronomic traits including insect resistance (Chapter 2). Tomato has been considered as a model system for genetic manipulation of vegetable crops. A number of transgenic phenotypes have been obtained to confer herbicide tolerance, disease and insect resistance, abiotic stress tolerance, superior quality traits in the harvested products and to improve shelf life of fruits (Chapter 3). Lettuce, chicory and spinach are the most commonly used leafy vegetables in which many agronomically useful genes have been transferred using gene transfer technology (Chapter 4).

Cauliflower is a common vegetable cultivated world wide. The combined production of India and China exceeds about 70% of the world production of this vegetable. By transferring proteinase inhibitor genes insect resistance have been developed in transgenic cauliflower. Chapter 5 describes the current status of *in vitro* cell culture techniques, and state of art of genetic transformation in cauliflower. Solanaceous vegetables, egg plant and chilli are two important vegetable crops cultivated in several countries. The progress made in genetic transformation of these plants for conferring viral and insect resistance has been presented in chapter 6. Garlic is one of the vegetatively propagated vegetable crop which has been recognised for its great therapeutic significance. The plant biotechnological techniques pave a way for genetic improvement in garlic for disease resistance and modification of the sulfur metabolism pathways in order to increase the pharmaceutical value of garlic bulbs (Chapter 7). Cucumber, one of the most important vegetables, is suffuring for yield instability due to biotic pathogens and abiotic stresses. The transformation systems for the production of transgenic cucumber for disease resistance have been developed (Chapter 8). Pea is a high protein containing legume vegetable which needs attention for its improvement for yield and quality using biotechnological tools and techniques (Chapter 9). The nutritional quality improvements in vegetable crops is one of the major focus of attention in the recent years. The genetic transformation provides great possibilities to modify the levels and compositions of minerals, vitamins, antioxidants, pigments, carbohydrates, proteins, nutraceuticals, anti-nutritive substances and shelf life of the various vegetables to enhance their nutritive and commercial value (Chapter 10).

This volume contains chapters on the diverse vegetable crops of great significance. The editors are grateful to the authors for providing authoritative and updated account in the knowledge of their interest. The students, academicians, scientists, planners and industrialists can find the current state of art, achievements and future directions on exploitation of gene transfer technology for the improvement of these vegetables at one place. We appreciate our family members, students and friends for their patience and understanding during planning and preparation of this book. We are thankful to the publishers especially Mr. Anil Jain and Mr. Arvind K Jain of Sci-Tech Pub. LCC Houston, USA and Ms. Reema and Mr. Ashok Datta of *LaPrints*, New Delhi, India for their personal care and euthusiasm which was maintained during the production of this title with quality output.

March, 2002
Rohtak, India

Rana P Singh
Pawan K Jaiwal

PLANT GENETIC ENGINEERING

Volume 5

Improvement of Vegetables

The Editors

Dr Rana P. Singh is a Reader in the Department of Biosciences, M. D. University, Rohtak, India. He is teaching courses in Plant Biology and Biotechnology to graduate and postgraduate students since 1986. He earned his first class M. Sc. in Botany from DDU Gorakhpur Univ., Gorakhapur and Ph.D. in Life Sciences (Plant Biochemistry) from Devi Ahilya University, Indore, India. He has more than 70 research publications including original research papers, reviews and book chapters in reputed International research journals and books. He has edited 5 books published by International publishers. He is fellow and member of various academic societies. Dr Singh has been awarded R. D. Asana Medal from Indian Soc. Plant Physiol., New Delhi and JEB Young Scientist award of The Academy of Environmental Biology, India for significant research contributions. He is recipient of INSA, New Delhi and UNESCO, Paris Biotech. Fellowships and visited twice at University of Guelph, Canada for advance research. He has visited Yunnan Agri. Univ., Kunming, China and hosted three Senegalese Scientists for research collaborations. He has guided many students for Ph.D., M. Phil. and M. Sc. degrees and completed many projects in areas of his interest. His group has contributed significant work in nitrogen relation to plants. Dr Singh is Editor-in-chief of an International research journal, Physiology and Molecular Biology of Plants and on editorial board of three other journals. His current research interests are metabolic engineering for nutrient use efficiency, stress tolerance and nutritional improvement in plants.

Dr Pawan K. Jaiwal is a Reader in the Department of Biosciences, M. D. University, Rohtak, India. He is teaching Molecular Biology, rDNA technology and Plant Biotechnology courses to graduate and postgraduate students since 1986. He has throughout first division academic career with Ph.D. in Botany (Plant Physiology) from Kurukshetra University, Kurukshetra, India. He has more than 70 research publications including original research papers, reviews and book chapters in reputed International research journals and books. He has edited three books published by International publishers. Dr Jaiwal has been awarded DST Young Scientist Project, INSA Visiting Fellowship, New Delhi and DBT Overseas Associateship (Long Term) for advance research at laboratory of Prof Ingo Potrykus, ETH, Switzerland. Dr Jaiwal is a member of several academic bodies and on the Editorial Board of four International research journals. His has guided many students for Ph.D., M. Phil. and M. Sc. degrees. He has completed several research projects on Biotechnology for improvement of important grain legumes. His group has developed gene transfer protocols in *Vigna radiata* and *V. mungo* for the first time. His current research interests are metabolic engineering for resistance to abiotic and biotic stresses, improvement of nutrient use efficiency and nutritional quality of grain legumes.

Other Volumes in this series:

Volume 1 **Applications and Limitations :** *Eds.* Rana P. Singh & Pawan K. Jaiwal

Volume 2 **Improvement of Food Crops :** *Eds.* Pawan K. Jaiwal & Rana P. Singh

Volume 3 **Improvement of Commercial Plants - I :** *Eds.* Rana P. Singh & Pawan K. Jaiwal

Volume 4 **Improvement of Commercial Plants - II :** *Eds.* Pawan K. Jaiwal & Rana P. Singh

Volume 6 **Improvement of Fruits :** *Eds.* Pawan K. Jaiwal & Rana P. Singh

CONTENTS

Contents of other volumes in the series

PLANT GENETIC ENGINEERING

SERIES ISBN 1-930813-17-1

ISBN : 1-930813-02-2

Volume 1 : **Applications and Limitations**
Editors : Rana P. Singh & Pawan K. Jaiwal

ISBN: 1-930813-03-1

Volume 2 : **Improvement of Food Crops**
Editors : Pawan K. Jaiwal & Rana P. Singh

ISBN: 1-930813-04-X

Volume 3 : **Improvement of Commercial Plants - I**
Editors : Rana P. Singh & Pawan K. Jaiwal

ISBN: 1-930813-14-7

Volume 5 : **Improvement of Vegetables**
Editors : Rana P. Singh & Pawan K. Jaiwal

ISBN : 1-930813-15-5

Volume 6 : **Improvement of Fruits**
Editors : Pawan K. Jaiwal & Rana P. Singh

Chapter 1

BIOTECHNOLOGICAL APPROACHES IN VEGETABLE CROPS : AN OVERVIEW

Major Singh★, Sanjeet Kumar, Sanjeev Kumar and G Kalloo

Indian Institute of Vegetable Research, 1, Gandhi Nagar (Naria) P.B. No. 5002, P.O. BHU, Varanasi - 221 005, India

Summary

In the historical background of the developmental phases of genetics as a scientific discipline and emergence of modern biotechnological approaches for vegetable improvement, this chapter describes various approaches, currently being utilized to change genotype of a vegetable species at genome, chromosome(s) and gene level. Recent examples of molecular tagging and mapping of desirable genes and QTLs in selected vegetables and their feasible applications through marker assisted selection have been briefly described. The section on genetic engineering deals with strategy and examples of developing genetically modified vegetables possessing transgene(s) expressing fungal resistance, viral resistance, parthenocarpic lines, and male sterility traits. Some successful examples of achievements made through anther/pollen culture, embryo rescue and protoplast culture/fusion in vegetable crops have been highlighted. Use of modern biotechnological tools for improvement of vegetables have been briefly highlighted.

Keywords : Disease resistance, molecular markers, male sterility, parthenocarpy, QTLs, transgene, tissue culture.

★Corresponding author : E-mail : majorsingh@satyam.net.in; pdveg@up.nic.in

1. INTRODUCTION

In the last century, scientific world has witnessed raise of genetics as a scientific discipline (1900), identification of DNA as hereditary material (1944), demonstration of double helix structure of the DNA (1953), deciphering of genetic code (1966), ability to isolate gene (1973), application of recombinant DNA techniques and molecular markers (from 1980 onwards) and the development of DNA chips and the success of plant genomics (1996 onwards). Similarly, methods of crop or vegetable improvement have also changed dramatically throughout the century (Oritz, 1998). Although application of biotechnology for the generation of food products is known from long time (Altman, 1999), the later three or four developmental phases coupled with the application of tissue culture techniques (1950s onwards) can be considered as current days modern biotechnology approaches for crop improvement. The most remarkable feature of utilization of these biotechnological approaches in crop improvement is that its application has changed the classical concept of primary, secondary and tertiary gene pool. Now the concept of gene pool has been extended to include trans gene (heterologous gene) and native exotic gene in the primary gene pool. We were asked to prepare an article focusing on overview of biotechnological approaches in vegetable improvement, which is a vast topic. As presentation of all areas of vegetable biotechnology in a single article may not be possible, we decided to restrict our discussion on some of the successful uses of biotechnology in selected important vegetables. Several reviews and books on the applications of biotechnology as a whole or its specific tools for specific breeding objectives in vegetable improvement are available (Palletier *et al.*, 1995; Lee, 1995; Joshi *et al.*, 1999; Mohan *et al.*, 1999; Singh *et al.*, 2000a,b). Therefore, in this article our focus will be on those modern tools, which are utilized to change the vegetable genotype at genome, chromosome or gene level. Obviously, certain techniques like micropropagation aiming to large-scale multiplication of selected vegetables, will not be described. However, applications of tissue culture based techniques such as protoplast fusion, embryo rescue etc. in selected vegetables have been described.

2. MOLECULAR MARKERS IN VEGETABLE IMPROVEMENT

In several vegetables, linkage map has been constructed containing loci for morphological marker, resistant genes and several mutants affecting physiological functions. In last two decades, however, attention has been paid on the construction of a detailed molecular map in several important vegetables including tomato, chilli, eggplant etc. A large number of DNA markers have been developed and many of them have been found to be linked with the gene of interest especially in tomato, one of the model crops for the molecular biologist (Hille *et al.*, 1989; Jhoshi *et al.*, 1999; Singh *et al.*, 2000a). Although this map has served well in genetic studies, its application in plant breeding is rather limited. Therefore, now attention is being shifting to construct detailed molecular map based on protein and DNA markers in many vegetables and use them in the breeding programmes.

Among the biochemical markers, isozyme markers have already been utilized in genetic studies and breeding programs of various crops including vegetables (Tanksley and Rick, 1980; Tanksley and Orton, 1983). For tomato, an isozyme linkage map has been constructed containing 41 isozymic genes corresponding to 15 unique enzyme reactions. Thirty-six of these genes have been mapped to their respective chromosomes. These available information in tomato have served many purposes, *e.g.*, verifying the purity of hybrid seed (Tanksley and Jones, 1979), tagging genes of economic importance (Rick and Fobes, 1977; Tanksley *et al.*, 1984), detecting genes and chromosomes from wild species following introgression (Rick *et al.*, 1986), identifying somatic hybrids and mapping genes underlying quantitative variation (Vallejos and Tanksley, 1983).

The use of isozyme loci has several limitations, which are not inherent to use of DNA markers. DNA markers directly reflect the variation at DNA level, which may be either coding or non-coding region. Molecular markers posses several distinct advantages over morphological or other conventional markers like they can be established at cellular or tissue level, they are not influenced by the environmental fluctuation, they usually show high degree of polymorphism etc. (Hille *et al.*, 1989; Singh *et al.*, 2000a). In vegetable crops, molecular markers are now increasingly being used for the germplasm management, gene or QTL tagging, marker assisted selection, identification of heterotic parents, hybrid purity test, comparative mapping etc. (Mohan *et al.*, 1997; Joshi *et al.*, 1999; Singh *et al.*, 2000a). Although detailed discussion of aforesaid applications is beyond the scope of this paper, some examples are being presented in coming text.

3. MOLECULAR TAGGING OF GENES

Tagging of desirable gene(s) is most interesting application of molecular markers. In plant breeding, gene tagging is a term commonly applied where a genetic marker is used as the selection criteria for a closely linked agronomic trait. A classic example resulted from the discovery of tight linkage between the isozyme marker *Aps1* and a gene (*Mi*) encoding root knot nematode resistance in tomato (Rick and Fobes, 1977). Many seed companies have successfully utilized the *Aps1* marker as a preferred method for the incorporation of this trait into elite tomato lines. Several disease resistance genes in vegetables have been tagged (Mohan *et al.*, 1997; Jhoshi *et al.*, 1999; Kumar *et al.*, 2000a). Some of the recent reports on identification of linkage between molecular markers and disease resistance gene in vegetable crops are depicted in table 1

Table 1. Identification of markers linked to disease resistance genes in vegetable crops

Vegetable	Pathogen	Gene	Markers type	Reference
Tomato	*Meloidogyne incognita*	*Mi*	RAPD	Williamson *et al., 1994*
	Meloidogyne incognita	*Mi3*	RAPD, RFLP	Yaghoobi *et al.*, 1995
	Phytophthora infestans	*Ph2*	RFLP	Moreau *et al.*, 1998
	Verticillium dahliae	*Ve*	RFLP	Diwan *et al.*, 1999
	Tomato mosaic virus	*Tm2*	SCAR	Sobir *et al.*, 2000
	Cucumber mosaic virus	*cmr*	RFLP	Stamova and Chetelat, 2000
	Yellow leaf curl virus	*Ty1*	RFLP	Zamir *et al.*, 1994
	Yellow leaf curl virus	*Ty2*	RFLP	Hanson *et al.*, 2000
	Tospovirus	*SW-5*	RFLP	Brommonschenkel and Tanksley, 1997
Bean	Common bean mosaic virus	*I*	RAPD	Melatto *et al.*, 1996
Pea	Pea common mosaic virus	*Mo*	RFLP	Dirlewanger *et al.*, 1994
	Erysiphe polygone	*Er*	RAPD	Dirlewanger *et al.*, 1994
	Fusarium oxysporum	*Fw*	USAT	Dirlewanger *et al.*, 1994

(Kumar *et al.*, 2000a). More than 1000 markers have been mapped onto the tomato molecular linkage maps and these markers cover a total of 1276 map units (Tanksley *et al.*, 1992). Many of the markers of the tomato map have also been mapped onto the potato and pepper molecular maps. All these solanaceous species share the same chromosome number and by having molecular maps based on the same RFLP probes, it has been possible to determine chromosomal homologies with fairly high degree of precision (Bonierbale *et al.*, 1988; Tanksley *et al.*, 1992).

Linkage drag is frequently encountered when wild species are used as a donor source. A study by Young and Tanskley (1989) demonstrates the power of RFLP in removing the linkage drag in breeding materials. The tomato cultivar 'Craigella' was found to carry approximately 50 cM piece of DNA derived from the disease resistant donor parents following 11 backcross generations. Young and Tanksley (1989) were able to reduce the size of this fragment to less than 7 cM, in a single generation of selection.

4. MOLECULAR TAGGING OF QTLs

Mapping of Quantitative trait loci (QTLs) is receiving much attention of plant breeders for the marker-assisted improvement of quantitative traits. Genetic variations in most traits of agricultural importance are attributed to allelic differences at a large but generally unknown number of loci having relatively small individual effects. These loci are termed quantitative trait loci (QTL), which control the expression of polygeneic traits (*e.g.*, grain yield, horizontal resistance etc.). Linkage map of several such polygenic traits with molecular markers have been constructed (Table 2).

Mapping and tagging of desirable genes or QTLs with molecular markers is a pre-requisite and forms the foundation for marker assisted selection (MAS) in plant breeding. Potential of marker assisted selection is very high but actual examples of the application of this approach in vegetable crops are limited to tomato (Joshi *et al.*, 1999).

5. ISOLATION OF GENE

One of the most challenging tasks to the advance molecular biology is isolation and cloning of genes with unknown protein product. Map based gene cloning and transposon tagging techniques provide opportunity to isolate gene of a particular phenotype. Construction of saturated molecular linkage map and ability to map any gene of economic

Table 2. Some recent examples of gene mapping in vegetable crops

Vegetables	Trait/Purpose	Molecular Marker	Reference
Tomato	*I-2 Fusarium oxysporum* resistance	RFLP	Segel et al., 1992
	ms-14 male sterile gene	RFLP	Gorman *et al.*, 1996
	Regulation of iron-metabolism	RFLP	Ling *et al.*, 1996
	Ripening inhibitor and non ripening	RFLP	Giovonnoni et al., 1995
	Seed weight	Several	Doganlar *et al.*, 2000a
	Earliness	RAPD	Doganlar *et al.*, 2000b
	Salt tolerance gene	RAPD	Fooled and Chen, 1997
	Golden nematode resistance gene (*Hero*)	RFLP	Ganal *et al.*, 1995
	Physical location of *Cf-2* gene	RFLP	Dixon *et al.*, 1995
	Carotenoid	RAPD	Zhang and Stommel, 1998
Cabbage	Linkage group alignment	RFLP	Hu *et al.*, 1998
	Agronomic trait	RFLP	Fulton *et al.*, 2000
	Location of self-incompatibility locus	RFLP	Camargo *et al.*, 1997
	Comparative genome mapping	RFLP	Lagercraniz *et al.*, 1996
	Linkage group alignment from four independent-RFLP maps	RFLP	Hu *et al.*, 1998
Cucumber	Plant habit, sex expression, lateral branches	RAPD	Serqueen *et al.*, 1997
	Construction of molecular map	AFLP	Wang *et al.*, 1997
Hot pepper	*Xanthomonas campestris*	RFLP, RAPD	Kang *et al.*, 1998

Table 2. Continued

Vegetables	Trait/Purpose	Molecular Marker	Reference
Lettuce	Detailed linkage map	RAPD, RFLP	Kesseli *et al.*, 1994
Eggplant	Bacterial wilt resistance gene	RAPD	Nunome and Hirai, 1998
Cowpea	Orthologous seed weight genes	RFLP	Fatokun *et al.*, 1992

importance to a defined chromosomal site opens the possibility of isolating genes via chromosome walking. This method is sometimes referred to as "map based cloning or the "top down" or "reverse genetics" approach, since it starts with an interesting phenotype observed at the whole plant level and proceeds to clone the gene via genetic mapping and chromosome walking.

A critical prerequisite for map-based gene cloning is that the target gene must be localized definitively with respect to markers in that region of the map. Since such localization is possible only with very small genetic distances (say 1 cM), analysis of very large progenies is often required for such fine mapping. Nevertheless, in vegetables like tomato, the availability of a high density molecular linkage map and a yeast artificial chromosomes (YACs) library potentially provide the foundation on which to initiate map-based gene cloning for genes underlying any trait that can be genetically mapped (Martin *et al.*, 1992). Map based cloning has been successfully utilized to tag several disease resistance genes in tomato, e.g. *pto* (Martin *et al.*, 1993), *fen* (Martin *et al.*, 1994), *prf* (Salmeron *et al.*, 1996) and *cf-2* (Dixon *et al.*, 1996). It is anticipated that, in the future, the availability of molecular linkage map will play a key role in cloning genes for a variety of economically valuable and/or biologically interesting characters including disease resistance, growth habit, fruit ripening behavior and even yield.

An inherent problem associated with the map based cloning is non-availability of tightly linked flanking markers to the gene of interest. Transposon tagging provides solution, for the isolation of useful genes not having of tightly linked flanking markers. The insertion of a transposable element in to a genetic locus can give rise to mutations. Using a transposable element from maize, the *Ac*-element, Fedoroff *et al.* (1984), were able to clone a locus which until then was only characterized genetically. Now it has been demonstrated that the maize transposable element *Ac* can transpose in dicots and using this technique, disease resistant gene (*Cf-9*) has been isolated from tomato (Jones *et al.*, 1994). This has opened the possibility to use this well characterized element for gene isolation in vegetables.

6. GENETIC ENGINEERING

Since the first success in the early 1980's, plant gene transfer has become a routine practice in crop plants and wide ranges of traits have been introduced in vegetables. In crops, transgenic herbicide resistance was the first to be field tested for an introduced novel trait (Daniell, 1999). Several hundred such field trials have been conducted and many field trials of transgenic plants carrying agriculturally significant new genes are in progress (Rai and Prasanna, 2000).

6.1. Engineering herbicide resistance

The criteria for an ideal herbicide are that it should have no toxicity towards humans, animals or soil organisms and, should kill selectively weeds and not crop species. Unfortunately, many herbicides effectively kill crop plants and weeds. Genetic engineering has enabled a new approach to this area by conferring resistance on crop plants, to an environmental manner safely. Unlike cereals, oil seeds or pulses, most of the economically important vegetable crops except certain leafy vegetables, are less densely planted and mechanized weeding is not a handicap. Therefore, developing herbicide resistant tomato, brinjal, chilli, cuiliflower etc., is not a priority agenda of transgenic vegetable research.

6.2. Engineering disease resistance

6.2.1. Viral resistance

Eversince the first report on the development of virus resistant transgenic tobacco plants using coat protein (*Cp*) gene by Powell and his associates (Powell *et al.*, 1986), several strategies have been described and executed to produce virus resistant transgenic plants. Presently, virus resistant transgenics are being produced utilizing following gene constructs : (i) virus resistance gene from one plant species to other, (ii) sequence corresponding to ameliorative satellite RNAs, (iii) gene(s) that encode antiviral proteins or compounds, (iv) antisense RNA gene corresponding to segments of viral genome and (v) viral genes like coat protein etc. Some of these approaches have been utilised to develop virus resistant transgenic plants in vegetable crops (Table 3). Recently, four transgenic tomato lines possessing single copy of four different *Cp* genes were evaluated under field conditions in various parts of Italy. In general, CMV resistance was confirmed, although it was less effective than that in growth chamber experiment (Tomassoli *et al.*, 1999).

6.2.2. Fungal resistance

Several molecular strategies have been developed and implemented to produce fungal disease resistant transgenic crops including vegetables (Kumar *et al.*, 2000a). Basically two approaches are being used to develop fungal disease transgenics. In the first approach, transgene encoding antifungal compound is used to develop transgenic plants. In this case the transgene product directly affects the growth and development of fungus. In the second approach, disease resistant transgenics are produced using transgene (under pathogen specific promoter), which encodes a product that leads to cell death upon pathogen

Table 3. Examples of virus resistant transgenics in vegetables

Crop/Trans gene	Origin of trans-gene	Resistant to	Reference
Tomato			
Antisense RNA	ToMV	Tomato mosaic virus (ToMV)	Beachy *et al.*, 1990
Satellite RNA	CMV	Cucumber mosaic virus (CMV)	Stommel *et al.*, 1998
N gene	TSWV	Tomato spotted wilt virus (TSWV)	Pang *et al.*, 1992
Truncated *Cp* gene	CMV	Cucumber mosaic virus (CMV)	Murphy *et al.*, 1998
Two *Cp* genes	CMV	Cucumber mosaic virus (CMV)	Kaniewski *et al.*, 1999
Cucumber			
Cp gene	CMV	Cucumber mosaic virus (CMV)	Gonsalves *et al.*, 1992

infection. Proteins with ability to inhibit the growth or even kill the fungi *in vitro* are abundantly present in nature as pathogenesis related (PR) proteins, ribosome-inactivating protein (RIP), cell wall degrading enzyme etc. (Kumar *et al.*, 2000a). One of the major approaches for engineering plants resistant to fungal disease includes use of chitinase genes of plant and/or bacterial origin. Chitinase enzyme degrades chitin rich cell walls of fungi and delays symptom development (Broglie *et al.*, 1993). Other antifungal protein genes include osmotin, glucanase, thionin, peroxidase etc. Recently, an effective antifungal protein from barley seeds called ribosome-inactivating protein (RIP) has been isolated. RIP acts by inhibiting protein synthesis via N-glycosidase modification of fungal 28 S rDNA. Transgenic crop plants expressing RIP proteins are now being tested against many fungi (Sharma and Kumar, 1995). In several cases, transgenic plants with PR protein gene showed enhanced resistance against fungus. In *in vitro* assays combinations of different PR protein genes displayed tremendous increase in fungal growth inhibition compare to single protein gene (Carnelissen *et al.*, 1996). Researchers have found that deliberately disarmed fungal pathogens can trigger the necessary response in susceptible (*i.e.* slow responding) plants before the virulent pathogen arrives. This approach is known as systemic acquired resistance (SAR), which uses pathogen derived bio-control agents to induce resistance in susceptible plants (Kumar *et al.*, 2000a).

6.2.3. Insect resistance

First report of transgenic insect resistant plants was published in 1987 (Fischhoff *et al.*, 1987; Hilder *et al.*, 1987; Vaeck *et al.*, 1987) and now this new technology is increasingly being utilized in many vegetables. Transformation strategy is an additional tool for the control of pests and offers certain advantages over conventional insecticides. These advantages include : more effective targeting of insects protected within plants, greater resilience to weather conditions, fast biodegradability, reduced operator exposure to toxins and financial savings (Gatehouse *et al.*, 1992).

The insect resistance genes transferred into plants to date mainly target the insect digestive system. Most of these have been derived from either a single species of bacterium (various versions of *cry* gene) or a range of higher plants. In addition some insect resistance genes from animals and other microorganisms have also been introduced into crop plants. The use of native insecticidal crystal protein (ICP) genes of *Bacillus thuringiensis* was not effective because the level of toxins expression was insufficient. Therefore, truncated versions/multiple cry gene construct are utilized, to achieve desirable level of gene expression (Kumar *et al.*, 1998a). In fact, this is the most successful application of genetic engineering in vegetables in India, because insect resistant transgenic are at the edge of field trial. A number of genetically engineered insect resistant transgenic tomato and eggplant lines have been developed (Singh *et al.*, 2000a;b), some of them are listed in table 4.

6.3. Engineering for parthenocarpy

In fruit vegetables like tomato, eggplant, sweet pepper etc., less seeded fruits increases the consumer acceptance. Now gene constructs are available, which can induce development of fruit without fertilization (parthenocarpic fruit). One such gene constructs, is derived from *Psuedomonas syringae*, which code for an enzyme involved in the auxin (IAA) biosynthesis. Expression of this gene under ovule specific promoter enhances the level of IAA in ovule and stimulates growth and development of unfertilized ovules and fruit set (Rotino *et al.*, 1997).

Table 4. Transgenic tomato and eggplant plants of vegetables expressing insect resistance genes

Trans gene	Target insects	Reference
Tomato		
Cry1Ab	Lepidoptera	Delannay *et al.*, 1989
Cry1Ac	Lepidoptera	Mandaokar *et al.*, 2000
CpTI (cowpea trypsin inhibitor)	Coleoptera Lepidoptera	Gatehouse *et al.*, 1992
Tomato proteinase inhibitor I	Lepidoptera	Mc. Gurl *et al.*, 1994
Tomato proteinase inhibitor II	Lepidoptera	Mc. Gurl *et al.*, 1994
α - Amylase inhibitors α *Al-Pv* (α - amylase inhibitor of the common bean)	Coleoptera	Schroeder *et al.*, 1995
Snowdrop lectins (gna)	Lepidoptera	Gatehouse *et al.*, 1992
Brinjal		
Cry1Ab	Lepidoptera	Kumar *et al.*, 1998b

Transgenic eggplants expressing the coding region of the *iaah* gene under the control of the ovule-specific *DefH9* promoter showed parthenocarpic fruit development (Rotino *et al.*, 1997).

The development and cultivation of parthenocarpy transgenics of fruit vegetables is having tremendous potential to harvest the fruits under adverse climatic conditions especially at extremely high and low temperatures.

6.4. Engineering male sterility

Transgenic male sterility systems provide an opportunity to exploit male sterility in wide range of crop species to produce economic hybrid seeds. From the beginning of 1990's molecular strategies have been proposed and implemented to develop male sterility through genetic transformation (Mariani *et al.*, 1990, 1992). The ability to design and execute molecular strategies has been possible because of the isolation, cloning and characterization of anther/pollen specific genes, anther/pollen specific promoters and chemically inducible promoters. These genes are expressed either in pollen themselves (gametophyte expression) or in cells and tissues (sporophytic expression) that directly or indirectly support pollen development, such as tapetum, filament, anther wall etc. (Williams *et al.*, 1997). There are several models to develop transgenic male sterility systems (Kumar *et al.*, 2000b), which are usually based on the molecular strategies to induce male sterility (for hybrid seed production) and fertility restoration (for seed increase of male sterile plant). Comparison of all the transgenic systems with respect to requirement of number of transgenic/gene constructs and possible risk involvement due to system failure is given in table 5 (Kumar *et al.*, 2000b). Internationally known seed companies are developing most of these male sterility systems and few of them are near the edges of utilization in hybrid seed production (Williams *et al.*, 1997).

7. TISSUE CULTURE IN VEGETABLE IMPROVEMENT

Plant tissue culture techniques, in combination with recombinant-DNA technology, are the essential requirements for development of transgenic plants. However, culture techniques like anther/pollen/ovule culture, meristem culture, protoplast fusion are also being used for improvement of different traits or as an aid to conventional plant breeding. In recent years, isolated microspore culture has been preferred as a breeding tool and an experimental system for various genetic manipulations. In addition, through the use of protoplast fusion and hybrid embryo rescue,

Table 5. Comparison of transgenic male sterility systems for requirement of number of transgenic(s) (with gene construct) and possible risk involved (Kumar et al., 2000b)

System	Number of transgenic(s) and trans gene construct(s) required	Possible risk in
Abolition-restoration	Two transgenics each with a distinct exogenous gene : (i) transgenic male sterile plants with disruptive gene under anther specific (usually tapetum) promoter and (ii) transgenic fertility restorer plants with inhibitor gene under same anther specific promoter★.	MS, HS
Abolition-reversible	One transgenic with two exogenous genes (co-transformed): reversible transgenic male sterile plants with (i) disruptive gene and (ii) inhibitor of disruptive gene under chemically inducible promoter.	MS, HS
Constitutive-reversible	One transgenic with one exogenous gene : reversible transgenic male sterile plants with endogenous gene for male sterility (*ms*) and exogenous male fertility gene (*mf*) for same *ms* gene under inducible promoter.	MS
Complementary-gene	Two transgenics, each with a complementary gene : transgenics with (i) gene *a* and (ii) gene *b*, under the control of anther specific promoter (gene *a* & *b* are complementary).	MS, HS
Gametocide-targetted	One transgenic with one gene : male sterile transgenic with disruptive gene under pollen specific chemically inducible promoter.	HS

MS and HS – Male sterile and hybrid seed production field, respectively
★In those crops where vegetative part is of economic importance, development of restorer transgenic will be not required.

hybridization between distant taxa/species have been successfully achieved in many vegetables (Singh *et al.*, 2000a). Protoplast fusion technique can be used for the transfer of cytoplasmic male sterility from one species to another in very short duration (Pelletier *et al.*, 1995). Mitochondrial recombination occurring after protoplast fusion has been of practical intact for the elimination of unfavorable traits resulting from nuclear cytoplasmic incompatibility after interspecific hybridization. There are successful examples of such genome transfer in *Brassica*, *Cichorium* and *Lycopersicon* (Melchers, 1992; Pelletier *et al.*, 1995; Petrova *et al*., 1998). In cauliflower and cabbage, male sterile cybrids are being utilized by seed companies in France to produce hybrid seeds (Pelletier *et al*., 1995).

Somaclonal variation generated during culture can also be utilized for recovery of variants showing disease resistance or other desirable traits. Some of the reports on utilization of tissue culture techniques in vegetable crops have been summarized (Table 6).

Table 6. Achievements through tissue culture based techniques in vegetables

Technique/Vegetables	Technique applied for	Reference
Anther/Microspore/Ovule Culture		
Brassica oleracea	Haploid production	Lillo and Hansen, 1987; Phippen and Ockendon, 1990
Capsicum annuum	Haploid production	Harn *et al*., 1975
Lycopersicon esculentum	Haploid production	Sharp *et al*., 1971; Greshoff and Doy, 1972
Solanum melongena	Haploid production	Raina and Iyer, 1973
Cucurbita pepo	Haploid Production	Metwally *et al*., 1998a
Cucurbita pepo	Haploid Production	Metwally *et al*., 1998b
Embryo Rescue		
L. esculentum × *L. peruvianum*	Hybrid embryo rescue	Thomas and Pratt, 1981
P. vulgaris × *P. angustissimus*	Hybrid embryo rescue	Belivanis and Dore, 1986
Vigna pubescens × *V. unguiculata*	Hybrid embryo rescue	Fatokun and Singh, 1987
S. melongena × *S. torvum*	Hybrid embryo rescue	Bletsos *et al*., 1998
Meristem Culture		
Allium sativum	Elimination of onion yellow dwarf virus	Walkey *et al*., 1987
Allium ascalonicum	Elimination of leek yellow dwarf virus	Walkey *et al*., 1987

Table 6. Continued

Technique/Vegetables	Technique applied for	Reference
Pisum sativum	Elimination of pea seed borne mosaic virus	Kartha and Gamborg, 1979
Solanum melongena	EMCV	Raj *et al.*, 1991
L. esculentum	Cryopreservation	Grout *et al.*, 1978
P. sativum	Cryopreservation	Kartha and Gamborg, 1979
Protoplast Fusion		
S. melongena + S. integrefolium	Resistance to *Pseudomonas*	Kameya *et al.*, 1990
S. melongena + S. saintwongesi	Resistance to *Pseudomonas*	Asao *et al.*, 1994
B. oleracea + B. campestris	Disease resistance	Itoh *et al.*, 1991
B. oleracea + B. napus	Atrazine resistance	Jourdan *et al.*, 1989
B. oleracea + Armoracia	Asymmetric hybridization	Navratilova *et al.*, 1997
L. esculentum × L. peruvianum + Lycopersicoides	Fertile somatic hybrids	Matsumoto *et al.*, 1997
L. esculentum + S. acaule	Male sterility induction	Melchers *et al.*, 1992
S. melongena + S. khasianum	Resistance to shoot and fruit borer	Sihachakr *et al.*, 1998
L. esculentum + L. peruvianum	Somatic hybrids	Chen *et al.*, 1998
Somaclonal Variation		
L. esculentum	Tolerance to *Fusarium oxysporum* culture filtrate	Scala *et al.*, 1984
	Resistance to *F. oxysporum*	Shahin and Spivey, 1986
	Tolerance to TMV	Cassells *et al.*, 1986
	Resistance to bacterial wilt	Toyoda *et al.*, 1989

8. CONCLUSIONS AND FUTURE PROSPECTS

The ultimate goal of molecular marker research is to develop plant varieties more efficiently and effectively through various described strategies. However, most of the techniques based on molecular markers are having certain handicaps in their direct utilization in plant breeding. In order to have cost and time effective molecular markers with increased efficiency of detecting polymorphism, development of various classes of markers is very fast, as a result now a large number of molecular markers are available. Therefore, selection of appropriate marker for applied plant breeding would depend on the crop, breeding objective(s), trait(s) under consideration, breeding material, cost effectiveness and reproducibility of molecular marker. Although PCR based marker like RAPD is cost effective and simple, its reproducibility has been questioned (Penner *et al.*, 1993). In breeding programme, even a 5% error precludes its applicability in MAS (Mohan *et al.*). Nevertheless, saturation of RAPD genetic map in vegetables will provide opportunity in future to extrapolate and use it in various ways. There is a need to test and identify reproducible RAPD marker so that its feasible use in hybrid seed production can be examined. Molecular marker tightly linked to male sterility gene can be used to identify male fertile sister plants at very early stage and this application of marker in hybrid seed production programs is seems to be feasible in vegetable crops with small population size (Kumar *et al.*, 2000b). Besides high cost, reproducible molecular markers like RFLP or AFLP may also have other drawbacks. For example, a marker developed in one cross for a gene may not work in other crosses for the same gene until the marker is from the gene itself (Mohan *et al.*, 1997; Joshi *et al.*, 1999). Therefore, before incorporating these markers in breeding program, its validity either in the same cross or other crosses should be tested. In vegetable crops, tomato is the right candidate to test such validity, since a large number of important genes and QTLs have been tagged with various classes of molecular markers.

In India, transgenic eggplant and tomato have been developed in many laboratories and few of them (especially insect resistant transgenic eggplant) are at the edge of field trails. It has been predicted that to obtain and stabilize improved transgenic genotype, selection would be needed (Jones and Cassells, 1995), because transgenic plants suffer from problems like adverse pleiotropy, co-suppression, position effects, gene silencing etc. These mechanisms are highly unstable because they are influenced by the environmental factors. For example, co-suppression can result in high frequency of environment specific gene expression.

Transfer of trans-gene into different genetic background can overcome these problems (Jones and Cassells, 1995). Obviously, at this stage such transfer is possible only through conventional back cross breeding. It is anticipated that the use of gene constructs with flanking scaffold-associated region (SAR) can solve the problem of trans gene silencing during transformation (insertion) process. Improved genotype(s) obtained through tissue culture techniques have already proven their worth in vegetable breeding. For instance, many seed companies in France are producing cabbage and cauliflower hybrids utilizing cytoplasmic male sterile cybrids (Pelletier *et al.*, 1995). Protoplast fusion has been successfully utilized to overcome problems associated with the sterile cytoplasm in *Brassica* vegetables *e.g.* yellowing of seedlings (Kumar *et al.*, 2000b). However, these techniques face novel problem of high frequency of unstable variation, resulting from methylation and sequence amplification (Jones and Cassells, 1995).

In the coming years, following research areas have been predicted to be very important in improvement of plant species: (i) apomixis to fix and exploit heterosis in hybrids, (ii) transgenic male sterility for hybrid seed production, (iii) parthenocarpy for seedless vegetables and fruits, (iv) short cycling for rapid insect of tolerant and fruit trees and (v) converting annual into perennial (Oritz, 1998). The first three areas are applicable to the most of the vegetable crops. In *Allium,* apomixis research utilizing apomictic accessions of *A. tuberosum* or *A. nutans* has not got attention by the molecular biologist. The development of transgenic male sterility systems and parthenocarpic vegetables along with the resistant transgenics to the biotic stresses will be the research thrust worldwide. Careful selection of breeding material(s) from breeder's germplasm and selection of biotechnological tool(s) for particular breeding material will be equally important and crucial for the success of any non-conventional tool to develop durable disease resistant varieties. In this regard, engineering resistant transgenics based on hypersensitive reaction strategy would be a better option than that of transgenics with gene for anti-fungal compounds (Kumar *et al.*, 2000a). In the proposed Indian sui-generis system (Plant Breeders and Farmers Rights) terminator technology is not permitted and farmers will have freedom to save and reuse seeds (Rai and Prasanna, 2000). Therefore, vegetable hybrid technology will be the priority research agenda, especially of the seed companies because this technology will provides opportunity to sale F_1 seeds every season. Hence, biotechnology will be increasingly utilized to develop very specific and potential parental lines like transgenic or cybrid male sterile, stresses resistant, parthenocarpic lines etc., and private sector may not directly release genetically modified plant varieties.

REFERENCES

Altman A (1999). Plant biotechnology in 21st century: the challenges ahead. *Elect. J. Biotech. http://www.ejb.org/content/vol2/issue2.*

Asao H, Arai S, Sato T and Hirai M (1994). Characteristics of a somatic hybrid between *Solanum melongena* L. and *Solanum sanitwongsei. Craib. Breed. Sci.,* **44** : 301-305.

Beachy RN, Loesh Fries S and Tumer NE (1990). Coat protein mediated resistance against virus infection. *Ann. Rev. Phytopath.,* **28** : 451-474.

Belivanis T and Dore C (1986). Interspecific hybridization of *Phaseolus vulgaris* L. and *Phaseolus angustissimus* using *in vitro* embryo culture XE "embryo culture". *Plant Cell Rep.,* **5** : 329-331.

Blestsos FA, Roupakias DG, Tsaktsira ML, Scaltsoyjannes AB and Thanassoulopoulos CC (1998). Interspecific hybrids between three eggplant (*Solanum melongena* L.) cultivars and two wild species (*Solanum torvum* Sw. and *Solanum sisymbriifolium* Lam.). *Plant Breed.,* **117** : 159-164.

Bonierbale M, Plaisted RL and Tanksley SD (1988). RFLP maps of potato and tomato based on a common set of clones reveal modes of chromosomal evolution. *Genetics,* **120** : 1095-1103.

Broglie R and Broglie K (1993). Production of disease resistant transgenic plants. *Curr. Opin. Biotech.,* **4** : 148-151.

Brommonschenkel SH and Tanksley SD (1997). Map based cloning of the tomato genomic region that spans the 100-seed weight-5 tospovirus resistance gene in tomato. *Mol. Gen. Genet.,* **256** : 121-126.

Camargo LE, Savides L, Jung G, Nienhuis J and Osborn TC (1997). Location of the self incompatibility locus in an RFLP and RAPD map of *Brassica oleracea. J. Hered.,* **88** : 57-60.

Carnelissen BJC, Does MP and Melchers LS (1996). Strategies for molecular resistance breeding (and transgenic plants). In: *Rhizoctonia Species: Taxonomy, Molecular Biology, Ecology, Pathology and Control,* (Eds Sneh B, Jabaji HS, Neale S and Deist G) Kluwer Acad. Pub., Netherlands, pp. 529-536.

Cassells AC, Coleman M, Farrell G, Long R, Goetz EM and Boyton V (1986). Screening for virus resistance in tissue culture adventitious regenerants and their progeny. In: *Genetic Manipulation in Plant Breeding,* (Eds Horn W *et al.*), W de Grujter, Berlin, pp. 535-545.

Chen LZ and Adachi T (1998). Protoplast fusion between *Lycopersicon esculentum* and *L. peruvianum* complex: somatic embryogenesis, plant regeneration and morphology. *Plant Cell Rep.,* **17** : 508-514.

Daniell H (1999). The next generation of genetically engineerd crops for herbicide and insect resistance. Containments of gene pollution and resistant insects. *AgBiotechNet,* **1** : 24-42.

Delannay X, La Vallee BJ, Proksh RK, Fuchs RL, Sims SR, Greenplate JT, Marrone PG, Dodson RB, Augustine JJ, Layton JG and Fischoff DA (1989). Field performance of transgenic tomato plants expressing the *Bacillus thuringiensis* var Kurstaki insect control protein. *Bio/Technology,* **7** : 1265.

Dirlewanger E, Isaac PG, Ranade S, Belajouza M, Cousin R and Vienne D De (1994). Restriction fragment length polymorphisin analysis of loci associated with disease resistance genes and developmental traits in *Pisum sativum* L. *Theor. Appl. Genet.,* **88** : 17-27.

Diwan N, Fluhr R, Eshed Y, Zamir D and Tansksley SD (1999). Mapping of *Ve* in tomato: a conferring resistance to the broad spectrum pathogen, *Verticillium dehliae* race 1. *Theor. Appl. Genet.,* **98** : 315-319.

Dixon MS, Jones DA, Hatzixanthis K, Ganal MW, Tanksley SD, Jones JD (1995). High resolution mapping of the physical location of the tomato *Cf-2* gene. *Mol. Plant Microbe. Interact.,* **8** : 200-206.

Dixon MS, Jones DA, Keddie JS, Thomos CM, Harrison K and Jones JDG (1996). The tomato *Cf-2* disease resistance locus comprises of two functional genes encoding leucin rich repeat proteins. *Cell,* **84** : 451-459

Doganlar S, Frary A and Tanksley SD (2000a). The genetic basis of seed-weight variation: tomato as a model system. *Theor. Appl. Genet.,* **100** : 1267-1273.

Doganlar S, Tanksley SD and Mutschler MA (2000b). Identification and molecular mapping of loci controlling fruit ripenig time in tomato. *Theor. Appl. Genet.,* **100** : 249-255.

Fatokun CA and Singh BB (1987). Interspecific hybridization between *Vigna pubescens* and *Vigna unguiculata* Walp. through embryo rescue. *Plant Cell Tiss. Org. Cult.,* **9** : 229-233.

Fatokun CA, Menancio-Hautea DI, Danesh D and Young ND (1992). Evidence for orthologous seed weight genes in cowpea and mungbean based on RFLP mapping. *Genome,* **132** : 841-846.

Federoff NV, Furtek DB and Nelson OE (1984). Cloning of the bronze locus in maize by a simple and generalizable procedure using the transposable controlling element Activator (Ac). *Proc. Natl. Acad. Sci. USA,* **81** : 3825-3829.

Fischhoff DA, Bowdish KS, Perlak FJ, Marrone PG, Mc Cormick SM, Niedermeyer JG, Dean DA, Kusano-Kretzmer K, Mayer EJ, Rochester DE, Rogers SG and Fraley RT (1987). Insect tolerant transgenic tomato plants. *Bio/Technology,* **5** : 807-813.

Foolad MR, Chen FQ (1997). RAPD markers associated with salt tolerance in an interspecific cross of tomato (*Lycopersicon esculentum* x *L. pinnellii*). *Plant Cell Rep.,* **17** : 306-312.

Fulton TM, Grandillo S, Bech-Bunn T, Fridman E, Frampton A, Lopez J, Petriard V, Uhlig J, Zamir D and Tanksley SD (2000). Advanced back cross QTL analysis of a *Lycopersicon esculentum* x *L. parviflorum* cross. *Theor. Appl. Genet.,* **100** : 1025-1030.

Ganal MW, Simon R, Brommonschenkel S, Arndt M, Phillips MS, Tanksley SD, Kumar A (1995). Genetic mapping of a wide spectrum nematode resistance gene (Hero) against *Globodera rostochiensis* in tomato. *Mol. Plant Microb. Interact.,* **8** : 886-891.

Gatehouse AMR, Boulter D and Hilder VA (1992). In. *Plant Genetic Manipulation for Crop Protection* (Eds Gatehouse AMR, Hilder VA and Boulter D), CAB International pp 155-181.

Giovannoni JJ, Noensie EN, Ruezinsky DM, Lu XH, Tracy SL, Ganal MW, Martin GB, Pillen K Alpert K and Tanksley SD (1995). Molecular and genetic analysis of the ripening inhibitor and non-ripening loci of tomato: a first step in genetic map-based cloning of fruit ripening genes. *Mol. Gen. Genet.,* **248** : 195-206.

Gonslaves D, Chee P, Provvidenti R, Seem R and Slightom JL (1992). Comparison of coat protein mediated and genetically derived resistance in cucumber to infection by cucumber mosaic virus under field conditions with natural challenge inoculations by vectors. *Bio/technology,* **10** : 1562-1570.

Gorman SW, Banasiak D, Fairley C and McCormick S (1996). A 610 kb YAC clone harbors 7 cM of tomato (*Lycopersicon esculentum*) DNA that includes the male sterile 14 gene and a hotspot for recombination. *Mol. Gen. Genet.,* **251** : 52-59.

Gresshoff PM and Doy CH (1972) Development of differentiation of haploid *Lycopersicon esculentum* (tomato). *Planta,* **107** : 161-170.

Grout BWW, West Cott RJ and Henshaw GG (1978). Survival of shoot meristems of tomato seedlings frozen in liquid nitrogen. *Cryobiology,* **15** : 478-483.

Hanson PM, Bernacchi D, Green S, Tanksley S, Muniappa V, Padmja AS, Kuo G, Fang D and Chen J (2000). Mapping of a wild tomato introgression associated with tomato yellow leaf curl virus resistance in a cultivated tomato line. *J. Amer. Soc. Hort. Sci.,* **125** : 15-20

Harn C, Kim MZ, Chov KT and Lee YI (1975). Production of haploid callus and embryoid from the cultured anther of *Capsicum annuum. SABRAO J.,* **7** : 71-77.

Hilder VA, Gatehouse AMR, Sheerman SE, Barker RF and Boulter D (1987). A novel mechanism of insect resistance engineered in tobacco. *Nature,* **330** : 160-163.

Hille J, Koornif M, Ramanna MS and Zabel P (1989). Tomato: A crop species amenable to improvement by cellular and molecular methods. *Euphytica,* **42** : 1-23

Hu J, Sadowski J, Osborn TC, Landry BS and Quiros CF (1998). Linkage group alignment from four independent *Brassica oleracea* RFLP maps. *Genome,* **41** : 226-235.

Itoh K, Iwabuchi M and Shinaonoto K (1991). *In situ* hybridisation with species-specific DNA probes gives evidence for asymmetric nature of *Brassica* hybrids obtained by X-ray fusion. *Theor. Appl. Genet.,* **81** : 356-362.

Jones DA, Thomos CM, Hammond Kosak KE, Balint Kurti PJ and Jones JDG (1994). Isolation the tomato *cf*-2 gene for resistance to *Cladosporium fulvum* by transposon tagging. *Science,* **266** : 789-793.

Jones PW and Cassells AC (1995). Crieteria for decision making in crop improvement programmes-technical consideration. *Euphytica,* **85** : 465-476.

Joshi SP, Ranjekar PK and Gupta VS (1999). Molecular Markers in plant genome analysis. *Curr. Sci.,* **77** : 230-240.

Jourdan PS, Earle ED and Mut Schler MA (1989). Atrazine resistant cauliflower obtained by somatic hybridisation between *Brassica oleracea* and *B. napus. Theor. Appl. Genet.,* **78** : 271-279.

Kameya T, Miyazawa N and Toki S (1990). Production of somatic hybrids between *Solanum melongena* L. and S. *integrifolium* POIR. *Jpn. J Breed.,* **40** : 429-434.

Kang BC, Nam SH, Huh JH, You HS and Kim BD (1998). Construction of a molecular linkage map and marker selection of hot pepper. *Plant and Animal Genome VI conference,* January 18-22, 1998, San Diego, CA.

Kaniewski W, Liardi V, Tamassoli L, Mitsky T, Layton J and Barba M (1999). Extreem resistance to cucumber mosaic virus (CMV) in transgenic tomato expressing one or two viral coat protiens. *Mol. Breed.,* **5** : 111-119.

Kartha KK and Gamborg OL (1979). Meristem culture techniques in the production of disease-free plants and freeze preservation of germplasm of tropical tuber crops and grain legumes. In: *Disease of Tropical Food Crops,* (Eds Maraite H and Mayer JA), Universite Catholique, Louvain-La-Neuve, pp 267-283.

Kesseli RV, Paran I, Michelmore RW (1994). Analysis of a detailed genetic linkage map of *Lactuca sativa* (lettuce) constructed from RFLP and RAPD markers. *Genetics,* **136** : 1435-1446.

Kumar PA, Mandaokar A, Sreenivasan K, Chakrabarti SK, Suman B., Sharma SR, Sarvjeet Kaur and Sharma RP (1998a). Insect resistant transgenic brinjal plants. *Mol. Breed.,* **4** : 33-37.

Kumar PA, Mandaokar AD, and Sharma RP (1998b). Genetic engineering for the improvement of egg plant (*Solanum melongena* L.). *AgBiotechNews and Information,* **10** : 329-332

Kumar Sanjeet, Banerjee MK and Kalloo G (2000a). Disease resistance: mechanisms and breeding in vegetable crops. In: *Emerging Scenario in Vegetable Research and Development,* (Eds Kalloo G and Singh Kirti), Researchco Publication, New Delhi, pp. 119-144.

Kumar Sanjeet, Banerjee MK and Kalloo G (2000b). Male sterility: mechanisms and idetification, chracterization and utilization in vegetables *Veg. Sci.,* **27** : 1-24.

Lagercrantz U and Lydiate DJ (1996). Comparative genome mapping in *Brassica. Genetics,* **144** : 1903-1910.

Lee M (1995). DNA markers and plant breeding programmes. *Adv. Agron.,* **55** : 265-344

Lillo C and Hansen M (1987). Anther culture of cabbage. Influence of growth temperature of donor plants and media composition on embryo yield and plant regeneration. *Norwegian J. Agric. Sci.,* **1** : 105-109.

Ling H, Pich A, Scholz G and Ganal MW (1996). Genetic analysis of two tomato mutants affected in the regulation of iron metabolism. *Mol. Gen. Genet.,* **252** : 87-92.

Mandaokar A, Goyal RK, Shukla A, Bhalla R, Chaurasia A, Raddy VS, Altosaar I, Sharma RP and Kumar PA (2000). Transgenic tomato plants resistant to shoot and fruit borer (*Helicoverpa armigera* Hubner). *Crop Protect.,* **19** : 307-312.

Mariani C, De Beuckelleer M, Truettner J, Leemans J and Goldberg RB (1990). Induction of male sterility in plant by a chimaeric ribonuclease gene. *Nature,* **347** : 737-741.

Mariani C, gossele V, De Beuckeleer M, De Block M, Goldberg RB, DeGreef W and Leemans J (1992). A chimeric ribonuclease inhibitor gene restores fertility to male sterile plants. *Nature,* **357** : 384-387.

Martin GB, Bromomnschenkel SH, Chunwongse J, Frary A, Ganal MW, Spivey R, Wu T, Earle ED and Tanksley SD (1993). Map based cloning of a protein kinase gene conferring disease resistance in tomato. *Science,* **262** : 1432-1436.

Martin GB, Frary A Wu T Bromomnschenkel SH, Chunwongse J, Earle ED and Tanksley SD (1994). A member of tomato *pto* gene family confers sensitivity to fenthion resulting in rapid cell death. *Plant Cell,* **6** : 1543-1552.

Martin GB, Ganal MW and Tanksley SD (1992). Construction of a yeast artificial chromosome library of tomato and identification of cloned segments linked to two disease resistance loci. *Mol. Gen. Genet.,* **233** : 25-32.

Matsumoto A, Imanishi S, Hossain M, Escalante A and Egashira H (1997). Fertile somatic hybrids between F_1 (*Lycopersicon esculentum* × *L. peruvianum* var. *humifusum*) and *Solanum lycc¨ersicoides*. *Breed. Sci.,* **43** : 327-333.

Mc Gurl B, Orozo-cardenas M, Pearce G and Ryan CA (1994). Over expression of the prosystemin gene in transgenic tomato plants generates a systemic signal that constitutively induces proteinase inhibitor synthesis. *Proc. Natl. Acad. Sci. USA,* **91** : 9799-9802.

Melchers G, Mohri Y, Wakabayashi S and Harada K (1992). One step generation of cytoplasmic male sterility by fusion of mitochondrial-inactivated protoplasts with nuclear-inactivated *Solanum* prtoplasts. *Proc. Natl. Acad. Sci. USA.,* **89** : 6832-6836.

Melotto M, Afanador L and Kelly JD (1996). Development of a SCAR marker linked to the *I* gene in common bean. *Genome,* **39** : 1216-1219.

Metwally I, Moustafa SA, El-Sawy BI, Haroun SA and Shalaby TA (1998a). Production of haploid plants from *in vitro* culture of unpollination ovules of *Cucurbita pepo*. *Plant Cell Tiss. Org. Cult.,* **52** : 117-121.

Metwally EI, Moustafa SA, El-Sawy BI and Shalaby TA (1998b). Haploid plantlets derived by anther culture of *Cucurbita pepo*. *Plant Cell Tiss. Org. Cult.,* **52** : 171-176.

Mohan M, Nair S, Bhagwat A, Krishna TG, Yaho M, Bhatia CR and Sasaki T (1997) Genome mapping, molecular markers and marker-assisted selection in crop plants. *Mol. Breed.,* **3** : 87-103.

Moreau P, Thoquet P, Olivier J, Laterrot H and Grimsley N (1998). Genetic Mapping of *Ph-2*, a single locus controlling partial resistance to *Phytophthora infestans* in tomato. *Mol Plant Microbe Interact,* **11** : 259-269.

Navratilova B, Bujek J, Siroky J and Havvanek P (1997). Construction of intergeneric somatic hybrids between *Brassica oleracea* and *Armocaria rusticana*. *Biol Plant.,* **39** : 531-541.

Nunome T and Hirai M (1998). A linkage map of egg plant (*Solanum melongena*) based on RAPD markers. *Plant and Animal Genome Conference,* Jan 18-22, 1998, San Diego, CA.

Oritz R (1998). Critical role of plant biotechnology for the genetic improvement of food crops: perspective for next millenium. *Elect. J. Biotech. http: //www.ejb.org/ content/vol1/issue3.*

Palletier G, Ferault M , Lancelin D , Baulidard L, Dore C Bonhomme S Grelon M and Budar F (1995). Engineering of cytoplasmic male sterility in vegetables by protoplast fusion. *Acta. Hort.,* **392** : 11-17

Pang SZ, Nagpala P, Wang M, Slightem JL and Gonsalves D (1992). Resistance to heterologous isolates of tomato spotted wilt virus in transgenic tobacco expressing its nucleo capsid protein gene. *Phytopathology,* **82** : 1223-1229.

Penner GA, Bush A, Wise R, Kim W, Domier L, Kasha K, Larochi, Sooles G, Moluav SJ and Fedak G (1993). Reproducibility of random amplified polymorfic DNA (RAPD) analysis from laboratories. *PCR Meth. and Appl.,* **2** : 341-345.

Petrova M, Yulkova Z, Gorinova N, Izhar S, Firon N, Jacquemin JM, Atanassove A and Stoeva P (1998). Characterization of cytoplasmic male sterile hybrid between *Lycopersicon peruvianum* Mill. X *Lycopersicon pennellii* Corr. and its crosses with tomato. *Theor. Appl. Genet.* **98** : 39-48.

Phippen C and Ockendon DJ (1990). Genotype, plant, bud size and media factors affecting anther culture of cauliflowers (*Brassica oleracea* var. *botrytis*). *Theor. Appl. Genet.*, **79** : 33-38.

Powell PA, Nelson RS, De BN, Hoffman N, Rogers SG, Fraly RT and Beachy RN (1986). Delay of disease development in transgenic plants that express the tobacco mosaic virus coat protein gene. *Science,* **232** : 738-743.

Rai Mangala and Prasanna BM (2000). *Trangenics in Agriculture.* Indian Council of Agricultural Research, New Delhi, India.

Raina SK and Iyer RD (1973). Differentiation of haploid plant from pollen callus in anther cultures of *Solanum melongena* L. *Z Pflazenzucht,* **70** : 275-280.

Rick CM and Fobes JF (1977). Association of an allozyme with nematode resistance. *Rep. Tomato Genet. Coop.,* **24** : 25.

Rick CM, Deverna JW, Chetelat RT and Stevens MA (1986). Meiosis in sesquidiploid hybrids of *Lycopersicon esculentum* and *Solanum lycopersicoides. Proc. Natl. Acad. Sci. USA,* **83** : 3580-3583.

Rotino GL, Perri E, Zottini M, Sommer and Spena A (1997). Genetic engineering of parthenocarpic plants. *Nature Biotech.,* **15** : 1398-1401.

Salmeron JM, Oldroyd EDG, Rommens CMT, Scoofield SR, Kim HS, Levelle DT, Dahlbech and Staskawicz BJ (1996). Tomato *prf* is a member of the leucin rich repeat class of plant disease resistance genes and lies embeded within the *pto* kinase gene cluster. *Cell,* **86** : 123-133

Scala A, Bethini P, Buiathi M, Bogani P, Pellegrini G and Tognoni F (1984). *In vitro* analysis of the tomato - *Fusarium oxysporum* system and selection experiments. In: *Plant Tissue and Cell Culture Application to Crop Improvement*, (Eds Novak FJ *et al.*), Czechoslovzk Acad Sci Progen, pp. 361-362.

Schroeder HE, Gollash S, Moore A, Tame LM, Craig S, Hardie DC, Chrispeels MJ, Spencer D and Higgins TJV (1995). Bean (-amylase inhibitor confers resistance to the pea weevil (*Bruchus pisorum*) in transgenic peas (*Pisum sativum* L.). *Plant Physiol.,* **107** : 1233-1239.

Segal G, Sarfatti M, Schaffer MA, Ori N, Zamir D and Fluhr R (1992). Correlation of genetic and physical structure in the region surrounding the *IZ Fusarium oxysporum* resistance locus in tomato. *Mol. Gen. Genet.,* **231** : 179-185.

Serquen FC, Bacher J and Staub JE (1997). Mapping and QTL analysis of horticultural traints in a narrow cross in cucumber (*Cucumis sativus* L.) using random-amplified polymorphic DNA markers. *Mol. Breed.*, 257-268.

Shahin EA and Spivey R (1986). A single dominant gene for *Fusarium* wilt resistance in protoplast derived tomato plants. *Theor. Appl. Genet.,* **73** : 164-169.

Sharma RP and Kumar PA (1995). Revolutionizing plant breeding through biotechnology. In : Sustaining Crop and Animal Productivity. In : *The Challenge of the Decade,* (Ed Deb DL), Associated Publishing Co, New Delhi, pp 147-156.

Sharp WR, Dougall DK and Paddock (1971). Haploid plantlets and callus from immature pollen grains of *Nicotiana* and *Lycopersicon. Bull. Torrey Bot. Club,* **98** : 219-222.

Sihachkr D, Serraf I, Chaput MH, Mussio I, RossiqnolL, and Dacreux G. (1995). Regeneration of plant from protoplasts of *Solanum khasianum* CB Clark and *Solanum laciniatum* Ait. In: *Biotechnology in Agriculture and Forestry,* (Ed Bajaj YPS), Springer Verlag Berlin, Heidelberg. **34** : 108-120.

Singh M, Chakraborty S, Srivastava K and Kalloo G (2000a). Biotechnological advances in vegetable crops. In: *Emerging Scenario in Vegetable Research and Development,* (Eds Kalloo G and Singh Kirti), Researchco Publication, New Delhi, pp. 59-85.

Singh M, Chakraborty S, Sanjeev Kumar and Kalloo G (2000b). Gentic engineering for insect resistance in vegetable crops. *Veg. Sci.,* **27** : 105-116.

Sobir OT, Murata M and Motoyoshi F (2000). Molecular characterization of the SCAR markers tightly linked to the *Tm-2* locus of genus *Lycopersicon. Theor. Appl. Genet.,* **101** : 64-69.

Stamova BS and Chetelat RT (2000). Inheritance and genetic mapping of cucumber mosaic virus resistance introgressed from *Lycopersicon chilense* into tomato. *Theor. Appl. Genet.,* **101** : 527-537

Stommel JR, Tousignant ME, Wai T, Pasini R and Kaper JM (1998). Viral satellite RNA expression in transgenic tomato confers field tolerance to cucumber mosaic virus. *Plant Disease,* **82** : 391-396.

Tanksley SD, Ganal MW, Prince JP, de Vicente MC, Boinerbale MW, Broun P, Fulton TM, Giovanonni JJ, Grandillo S, Martin GB, Messenguer R, Miller JC, Miller L, Paterson AH, Pierda O, Roder M, Wing RA, Wu W and Young ND (1992). High density molecular linkage maps of the tomato and potato genomes: biological inferences and practical applications. *Genetics,* **132** : 1141-1160.

Tanskley SD and Jones RA (1979). Application of alcohol dehydrogenase allozymes in testing the purity of F_1 hybrids of tomato. *HortScience,* **16** : 179-181.

Tanskley SD and Orton TJ (1983). *Isozymes in Plant Genetics and Breeding,* Elsevier, Amsterdam.

Tanskley SD and Rick CM (1980). Isozymic gene linkage map of the tomato. Applications in genetics and breeding. *Theor. Appl. Genet.,* **57** : 161-170.

Tanskley SD, Rick CM and Vallejos CE (1984). Tight linkage between a male sterile locus and an enzyme marker in tomato. *Theor. Appl. Genet.,* **68** : 109-113.

Thomas BR and Pratt D (1981). Efficient hybridisation between *Lycopersicon esculentum* and *L. peruvianum* via embryo callus. *Theor. Appl. Genet.,* **59** : 215-219.

Tomassoli L, Liardi V, Barba M and Kaniewski W (1999). Resistance of transgenic tomato to cucumber mosaic cucumovirus under field conditions. *Mol. Breed.,* **5** : 121-130.

Toyoda H, Shimizu K, Chatani K, Kita N, Matsuda Y and Onchi S (1989). Selection of bacterial wilt resistant tomato through tissue culture. *Plant Cell Rep.,* **8** : 317-320.

Vaeck M, Reynaerts A, Hofte H, Jansens S, De Beuckeleer M, Dean C, Zabeau M, Var Montagu M and Leemans J (1987). Transgenic plants protected from insect attack. *Nature,* **327** : 33-37.

Vallejos CE and Tanksley SD (1983). Segregation of isozyme markers and cold tolerance in an interspecific backcross of tomato. *Theor. Appl. Genet.,* **66** : 241-247.

Walkey DGA, Webb MJW, Bolland CJ and Miller A (1987). Production of virus free garlic (*Allium sativum* L.) and shallot (*A. ascalonicum* L.) by meristem-tip culture. *J. Hort. Sci.,* **62** : 211-220.

Wang Y, Sun C, Wang C and Chien N (1997). Induction of pollen plantlets of *Triticale* and *Capscium annuum* fron anther culture. *Sci. Sin.,* **16** : 147-157.

Williamson VM, Ho JY, Wu FF, Miller N and Kaloshian I (1994). A PCR based marker tightly linked to the nematode resistance gene, *Mi* in tomato. *Theor. Appl. Genet.,* **87** : 757-763.

Williams ME, Lecmans J and Michiels F (1997). Male sterility through recombinant DNA technology. In: *Pollen Biotechnology for Crop Production and Improvement,* (Eds Shivanna KR and Sawhney VK), Cambridge Univ. Press, pp. 237-257.

Yaghoobi J, Kaloshian I, Wen Y and Williamson VM (1995). Mapping a new nematode resistance locus in *Lycopersicon peruvianum. Theor. Appl. Genet.,* **91** : 457-464.

Young ND and Tanksley SD (1989). RFLP analysis of the size of chromosomal segments retained around the *Tm-2* locus of tomato during backcrossing breeding. *Theor. Appl. Genet.,* **77** : 353-359.

Zamir D, Ekstein-Michelson I, Zakay Y, Navot N, Zeidan M, Sarfatti M, Eshed Y, Harel E, Pleban T, Oss H Van, Kedar N, Rabinowitch HD and Czoshek H (1994). Mapping and introgression of a tomato yellow leaf curl virus tolerance gene, *TY-1. Theor. Appl. Genet.,* **88** : 141-146.

Zhang Y and Stommel JR (1998). Identification of carotenoid loci in tomato with RAPD and AFLP markers. *Plant and Animal Genome Conference,* January 18-22, 1998, San Diego, CA.

Chapter 2

GENETIC ENGINEERING OF VEGETABLE CROPS FOR INSECT RESISTANCE

P Ananda Kumar

National Research Centre for Plant Biotechnology, Indian Agricultural Research Institute, New Delhi - 110 012, India

Summary

Vegetables are the most important constituents in the diet of people in the developing world, especially Indian subcontinent. Cultivation of vegetable crops is bedeviled by many pest and disease problems. Over the past five decades, plant breeders have done yeoman service towards the improvement of vegetable productivity. However, incorporation of resistance to several pests and diseases still remains elusive which is the major limiting factor for productivity. Excessive use of synthetic, organic insecticides for insect pest management in horticulture resulted in the degradation of environment, adverse effects on human health and development of resistant insects. Recent advances in genetic engineering have clearly demonstrated the possibility of incorporating foreign genes for desired characters while preserving the existing traits of improved genotypes. Reduction in the consumption of insecticides is only possible by introducing the genes encoding insecticidal proteins such as δ-endotoxins of Bacillus thuringiensis (Bt), protease inhibitors, alpha-amylase inhibitors, lectins, enzymes such as chitinase and peroxidase. Genetically modified vegetable crops carrying insect resistance genes have been shown to exhibit considerable protection against the target insects. It is possible to incorporate the resistance against more than one insect pest in the transgenic plants through genetic engineering. This can be achieved by introducing fusion genes or

*Corresponding author : E-mail : polumetla@hotmail.com

multiple genes in combination encoding the insecticidal proteins which are toxic against a wide range of target pests. Development of resistance in insects is a distinct possibility that can be prevented by deploying suitable resistance management strategies. Such approaches include the usage of multiple toxins, crop rotation practices, high dosage of toxins and refugia. Use of δ-endotoxins of Bacillus thuringiensis in conjunction with other bioinsecticides in an IPM mode can drastically reduce the consumption of chemical pesticides and pave the way for safe and sustainable horticulture.

Keywords : Amylase inhibitors, *Bacillus thuringiensis*, insect resistance, insect pest management, lectins, protease inhibitors, transgenic vegetables

Abbreviations : Bt-*Bacillus thuringiensis*, ICP-Insecticidal crystal Protein, CpTi -Cowpea Trypsin inhibitor, CPB-Colorado Potato beetle, IPM -Integrated Pest Management.

1. INTRODUCTION

Fruits and vegetables constitute an important part of Indian diet which is predominantly vegetarian in nature. Vegetables are rich and comparatively cheaper source of vitamins and minerals. The beneficial effects of vegetables on human health and well being are well documented by various sources. Considering the present area and yield, the production of vegetables in India is inadequate to meet the needs of the country (Bose *et al.*, 1993). In near future, the burgeoning population will make additional demands for further increases in the yield and productivity of vegetable crops.

Over the past few decades, plant breeders have rendered yeoman service to agriculture and horticulture by developing high yielding hybrids and cultivars of different vegetable crops (Bose *et al.*, 1993). Another quantum jump in the productivity of these crops can only be realized by adopting the novel methods and techniques of modern biology. Fortunately, advancements made in the recent years in the areas of molecular biology, genetic engineering and plant tissue culture provide an altogether new dimension to crop improvement. It is now possible to harness economically important genes from any biological source and introduce into a plant of our choice with precision. Using such tools many important crop plants have been genetically engineered over the past one decade to confer agronomically valuable traits such as herbicide tolerance, viral resistance, insect resistance, fungal resistance, protein and oil quality, metabolic changes, production of biodegradable plastics, edible vaccines etc., (Dale *et al.*, 1993). Many vegetable crops such as potato, tomato, cauliflower, cabbage and eggplant have been genetically transformed for important characters like pest resistance. Some of these crops are already under commercial cultivation (Schuler *et al.*, 1998). Considerable progress has been made in India too, to develop transgenic vegetable crops, especially for insect resistance (Kumar *et al.*, 1998). In the present account I describe the progress made so far with respect to the genetic engineering of vegetable crops and discuss the prospects for future.

2. INSECT PESTS OF VEGETABLE CROPS

India is predominantly tropical and sub-tropical in its climate. Because of the warm temperatures prevailing all over the country the insect prevalence is permanent and the biodiversity of the insect pests is astounding. Vegetable crops in such a situation are damaged by a plethora of insect pests. The most damaging insects are usually lepidopterans followed by coleopterans and dipterans. Some of the major insect pests of vegetable crops of India are enumerated in Table 1. In addition to these major pests, insects such as aphids spread debilitating viral diseases (*e.g.*, Tomato Leaf Curl Virus). Thus, insect pest management in vegetable crops is of paramount importance. Over the past few decades usage of synthetic, organic insecticides has resulted in considerable control of pests in horticulture. On a global scale, the major share of insecticide consumption is taken by fruits and vegetables (Fig. 1). The consequences of extensive and very often indiscriminate usage of pesticides have been the adverse effects on human health, degradation of the environment, extermination of the beneficial insects (pollinators and predators) and development of resistance in insects. Hence, a drastic reduction in the consumption of insecticides in horticulture is imperative to safeguard the environment and human health.

3. BIOTECHNOLOGY AND INSECT PEST MANAGEMENT

A durable and eco-friendly alternative for insect pest management in horticulture is to genetically transform the elite genotypes and breeding lines of vegetable crops using genes that encode insecticidal proteins. Such insecticidal proteins should be innocuous to all other organisms except the target pests and can be taken from any biological source. The transgenic plants carrying foreign genes for insect resistance have

Table 1. Major insect pests of vegetable crops of India

Crop species	Insect Pest	
Cabbage/Cauliflower	Diamondback Moth	(*Plutella xylostella*)
	Webworm	(*Hellula undalis*)
	Hairy Caterpillar	(*Spilosoma obliqua*)
Tomato	Fruit borer	(*Helicoverpa armigera*)
	Epilachna beetle	(*Epilachna sps.*,)
	Tobacco caterpillar	(*Spodoptera litura*)
Brinjal	Shoot and fruit borer	(*Leucinodes orbonalis*)
	Epilachna beetle	(*Epilachna sps.*,)
	Lacewing bug	(*Urentius echinus*)
	Jassids	
Chilli	Thrips	
	Aphids	(*Aphis gossypi*)
Potato	Aphid	(*Myzus persicae*)
	Tubermoth	(*Phthorimaea operculella*)
Pea	Pea aphid	(*Macrosiphum pisi*)
	Podborer	(*Helicoverpa armigera*)
	Pea weevil	(*Bruchus pisorum*)
Onion	Thrips	(*Thrips tabaci*)
	Headborer	(*Helicoverpa armigera*)
Okra	Spotted Bollworm	(*Earias sps.*,)
	Jassid	
Spinach	Aphids	
Cucurbits	Red pumpkin beetle	(*Aulacophora sps.,*)
	Fruitfly	(*Dacus cucurbitae*)
	Aphids	

many advantages which include decreased pesticide usage, environment-friendly nature, decreased input costs to the farmer, season-long protection independent of weather conditions, effective control of burrowing insects difficult to reach with sprays and control at all stages of insect development (Kumar *et al.*, 1996). Nature has in its bounty many insecticidal molecules and proteins which can be safely expressed in transgenic plants for insect control. Some examples are protease inhibitors, alpha-amylase inhibitors, lectins, enzymes such as chitinase and peroxidase and delta-endotoxins of *Bacillus thuringiensis* (Bt). A list of the insecticidal protein genes and transgenic horticultural crops expressing these genes is given in Table 2. A brief description of these insecticidal proteins is given below.

Table 2. Transgenic vegetable crops engineered for insect resistance

	Proteinase inhibitors	
1.	Soybean serine-proteinase inhibitor	Potato
2.	Cowpea trypsin inhibitor	Potato, Sweet potato, Tomato
3.	Tomato proteinase inhibitor I	Tomato
4.	Tomato proteinase inhibitor II	Tomato
	Amylase inhibitors	
1.	Amylase inhibitor of the common bean	Pea, Potato
	Lectins	
1.	Snowdrop lectin	Potato, Sweet potato, Tomato
2.	Pea lectin	Potato
	Others	
1.	Bean chitinase	Potato
2.	Tobacco anionic peroxidase	Tomato

Ref : Schuler *et al.* (1998).

3.1. Insecticidal proteins of *Bacillus thuringiensis* (Bt)

The leading biorational pesticide, *Bacillus thuringiensis* (Bt), is a ubiquitous gram-positive, spore-forming bacterium that forms a parasporal crystal during the stationary phase of its growth cycle. Bt has a very wide insecticidal spectrum ranging from Lepidoptera, Diptera, Coleoptera, Hymenoptera, Homoptera, Orthoptera, Mallophaga and extending up to nematodes, mites and protozoa (de Maagd *et al.*, 2000; Kumar *et al.*, 1996; Schnepf *et al.*, 1998). Bt produces two different types of insecticidal proteins which are agronomically important, the most widely known one being called delta-endotoxin or insecticidal crystal protein (ICP).

The ICPs usually exist as protoxins of high molecular weight (135-

138 kD). Upon ingestion by insect larvae these are proteolytically processed into small molecular weight toxins in the highly alkaline midgut of the larvae. The toxin molecules bind to specific receptors present in the membranes of midgut epithelium and cause pore formation. This causes disruption of the electrical, K+ and pH gradients across the membrane leading to the death of the larvae (Kumar *et al.*, 1996). The presence of specific receptors in the insects and insect classes is what determines the lack of activity of Bt delta-endotoxins towards mammals and other organisms including beneficial insects. The advent of molecular biology and plant genetic engineering has facilitated stable expression of Bt genes in crop plants with considerable success (Schuler *et al.*, 1998). Bt-transgenic vegetable crops are dealt with in the section 4.

3.2. Protease inhibitors

Currently, there are two major groups of plant-derived genes used to confer insect resistance on crops: inhibitors of insect's digestive enzymes (proteinase and amylase inhibitors) and lectins. Plant protease/proteinase inhibitors are polypeptides or proteins that occur naturally in a wide range of plants and are a part of plant's natural defense system against herbivory. Fourteen different plant proteinase-inhibitor genes have been introduced into crop plants. Table 3 gives a list of the vegetable crop species transformed with different proteinase inhibitor genes. The most active inhibitor identified to date is the cowpea trypsin inhibitor (CpTi),

Table 3. Bt-Transgenic vegetable crops.

	Bt ICP	Crop	Reference
1.	Cry1Aa	Rutabaga	Li *et al.*, 1995
2.	Cry1Ab	Potato	Jansens *et al.*, 1995
		Tomato	Delannay *et al.*, 1989
		Brinjal	Kumar *et al.*, 1998
		Potato	Chakrabarti *et al.*, 2000
3.	Cry1B	Brinjal	Kaushik *et al.*, 2002
4.	Cry1Ac	Broccoli	Metz *et al.*, 1995
		Cabbage	Metz *et al.*, 1995
		Tomato	Mandaokar *et al.*, 2000
5.	Cry1C	Broccoli	Cao *et al.*,1999
		Chinese cabbage	Cho *et al.*, 2001
6.	Cry3Aa	Eggplant	Jelenkovic *et al.*, 1997
		Potato	Perlak *et al.*, 1993

which has been transferred into ten different crop species. Experiments with transgenic plants and artificial diets have shown that CpTi affects a wide range of lepidopteran and coleopteran species (Gatehouse and Hilder, 1994). The serine-proteinase inhibitors (from soybean) when expressed in transgenic tobacco and potato resulted in considerable larval mortality of *Spodoptera littoralis.* In addition to serine-proteinase inhibitors, one cysteine-proteinase inhibitor from rice has been introduced into several other crops. To date, no crop expressing proteinase inhibitor transgenes has been commercialized.

3.3. Amylase inhibitors

Inhibitors of alpha-amylases are the second type of enzyme inhibitors used to modify crop plants. Genes for three alpha amylase inhibitors have been expressed in tobacco, pea and Azuki bean (Table 3). Alpha-amylase inhibitor from the common bean (*Phaseolus vulgaris*) forms a complex with and inhibits alpha-amylases in the midgut of coleopteran and storage pests of the genera *Callosobruchus* and *Bruchus* and blocks larval development (Ishimoto *et al.*, 1996).

3.4. Lectins

Lectins are carbohydrate-binding proteins, some of which are toxic to insects. Various lectins have shown some toxicity against species of the insect orders, Homoptera, Coleoptera, Lepidoptera and Diptera (Czapla, 1997). The mode of action of lectins against insects remains unclear, but it has been shown that at least some bind to midgut epithelial cells (Gatehouse and Gatehouse, 1998). However, some insecticidal lectins also show significant mammalian toxicity, including lectins from *P. vulgaris*, winged bean, soybean and wheat germ. Other lectins, for example those from pea and snowdrop have demonstrated insecticidal activity and are inoccuous to mammals (Gatehouse and Hilder, 1994).

Lectin from snowdrop (*Galanthus nivalis*) has been shown to be very effective against aphids and rice brown planthopper (Powell *et al.*, 1995). It has been expressed in nine different crops including potato and tomato (Table 3). Laboratory tests with engineered potatoes showed that snowdrop lectin did not increase the mortality or development time of potato aphid but considerably reduced fecundity (Down *et al.*, 1996). Results of experiments with the potato peach aphid were similar but in addition, the establishment of aphids on transgenic potatoes was reduced (Gatehouse *et al.*, 1996). Snowdrop lectin also enhanced the resistance of potato to larvae of tomato moth (*Lacanobia oleracea*). Gatehouse and others concluded that the effect of snowdrop lectin is antifeedant rather than insecticidal (Gatehouse *et al.*, 1997).

3.5. Other insecticidal proteins

In addition to the insecticidal proteins described above, there are examples of other toxins derived from microorganisms, plants and animals which may be valuable for plant expression.

4. BT-TRANSGENIC VEGETABLE CROPS

Although many insecticidal genes have been transferred into different crop species, the most satisfactory system so far in terms of field resistance is the one based on Bt. Bt toxins have been transferred and expressed in at least thirty different plant species (Schuler *et al.*, 1998). However, the level of resistance they confer will depend on whether native (wild-type) or truncated, modified genes have been used (Kumar *et al.*, 1996). The prokaryotic codon usage in Bt genes needs to be modified towards that of higher plant genes. In addition, features that can destabilize the transcripts in higher plant cells need to be removed. A pioneering study of this kind has been made by the scientists at Monsanto, who modified the Bt genes and expressed in cotton with considerable success (Perlak *et al.*, 1990, 1991). The performance of cotton plants transgenic for *cry1Ab or cry1Ac* against cotton bollworm and pink bollworm was very satisfactory. This was followed by the expression of a coleopteran-specific gene *cry3Aa* from Bt subspecies *tenebrionis* in transgenic potato plants (Perlak *et al.*, 1993). The plants were resistant to Colorado potato beetle (CPB; *Leptinotarsa decemlineata*) under high levels of natural field infestation. Improved plant expression of the *cry3Aa* was achieved by increasing its overall G/C content from 36% to 49%. Potato is also infested by a lepidopteran insect called Potato tuberworm (*Phthorimaea operculella*), especially in storage. Jansens *et al.*, (1995) modified a *cry1Ab* gene and expressed it in potato (Kennebec and Bintje) under the control of a TR2 promoter. Selected transgenic potato lines showed significantly less damage than the control lines in greenhouse trials. Transgenic potato expressing Cry3Aa was first commercialized in 1996 under the trade name New Leaf by Monsanto. Transgenic potatoes (Bt) are also being developed and tested by AgrEvo (PGS), Hettema ZK and Fritolay companies. Recently, scientists at Central Potato Research Institute (Shimla) developed transgenic potatoes using 5 elite genotypes and introducing a synthetic *cry1Ab* gene into them (Chakrabarti *et al.*, 2000).

Bt-transgenic tomato was first developed and tested by Monsanto in 1989 (Delannay *et al.*, 1989). Significant protection against the larvae of tobacco hornworm (*Manduca sexta*), tomato fruitworm (*Helicoverpa zea*) and tomato pinworm (*Keiferia lycopersicella*) was observed. Field trials of tomatoes carrying *cry1Ab* gene are being carried out by Monsanto, Novartis and Mycogen corporations in USA (Krattiger, 1997). Transgenic tomato carrying a synthetic *cry1Ac* gene was developed in my laboratory. Insect bioassays using the leaves and fruits of transgenic plants demonstrated complete protection against the larvae of *Helicoverpa armigera* (Mandaokar *et al.*, 1999). The transgenic plants are currently under field evaluation.

One of the major insect pests of eggplant in Europe and North America is the Colorado potato beetle. Attempts have been made to introduce a native version of *cry3B* into Blackjack variety (Chen *et al.*, 1995). However, the expression of the mRNA and Cry3B protein were not sufficient enough to kill the larvae of CPB. A mutagenized version of *cry3B* gene was developed by Rotino and his group (Iannacone *et al.*, 1995). The modified gene was introduced into the female parent of the eggplant hybrid Rimina (Arpaia *et al.*, 1997). Twentythree out of fortyfour transformants showed significant insecticidal activity on neonate larvae of CPB. In a similar study, Jelenkovic *et al.*, (1998) introduced a synthetic, fully modified gene into eggplant. A population of three hundred transformants was produced out of which 69% of the plants were found to be resistant to neonate larvae and adult CPB. In Indian sub-continent and South Asia, Eggplant is infested by a lepidopteran insect namely Brinjal Shoot and Fruit Borer. A synthetic *cry1Ab* gene modified for plant codon usage has been introduced in eggplant (Cultivar Pusa Purple Long) in my laboratory (Kumar *et al.*, 1998). Fruits of transgenic plants have been found to be totally protected from the larval attack. Limited field trials of the transgenic lines over a period of three years showed 70% resistance to BSFB.

Cruciferous vegetables such as cauliflower, cabbage and broccoli face a host of insect pests all over the world, the most serious of which is Diamondback moth (DBM). DBM has developed resistance to virtually all insecticides in many countries. Earle and coworkers at Cornell University have developed transgenic cabbage and broccoli by introducing a synthetic *cry1Ac* gene (Metz *et al.*, 1995a, b). The insect-resistant broccoli plants have uniform levels of toxin expression in leaves, based on the comparisons of insect mortality on leaves taken from the crown, middle and lower portions of the plants. Rutabaga (*Brassica napobrassica*) is a popular cruciferous vegetable in China. Efforts have been made to introduce a *cry1Aa* gene to confer resistance to cabbage caterpillar (*Pieris rapae*). Mortality of the larvae was complete on some of the transformants (Li *et al.*, 1995).

5. RESISTANCE MANAGEMENT

The fundamental purpose of the deployment of resistance genes in transgenic plants is to manage the insect pest population and to prevent the development of resistance in insects. Resistance management strategies try to prevent or diminish the selection of rare individuals carrying resistance genes and hence to keep the frequency of resistance genes sufficiently low for insect control. Strategy development generally relies on theoretical assumptions and on computer models simulating insect population growth under various conditions (Tabashnik, 1994). Proposed strategies include the use of multiple toxins, crop rotation, high or ultra high dosages, and spatial or temporal refugia. The most promising and currently practical strategy is that of using refugia. The strategy would reduce the possibility of resistant insects from mating with other resistant insects, thereby preventing the creation of a resistant population. This is achieved by ensuring that there are always plenty of susceptible insects nearby for the few resistant ones to mate with. The basic principle of high dose strategy is to deploy plants with high levels of expression of the toxin with the expectation that it would take a long time for insects to overcome the toxin. It assumes that most or all resistance is recessive and that most resistance carriers would be heterozygous. A viable complementary strategy that is best adopted with the above two strategies is the deployment of multiple resistance or pyramiding of resistance genes. This strategy requires more than one resistant gene with different modes of action. It could be achieved either with additional *cry / vip* genes or with novel methods of insect resistance, but requires the use of refugia (Gould, 1997). Targeted expression is also complementary to the above-described strategies and will become possible in the near future. A toxin gene is expressed only specifically in a certain vulnerable tissue / part of the plant or is expressed both in a certain part of the plant as well as at a particularly critical time in the development of the plant. This strategy would allow plenty of susceptible insects to breed normally, thus increasing their predator and parasitic populations, while at the same time be prevented from causing damage in the critical plant parts or life cycles.

One of the most important tools of resistance management is to apply integrated pest management (IPM) principles is transgenic crop cultivation. Use of biological control methods (predators, viruses, fungi etc.,), botanical pesticides (Neem and Pyrethrum), crop rotation and sanitation, traditional methods coupled with minimal application of chemical insecticides will prolong the "life" of transgenic technology.

6. CONCLUSIONS AND FUTURE PERSPECTIVES

Vegetable crops suffer from a variety of insect pests. It is possible to implement biotechnological approaches to manage insect pests in a rational, durable and eco-friendly manner. Insecticidal proteins / genes of Bt will be the backbone of transgenic crop cultivation in the 21st century. However, novel insecticidal proteins and respective genes need to be identified and used in conjunction with Bt to prevent the development of resistant insects. In addition, integrated pest management will have to play a central role in sustainable horticulture. Synthetic insecticides will continue to have a minimal role in combating minor pests and those that cannot be managed by biotechnological tools. Some of the horticultural crops are recalcitrant in tissue culture and it may be difficult to develop genetic transformation procedures for such species. Bt and other biocides in combination with IPM measures can reduce the consumption of chemical pesticides in such instances.

REFERENCES

Arpaia S, Mennella G, Onofaro V, Perri E, Sunseri F and Rotino GL (1997) Production of transgenic eggplant (*Solanum melongena* L.) resistant to Colorado potato beetle (*Leptinotarsa decemlineata* Say). *Theor. Appl. Genet.*, **95** : 329-334

Bose TK, Som MG and Kabir J (1993) Vegetable Crops. Naya Prokash, Calcutta

Boulter D (1993) Insect pest control by copying nature using genetically engineered crops. *Phytochem.*, **34** : 1453-1466

Cao J, Tang JD, Strizhov N, Shelton AM and Earle ED (1999) Transgenic broccoli with high levels of *Bacillus thuringiensis* Cry1C protein control diamondback moth larvae resistant to Cry1A or Cry1C. *Mol. Breed.*, **5** : 131-141.

Chakrabarti SK, Mandaokar A, Pattanayak D, Shukla A, Naik PS, Sharma RP and Kumar PA (2000) *Bacillus thuringiensis cry1Ab* gene confers resistance to potato against *Helicoverpa armigera* Hubner. *Potato Res.,* **42** : 227-238.

Chen Q, Jelenkovic G, Chin CK, Billings S, Eberhardt J, Goffreda JC & Day P (1995) Transfer and transcriptional expression of coleopteran *cryIIIB* endotoxin gene of *Bacillus thuringiensis* in eggplant. *J. Amer. Soc. Hort. Sci.*, **120** : 921-927.

Cho HS, Cao J, Ren JP and Earle ED (2001) Control of lepidopteran insect pests in transgenic Chinese cabbage (*Brassica rapa* ssp. *Pekinensis*) transformed with a synthetic *Bacillus thuringiensis cry1C* gene. *Plant Cell Rep.*, **20** : 1-7.

Choudhary B (1979) *Vegetables.* National Book Trust, New Delhi.

Czapla TM (1997) Plant lectins as insect control proteins in transgenic plants. In: *Advances in Insect Control: The Role of Transgenic Plants* (Eds Carozzi N and Koziel M), Taylor and Francis, London. pp. 123-138.

Dale PJ, Irwin JA and Scheffler JA (1993) The experimental and commercial release of transgenic crop plants. *Plant Breed.*, **111** : 1-22.

Delannay X, Lavallee BJ, Proksch RK, Fuchs RL, Sims SR, Greenplate JT, Morrone PG, Dodson RB, Augustine JJ, Layton JG and Fischhoff DA (1989) Field performance of transgenic tomato plants expressing the *Bacillus thuringiensis* var. *kurstaki* insect control protein. *Biotechnol,* **7** : 1265-1269.

De Maagd RA, Bosch D and Stiekema W (1999) *Bacillus thuringiensis* toxin-mediated insect resistance in plants. *Trends Plant Sci*, **4** : 9-13.

Down RE, Gatehouse AMR, Hamilton WDO and Gatehouse JA (1996) Snowdrop lectin inhibits development and decreases fecundity of the glasshouse potato aphid (*Aulacorthum solani*) when administered *in vitro* and via transgenic plants both in laboratory and glass house trials. *J. Insect Physiol.*, **42** : 1035-1045

Dowd PF and Lagrimini LM (1997) Role of Transgenic Plants. In: *Advances in Insect Control* (Eds Carozzi N and Koziel M), Taylor and Francis, London. pp. 195-223.

Gatehouse AMR, Down RE, Powell KS, Sauvion N, Rahbe Y, Newell CA, Merryweather A, Hamilton WDO and Gatehouse JA (1996) Transgenic potato plants with enhanced resistance to the peach-potato aphid *Myzus persicae*. *Entomol. Exp. Appl.*, **79** : 295-307

Gatehouse AMR, Davison GM, Newell CA, Marryweather A, Hamilton, WDO, Burgess EPJ, Gilbert RJC and Gatehouse JA (1997) Transgenic potato plants with enhanced resistance to the tomato moth, *Lacanobia oleracea*: growth room trials. *Mol. Breed.*, **3** : 49-63

Gatehouse AMR & Hilder VA (1994) Genetic manipulation of crops for insect resistance. In: *Molecular Perspectives: Crop Protection* (Eds. Marshall G and Walters D), Chapman and Hall, London. pp 177-201.

Gould F (1997) Integrating pesticidal engineered crops into mesoamerican agriculture. In: *Transgenic Plants: Bacillus thuringiensis in Mesoamerican Agriculture* (Eds. Hruska AJ and Pavon ML), Zamorono, Honduras, pp 6-36,

Iannacone R, Flore MC, Macchi A, Grieco PD, Arpaia S, Perrone D, Mennella G, Sunseri F, Cellini F and Rotino GL (1995) Genetic engineering of eggplant (*Solanum melongena* L.). *Acta. Hortic.*, **392** : 227-233

Ishimoto M, Sato T, Chrispeels MJ and Kitamura K (1996) Bruchid resistance of transgenic azuki bean expressing seed amylase inhibitor of common bean. *Entomol. Exp. Appl.*, **79** : 309-315.

Jansens S, Cornelissen M, De Clercq R, Reynaerts A and Peferoen M (1995) *Phthorimaea operculella* (Lepidoptera: Gelechiidae) resistance in potato by expression of the *Bacillus thuringiensis* Cry1Ab insecticidal crystal protein. *J. Econ. Entomol.*, **88** : 1469-1476.

Jelenkovic J, Billings S, Chen Q, Lashomb J, Hamilton G and Ghidiu G (1998) Transformation of eggplant with synthetic *cryIIIA* gene produces a high level of resistance to the Colorado potato beetle. *J. Amer. Soc. Hort. Sci.*, **123** : 19-25.

Krattiger AF (1997) *Insect Resistance in Crops: A Case Study of Bacillus thuringiensis* ISAAA: Ithaca, NY.

Kumar PA, Sharma RP and Malik VS (1996) Insecticidal proteins of *Bacillus thuringiensis*. *Adv. Appl. Microbiol.*, **42** : 1-43.

Kumar PA, Mandaokar A, Sreenivasu K, Chakrabarti SK, Bisaria S, Sharma SR, Kaur S and Sharma RP (1998) Insect-resistant transgenic brinjal plants. *Mol. Breed.*, **4** : 33-37.

Kumar PA, Mandaokar A and Sharma RP (1998b) Genetic engineering for the improvement of eggplant (*Solanum melongena*). *Ag. Biotech. News and Info.*, **10** : 329-332.

Li N, Mao HZ and Bai YY (1995) Transgenic plants of rutabaga (*Brassica napobrassica*) tolerant to insect pests. *Plant Cell Rep.*, **15** : 97-102.

Mandaokar A, Goyal RK, Shukla A, Bhalla R, Chaurasia A, Reddy VS, Altosaar I, Sharma RP and Kumar PA (2000). Transgenic tomato plants resistant to fruitborer (*Helicoverpa armigera* Hubner). *Crop Protection,* **19** : 307-312.

Metz TD, Dixit R and Earle ED (1995a) *Agrobaterium*-mediated transformation of broccoli (*Brassica oleracea* var. *italica*) and cabbage (*B. oleracea* var. *capitata*). *Plant Cell Rep.*, **15** : 287-292.

Metz TD, Roush RT, Tang JD, Shelton AM and Earle E (1995b) Transgenic broccoli expressing a *Bacillus thuringiensis* insecticidal crystal protein: implications for pest resistance management strategies. *Mol. Breed.*, **1** : 309-317.

Perlak FJ, Deaton RW, Armstrong TA, Fuchs RL, Sims SR, Greenplate JT & Fishhoff DA (1990) Insect resistant cotton plants. *Biotechnology*, **8** : 939-943.

Perlak FJ, Fuchs RL, Dean DA, McPherson S and Fischhoff DA (1991) Modification of the coding sequence enhances plant expression of insect control protein genes. *Proc. Natl. Acad. Sci. USA,* **88** : 3324-3328

Perlak FJ, Stone TB, Muskopf YM, Petersen LJ, Parker GB, McPherson SA, Wyman J, Love S, Reed G, Biever D and Fischhoff DA (1993) Genetically engineered potatoes: protection from damage by Colorado potato beetles. *Plant Mol. Biol.*, **22** : 313-321

Powell KS, Gatehouse AMR, Hilder VA and Gatehouse JA (1995) Antifeedant effect of plant lectins and an enzyme on the adult stage of the rice brown planthopper. *Nilaparvata lugens. Entomol. Exp. Appl.*, **75** : 51-59

Schnepf E, Crickmore N, Van Rie J, Lereclus D, Baum J, Feitelson J, Zeigler DR and Dean DH (1998) *Bacillus thuringiensis* and its pesticidal crystal proteins. *Microbiol. Mol. Biol. Rev.*, **62** : 775-806

Schuler TH, Poppy GM, Kerry BR and Denholm I (1998) Insect-resistant transgenic plants. *TIBTECH,* **16** : 168-175

Tabashnik BE (1994) Evolution of resistance to *Bacillus thuringiensis*. *Annu. Rev. Entomol.*, **39** : 47-79

Thomas JC, Wasmann CC, Echt C, Dunn RL, Bohnert HJ and McCoy TJ (1994) Introduction and expression of an insect proteinase inhibitor in alfalfa (*Medicago sativa* L.). *Plant Cell Rep.*, **14** : 34-36.

Chapter 3

TRANSGENIC TOMATO

Wagner C Otoni[1]*, Edgard AT Picoli[1], Marcio GC Costa[1], Fábio TS Nogueira[1], and Francisco M Zerbini[2]

[1]*Departamento de Biologia Vegetal/BIOAGRO,*
[2]*Departamento de Fitopatologia/BIOAGRO, Universidade Federal de Viçosa, 36571-000, Viçosa, MG, Brazil*

Summary

Plant genetic engineering has experienced great and important achievements along the last 15-20 years. Tomato has been one of the major vegetable crops in which the amenability for in vitro responses has enabled several progresses in many areas of knowledge. In tomato, genetic transformation techniques have been used for a wide array of purposes, from protocol optimization to gene identification, isolation, and further structural and functional characterization and expression. Due to its worldwide importance, future efforts will most likely be directed to the improvement of nutritional quality, characterization of novel disease and insect resistance genes, and their incorporation in elite cultivars. In the present review we will focus on past and recent advances on tomato biotechnology with emphasis on genetic transformation.

Keywords : Genetic transformation, *Lycopersicon* sp., tomato, tissue culture.

Abbreviations : BDMV, *Bean dwarf mosaic virus*; BiP, Binding protein; CaMV, *Cauliflower mosaic virus*; CMV, *Cucumber mosaic virus*; GRSV, *Groundnut ring spot virus*; *ipt*, isopentenyltransferase; MJ, methyl

*Corresponding author : E-mail : wotoni@mail.ufv.br
Plant Biology Department, University Campus, Federal University of Viçosa, 36571-000, Viçosa, Minas Gerais, Brazil.

jasmonate; PE, pectin esterase; PG, polygalacturonase; PDR, pathogen-derived resistance; PTGS, post-transcriptional gene silencing; SPS, sucrose phosphate synthase; PhMV, *Physalis mottle virus*; TAV, *Tomato aspermy virus*; TCSV, *Tomato chlorotic spot virus*; TMV, *Tobacco mosaic virus*; TGMV, *Tomato golden mosaic virus*; ToMV, *Tomato mosaic virus*; TSWV, *Tomato spotted wilt virus*; TYLCV, *Tomato yellow leaf curl virus*; ToMoV, *Tomato mottle virus*, NBS-LRR, nucleotide binding site-leucine rich repeats; MP, movement protein; SPS, sucrose phosphate synthase.

1. INTRODUCTION

It has been nearly fifteen years since the first transgenic tomato plant was regenerated. Since then a tremendous scientific progress in tomato genetic engineering has been noticed. Tomato is a worldwide-consumed vegetable crop, for which successful applications of *in vitro* regeneration and genetic transformation have already been implemented for genetic improvement, and is considered a favorable experimental material for tissue culture. This favorable situation enabled the widening of different biotechnological approaches in tomato. Although other methods for introducing foreign DNA into tomato have been proposed and effectively used, none has proven to be as popular as the *Agrobacterium*-mediated method (Frary, 1995). The popularity of this particular technique might be best explained by the fact that all the other methods have distinct disadvantages that make them less practical and less effective for the production of stable transformed plants.

A number of important phenotypes have been obtained in transgenic tomato. Many of the phenotypes improve agronomic traits by conferring herbicide tolerance, disease resistance, insect resistance, and superior quality traits in the harvested product. Also a number of genes that are development-, wounding- or stress-related have been examined in transgenic tomato (Houck *et al.*, 1993). Despite the great importance of tomato biotechnology, there is a lack of literature reviews on transgenic tomatoes, especially on the molecular biology of ripening-related processes. One excellent review is the one by Houck *et al.* (1993), which discusses aspects related to transgenic solanaceous plants including tomato. We found several thesis works on the subject: Latif (1993), Patil (1994), Frary (1995), Gisbert (1997), Costa (1999) and Nogueira (2000), among others, but with limited access by the scientific community worldwide.

2. THE ROLE OF TISSUE CULTURE IN GENETIC MANIPULATION OF TOMATO

Since White's pioneer work (White, 1934) several new possibilities were raised in tomato tissue culture. Long time ago, tissue culture techniques such as embryo culture had already made possible the rescue of *Lycopersicon esculentum* and *L. peruvianum* hybrids, which are cross-sterile, enabling the development of normal and vigorous plants from these embryos (Smith, 1944). *In vitro* regeneration of tomato has been a subject of consistent research because of the commercial value of the crop and the potential for further improvement via genetic manipulation (Evans, 1989).

Tomato *in vitro* regeneration was established by Padmanabhan *et al.* (1974). Since then, other protocols involving regeneration were developed (Kut *et al.*, 1984; Sink and Reynolds, 1986; Wijbrandi and De Both, 1993; McCormick, 1991; Costa *et al.*, 2000c; Fári *et al.*, 2000). As a result, numerous studies on plant regeneration from a wide variety of tissues and organs of wild and cultivated tomato germplams have been conducted either through organogenesis (Ohki *et al.*, 1978; Locy, 1983; Zelcer *et al.*, 1984; Kartha *et al.*, 1976; Tan *et al.*, 1987; Uddin *et al.*, 1988; Branca *et al.*, 1991; Koornneef *et al.*, 1993; Düzyaman *et al.*, 1994; Ichimura *et al.*, 1995; Newman *et al.*, 1996; Pratta *et al.*, 1997; Costa, 1999; Costa *et al.*, 2000a, c; Fári *et al.*, 2000; Nogueira, 2000) or in a small scale via somatic embryogenesis (Young *et al.*, 1987; Lamproye *et al.*, 1990; Chen and Adachi, 1994; Gill *et al.*, 1995; Newman *et al.*, 1996; Chen and Adachi, 1998). The primary mode of regeneration in tomato is via shoot organogenesis from calli that originate following dedifferentiation of leaf or cotyledon explants, or directly from inflorescence-derived thin layers (Compton and Veilleux, 1991).

For tomato, in terms of basic regeneration studies, several different cell and tissue culture approaches have been used such as protoplast isolation and somatic hybridization (Adams *et al.*, 1985; Niedz *et al.*, 1985; Kinsara *et al.*, 1986; Tan *et al.*, 1987a; Tan *et al.*, 1987b; Wijbrandi *et al.*, 1988; Wijbrandi *et al.*, 1990; Latif *et al.*, 1993; Patil, 1994; Chen and Adachi, 1998; Kochevenko *et al.*, 2000), embryo culture (Patil *et al.*, 1993; Chen and Adachi, 1994), somaclonal variation (Gavazzi *et al.*, 1987; Uddin *et al.*, 1988; Koornneef *et al.*, 1989; Hanus-Fajerska *et al.*, 2000), and androgenesis (Gresshoff and Doy, 1972; Chlyah *et al.*, 1990; Summers *et al.*, 1992; Shtereva *et al.*, 1998; Zagorska *et al.*, 1998). The applications of tomato cell and tissue culture have been reviewed elsewhere (Sink and Reynolds, 1986; Koblitz, 1991). Presently, tomato *in vitro* morphogenesis coupled with genetic manipulation processes is the current goal of tomato tissue culture.

2.1. Regeneration capacity is under genetic control

Studies on the genetics of regeneration in tomatoes have demonstrated that this trait is highly heritable (Frankenberger *et al.*, 1981) with a dominance effect (Ohki *et al.*, 1978; Adams and Quiros, 1985; Wijbrandi *et al.*, 1988). The importance of the genetic background on morphogenesis may be visualized on the screening for the regeneration ability of fifteen tomato lines performed by McCormick *et al.* (1986). Regardless of the medium, it was observed that the lines had in general distinct regeneration abilities.

Regeneration dependence on genotype and inheritance of regeneration ability in tomato has been studied by several workers (Tal *et al.*, 1977; Kurtz and Lineberger, 1983; Tan *et al.*, 1987; Uddin *et al.*, 1988; Koornneef *et al.*, 1993; Zelcer *et al.*, 1984). According to Koornneef *et al.* (1993), the genetic component associated with regeneration determines not only the tissue sensibility to growth regulators but also the morphogenic competence. Pratta *et al.* (1997) detected significant differences in the regeneration frequency and number of shoots per explant among species and genotypes of *Lycopersicon,* even though genotypes of different species displayed the same morphogenic response *in vitro*.

Based on reciprocal crosses in parental lines and analysis of the progenies regarding regeneration capacity, Ohki *et al.* (1978) pointed out the importance of the cytoplasmic inheritance effect, suggesting a heterotic effect, and also that hybrid performance was influenced by media and explant effects. Frankenberger *et al.* (1981) demonstrated genotype influence on regeneration, as well as effects of explant type and source, time of the year in which the source plant was grown, and the environment in which the donor plant was grown. The authors examined heritability of shoot-forming capacity of six genotypes. There was significant general combining ability (CGA) of the parental lines suggesting that crosses between high regenerative lines would result in progeny with high shoot-forming capacities. Recessive gene effects on regeneration capacity and additive gene action did also account for a large part of the shoot forming variation between tomato F_1s in a diallel cross.

Segregation analysis revealed that the favorable cell culture traits and regeneration capacity of *L. peruvianum* were dominant and controlled by two genes. A gene controlling shoot regeneration was further mapped on chromosome 3 of *L. peruvianum* (Koornneef *et al.*, 1987a, 1993). Interestingly, an especial tomato lineage, MsK93, with marked regenerability, was bred by Koornneef *et al.* (1986), resulting from a backcross of hybrids of *L. peruvianum* × *L. esculentum* with *L. esculentum* cv. VF11. Genotypes MKS8 and MKS9 were selected from a further backcross of MKS93 with VF11. These genotypes were selected based on their efficient shoot regeneration either from protoplasts or established callus cultures. This material proved to be extremely important for the success of genetic transformation experiments, including those involving direct DNA uptake from protoplasts or *Agrobacterium*-mediated leaf-disk transformation, thus reinforcing the importance of the genotype for genetic manipulation.

Studying the inheritance of *in vitro* plant regeneration ability in tomato, Faria and Illg (1996) used *Lycopersicon pimpinellifolium* (genotype WV-700), with high regeneration ability, as a pollen donor, and crossed it with five recalcitrant *L. esculentum* genotypes. Statistical analysis was based upon the frequency of calli that regenerated plants (R) in F_1, $F_{2,}$ BC_1 and on reciprocal crosses with genotypes that did not regenerate plants (NR). F_1 hybrids showed a dominance effect, and no differences were observed in reciprocal crosses. The F_2 rate was close to 9:7 (R:NR) and BC_1 expressed a 1:3 (R:NR) ratio. The results suggested an interaction of two dominant genes, indicating that the transfer of high regeneration ability from wild to commercial varieties would be relatively easy, since a qualitatively inherited trait is involved.

Tomato anthocyaninless mutants that have increased germination potential under optimal and large scale stress conditions (Shtereva *et al.*, 2000) did not present significant differences in hypocotyl and root lengths growth under 100 mM NaCl treatment when compared to the wild-type line. Conversely, significant differences in callus growth and number of shoots were detected, depending on the media. The anthocyaninless mutants ah- (Hoffmann's anthocyaninless) and *aw*- (anthocyanin without), and in some cases *bls*- (baby lea syndrome), exhibited higher callogenic and shoot-forming capacity. Although more detailed studies are necessary to evaluate the effect of these three genes in plant development, the results suggest the influence of specific genes in tomato morphogenesis.

Bearing in mind the marked genotype effect, the use of material with proved *in vitro* regeneration ability should be taken into account, extending the chances of increased *in vitro* regeneration of plants and production of transgenic plants with desired traits.

3. GENETIC TRANSFORMATION IN TOMATO

Horsch *et al.* (1985) and McCormick *et al.* (1986) were the first to report, in independent works, the genetic transformation of tomato. Following these first reports, it was further achieved by using different methodologies such as *Agrobacterium tumefaciens* (Fillatti *et al.*, 1987; Chyi and Phillips, 1987; Koornneef *et al.*, 1987b; Bird *et al.*, 1988), *Agrobacterium rhizogenes* (Morgan *et al.*, 1987; Sukhapinda *et al.*, 1987; Urwin *et al.*, 1995; Hashimoto *et al.*, 1999), microinjection (Toyoda *et al.*, 1988), electroporation (Nakata *et al.*, 1992), PEG-mediated transformation in protoplasts (Koornneef *et al.*, 1986), and biolistics (Xue *et al.*, 1994; Van Eck *et al.*, 1995; Baum *et al.*, 1997; Meier *et al.*, 1997).

Agrobacterium-mediated transformation of tomato has been used for a wide array of purposes: to optimize protocols (Fillati *et al.*, 1987; Chyi and Phillips, 1987; Hamza and Chupeau, 1993; Lipp João and Brown, 1993; van Roekel *et al.*, 1993; Frary, 1995; Frary and Earle, 1996; Ling *et al.*, 1998; Costa, 1999; Costa *et al.*, 2000b; Nogueira, 2000; Hu and Phillips, 2001), to characterize the roles of genes involved in plant growth and development (Martineau *et al.*, 1994; Ohyama *et al.*, 1995; Janssen *et al.*, 1998), to analyze mutants (Goldsbrough *et al.*, 1993; Groot *et al.*, 1995), for tomato improvement (Nelson *et al.*, 1988; Xue *et al.*, 1994; Martineau *et al.*, 1995; Jongedijk *et al.*, 1995; Rhim *et al.*, 1995; Ultzen *et al.*, 1995; Brunetti *et al.*, 1997; Costa *et al.*, 1998; Rhim, 1998a, b; Stommel *et al.*, 1998; Tomassoli *et al.*, 1999; Costa *et al.*, 2000b; Mandaokar *et al.*, 2000), in physiological studies (Klee *et al.*, 1991; Worrell *et al.*, 1991; Oeller *et al.*, 1991; Peñarrubia *et al.*, 1992; Klee, 1993; Gray *et al.*, 1994; Watson *et al.*, 1994; Galtier *et al.*, 1995; Micallef *et al.*, 1995; Ohyama *et al.*, 1995; Picton *et al.*, 1995a; Picton *et al.*, 1995b; Klann *et al.*, 1996; Beaudoin and Rothstein, 1997; Laporte *et al.*, 1997; Aloni *et al.*, 1998; Arrilaga *et al.*, 1998; Kuntz *et al.*, 1998; Brüggemann *et al.*, 1999; Mansouri *et al.*, 1999; Murchie *et al.*, 1999; Simons and Tucker, 1999; Griffiths *et al.*, 1999; Gisbert *et al.*, 2000;

Grichko *et al.*, 2000; Hong *et al.*, 2000), and for the isolation of disease resistance and other genes by map-based cloning (Martin *et al.*, 1993; Jones *et al.*, 1994; Martin *et al.*, 1994; Brommonschenkel, 1996; Brommonschenkel and Tanksley, 1997; Brommonschenkel *et al.*, 1998; Ling *et al.*, 1999; Brommonschenkel *et al.*, 2000). A general procedure adopted in our laboratory is described in Figure 1, which is used in transformation experiments for processing and fresh market tomato cultivars (Costa, 1999; Costa *et al.*, 2000b; Nogueira, 2000).

However, tomato is still considered more difficult to transform than other solanaceous species like *Petunia hybrida* and *Nicotiana tabacum,* and can show widely discrepant success rates can be obtained, possibly depending on the cultivar, *Agrobacterium* strain, antibiotic selection, and/or the technical expertise of personnel (McCormick, 1991). Table 1 reviews in chronological order the different tomato species that have been transformed and regenerated. As already stated by Frary (1995), a review of the past and recent literature along with a significant number of publications illustrates the importance of *Agrobacterium*-mediated transformation in developmental, molecular, genetic and breeding work with cultivated tomato.

3.1. Factors affecting transformation

A prerequisite for transferring genes into plants is the availability of efficient transformation and regeneration systems. Although the transfer of genes into tomato by means of *Agrobacterium*-mediated transformation systems has been reported previously, it is far from being routine. Moreover, transformation efficiency varies within and between species and can be very low (Frary and Earle, 1996). Since 1986, several reports have been published describing the use of *A. tumefaciens*-mediated transformation of the cultivated tomato (*L. esculentum* Mill.) (Hamza and Chupeau, 1993; van Roekel *et al.*, 1993; Frary and Earle, 1996; Nogueira *et al.*, 1998 a,b; Brüggemann *et al.*, 1999; Costa *et al.*, 2000b) or wild tomato species (Koornneef *et al.*, 1987b; Morgan *et al.*, 1987; Hamza and Chupeau, 1993; Agharbaoui *et al.*, 1995).

Tomato transformation may yield widely variable success rates, depending on several factors. One of the most studied factors is the genotype (McCormick *et al.*, 1986). Other factors related with the plant material have been analyzed, such as the age of the plant material (Fillatti *et al.*, 1987; Davis *et al.*, 1991a; Hamza and Chupeau, 1993; Frary and Earle, 1996), the size and position of the explants (Davis *et*

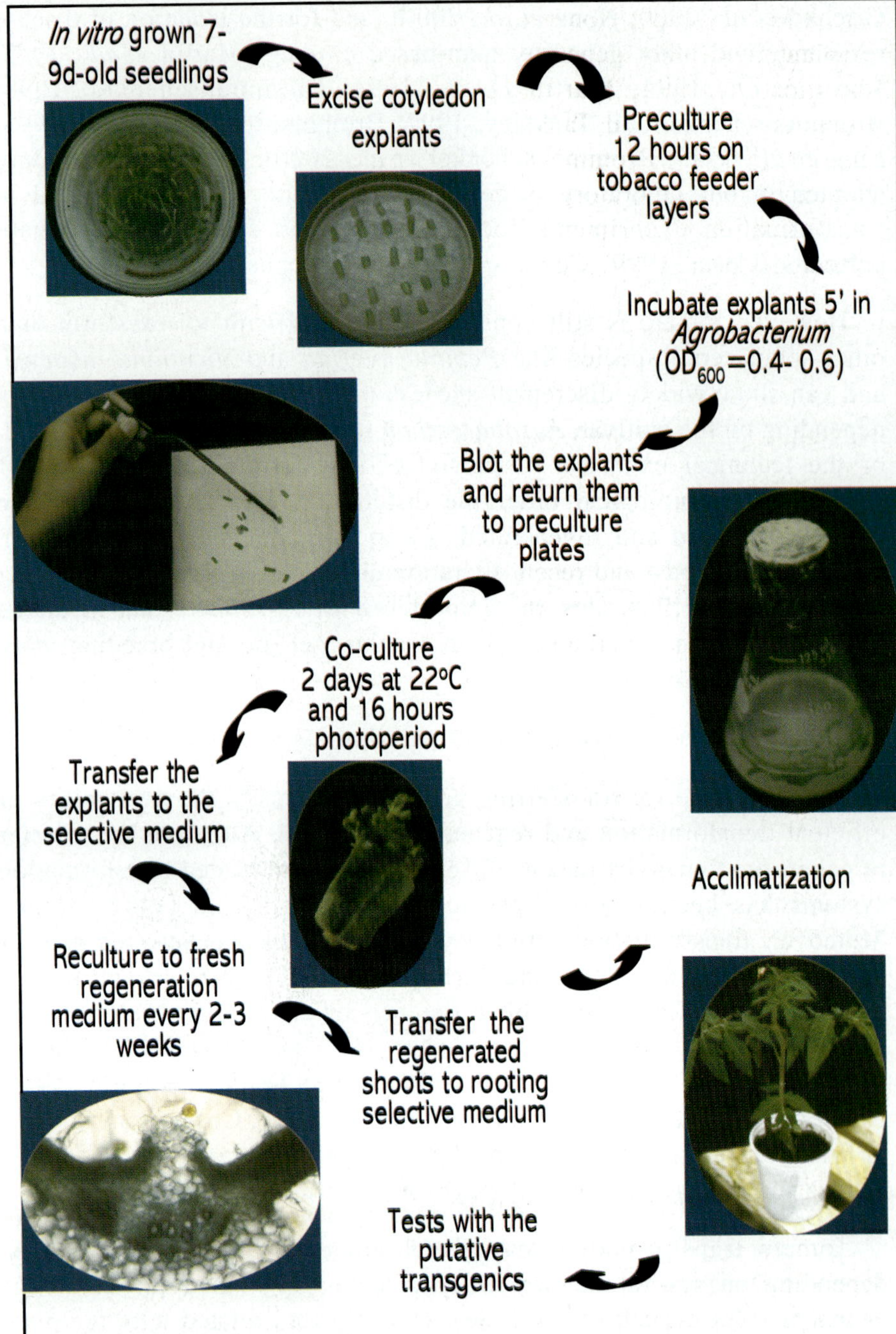

Figure 1 : Outline of the protocol used for transformation of some Brazilian processing tomato cultivars (Modified from Frary and Earle, 1996, according to Costa *et al.*, 2000b).

al., 1991b; van Roeckel *et al.*, 1993; Frary and Earle, 1996), and the explant source (Van Hoekel *et al.*, 1993; Frary and Earle, 1996). In addition, factors related with the virulence of *Agrobacterium tumefaciens* such as the concentration of the bacterial suspension used in the co-cultivation phase (Filatti *et al.*, 1987; Davis *et al.*, 1991a), presence of acetosyringone (Lipp João and Brown, 1993), co-cultivation temperature (Costa, 1999; Costa *et al.*, 2000b; Nogueira, 2000) and the presence of different antibiotics (carbenicilin, cefotaxime, timentin) in the selective culture medium for elimination of *Agrobacterium* were also investigated (Nauerby *et al.*, 1997; Ling *et al.*, 1998; Costa, 1999; Costa *et al.*, 2000a; Nogueira, 2000; Hu and Phillips, 2001). The use of different genotypes, explant sources and DNA delivery systems for tomato transformation results in a wide range of transformation efficiencies, reinforcing the range of factors influencing tomato transgenesis and the need of optimization for any transformation protocol. It is therefore not surprising that several different protocols have been described for *Agrobacterium*-mediated tomato transformation, using stems, leaves or cotyledons as explant sources (Frary, 1995, See Table 1). Considering the large number of surveyed publications that dealt with tomato transformation, few of them have focused on examining the factors that affect transformation efficiency, in order to develop an optimized transformation protocol. The most comprehensive evaluations of factors affecting tomato transformation were carried out by Hamza and Chupeau (1993) and Frary and Earle (1996).

Since the development of the first protocols for tomato transformation, the importance of the genotype has often been emphasized. As observed for others species (Dobigny *et al.*, 1996; Bec *et al.*, 1998), the genotype is one of the main factors influencing genetic transformation. Considering this matter, several researchers have pointed out the need for optimization of tomato regeneration protocols (Xue *et al.*, 1994; Frary, 1995; Frary and Earle, 1996; Costa, 1999; Costa *et al.*, 2000b; Nogueira, 2000). McCormick *et al.* (1986) observed variation in time and ability for shoot morphogenesis in fifteen tomato cultivars. Different transformation efficiencies of four tomato cultivars were also detected by Frary (1995), which corroborates the state of the art of plant tissue culture and attests its peculiarity as a meaningful tool for plant breeding and genetic manipulation. In addition to genotype, the explant source (Chyi and Phillips, 1987; Bird *et al.*, 1988; Patil, 1994; McCormick, 1991; Frary and Earle, 1996), age of the donor plants (Fillatti *et al.*, 1987; Frary, 1995) and medium components (Fillatti *et al.*, 1987; McCormick, 1991; João and

Table 1. Summary of tomato transformation publications in chronological order, including only species that have been transformed and regenerated.

Species	Genotype	Explant Source	Transformation Method	Selective Antibiotic	Killing *Agrobacterium* Antibiotic	Protocol	Main Gene (s) Involved	Reference
L. esculentum	L2	Leaf discs from greenhouse-derived plants	AT GV3Ti11SE	Kanamycin	Carbenicillin	Protocol development	Octopine and NOS/ *npt II*	Horsch *et al.*, 1985
L. esculentum	15 cultivars and hybrids surveyed	Leaves	AT GV3111SE AT 208	Kanamycin	Carbenicillin	Protocol development	*nptII*	McCormick *et al.*, 1986
L. esculentum	Red Cherry, Improved Pearson, UC82, Heinz 2152, ONT 7710	Cotyledons	AT LBA4404 AR A4	Kanamycin	Cefotaxime	Protocol development	*nptII*	Shahin *et al.*, 1986
L. esculentum × *L. pennellii*	VF36	Stem internodes	AT C58C1/pGV 3850::kanR	Kanamycin	Cefotaxime Carbenicillin	Protocol optimization	*nos* and *nptII*	Chyi and Phillips, 1987

Table 1. Continued

Species	Genotype	Explant Source	Transformation Method	Selective Antibiotic	Killing *Agrobacterium* Antibiotic	Protocol	Main Gene (s) Involved	Reference
L. peruvianum and *L. esculentum*	PI 128650; LA292, LA1166, LA1179, LA1182, LA1444, MsK93	Stem and leaf expalnts derived from *in vitro* grown plants and leaf callus-derived protoplasts	AT C58, AT LBA4404 and for IT (plasmids pCTW22 and PCTW78)	Kanamycin	Carbenicillin	Protocol development		Horsch *et al.*, 1985; Koornneef *et al.*, 1986; Korneeff *et al.*, 1987
L. peruvianum × *L. esculentum hybrid*	MsK93	Leaf-derived callus protoplasts	Ca^{++}phosphate-mediated combined with PEG or PVA	Kanamycin	NU	Protocol development	*nptII* and putative tomato origins of replication (ars)	Jongsma *et al.*, 1987
L. esculentum	Heinz 2152 and UC82	Cotyledons	AR A4 AT LBA4404	Kanamycin	Cefotaxime	Shahin *et al.*, 1986	*npt*	Sukhapinda *et al.*, 1987
L. esculentum	UC82B	Cotyledons	AT LBA4404	Kanamycin	Carbenicillin	Protocol developement	*nptII* and *aroA*	Fillatti *et al.*, 1987

Table 1. Continued

Species	Genotype	Explant Source	Transformation Method	Selective Antibiotic	Killing *Agrobacterium* Antibiotic	Protocol	Main Gene (s) Involved	Reference
L. esculentum	Lukullus	Leaf discs	AT C58C1	Kanamycin	Cefotaxime	Deblaere *et al.,* 1987	*bar*	De Block *et* al., 1987
L. esculentum, L. peruvianum, L. hirsutum × *L. esculentum*	TinyTim, LA111 and rootstock KNVF	Decapitated seedlings or stem explants	AR A4T	Kanamycin	Cefotaxime	Protocol developement	*nptII*	Morgan *et al.,* 1987
L. esculentum	VF36	Cotyledons	AT (NA)	Kanamycin	Carbenicillin	McCormick *et al.* 1986	Bt. var. kustaki (B.t.k.) *HD-1* gene	Fischhoff *et al.*, 1987
L. peruvianum *L. esculentum*	LA2172 and Manapal	Greenhouse grown leaves-derived protoplasts	Electroporation	Kanamycin	NU	Protocol developement	Plasmids pABD1 and PGH1 with kan^r and chlorsulfuron^r respectively, and *CAT*	Bellini *et al.*, 1989

Table 1. Continued

Species	Genotype	Explant Source	Transformation Method	Selective Antibiotic	Killing *Agrobacterium* Antibiotic	Protocol	Main Gene (s) Involved	Reference
L. esculentum	Ailsa Craig	Stem internodes	AT LBA4404	Kanamycin	Carbenicillin	Protocol development	PG and *CAT*	Bird *et al.*, 1988
L. esculentum	VF36	Cotyledons	AT GV3111	Kanamycin	Carbenicillin	Koornneeff *et al.*, 1986; Fillati *et al.*, 1987	Ac7 and Ds1 transposable elements from maize	Yoder *et al.*, 1988
L. esculentum	UC82B	Cotyledons	AT (NA)	Kanamycin	Carbenicillin	Filatti *et al.*, 1987	Antisense PG cDNA	Sheehy *et al.*, 1988
L. esculentum	VF36	Leaves and cotyledons	AT (NA)	Kanamycin	Carbenicillin	McCormick *et al.*, 1996	TMV coat protein	Nelson *et al.*, 1988
L. esculentum	Ailsa Craig and mutant rin	Cotyledons	AT pGV 3850	Kanamycin	Carbenicillin	Filatti *et al.*, 1987	Chimeric E8-PG	Giovannoni *et al.*, 1989

Table 1. Continued

Species	Genotype	Explant Source	Transformation Method	Selective Antibiotic	Killing *Agrobacterium* Antibiotic	Protocol	Main Gene (s) Involved	Reference
L. esculentum	VF36	Cotyledons	AT LBA4404	Kanamycin	Carbenicillin	McCormick *et al.*, 1986	Oat phytochrome cDNA	Boylan and Quail, 1989
L. esculentum	Ailsa Craig	Stem segments	AT LBA 4404	Kanamycin	Carbenicillin	Bird *et al.*, 1988	PG (pT OM6) antisense cDNA	Smith *et al.*, 1990
L. esculentum	VF36	Cotyledons	AT LBA 4404	Kanamycin	Carbenicillin	McCormick *et al.*, 1986	pTACC2 cDNA in antisense orientation	Oeller *et al.*, 1991
L. esculentum	Ailsa Craig	Stem segments	AT LBA 4404	Kanamycin	Carbenicillin	Bird *et al.*, 1988	pTOM5 antisense gene (pJR15B)	Bird *et al.*, 1991
Lycopesicon	NA	Cotyledons and hypocotyls	AT (NA)	Kanamycin	Carbenicillin or cefotaxime	McCormick *et al.*, 1986	*nptII*	McCormick, 1991

Table 1. Continued

Species	Genotype	Explant Source	Transformation Method	Selective Antibiotic	Killing *Agrobacterium* Antibiotic	Protocol	Main Gene (s) Involved	Reference
L. esculentum	UC82B	Cotyledons	AT (NA)	Kanamycin	Carbenicillin	Fillati *et al.*, 1987	Maize SSP	Worrell *et al.*, 1991
L. esculentum	UC82B	Leaf discs	AT ABI (A208)	Kanamycin	Carbenicillin	McCormick *et al.*, 1986	*Arabidopsis* acid chitinase and *gus*	Samac and Shah, 1991
L. esculentum	NA	Leaves	AT (NA)	Kanamycin	Carbenicillin	McCormick *et al.*, 1986	*cryI*A(b) and modified *cryI*A(b)	Perlak *et al.*, 1991
L. esculentum	Better Boy	Cotyledons	AT LBA4404	Kanamycin	Carbenicillin and cefotaxime	Narvaez-Vasquez, 1991	Antisense prosystemin	McGurl *et al.*, 1992
L. esculentum	Ailsa Craig	Cotyledons	AT LBA4404	Kanamycin	Carbenicillin	Fillati *et al.*, 1987	E-8 cDNA antisense gene	Peñarrubia *et al.*, 1992

Table 1. Continued

Species	Genotype	Explant Source	Transformation Method	Selective Antibiotic	Killing *Agrobacterium* Antibiotic	Protocol	Main Gene (s) Involved	Reference
L. esculentum	JTM-7	Cotyledons	AT pGV3850	Kanamycin	Cefotaxime	Masuta *et al.*, 1989 modified	CMV T73-satellite RNA	Saito *et al.*, 1992
L. esculentum	UC82B	NA	AT (NA)	Kanamycin	NA	NA	TMV or ToMV CP	Sanders *et al.*, 1992
L. esculentum	UC82B	Leaf	AT ABI	Kanamycin	Carbenicillin	McCormick *et al.*, 1986	*ACCd* gene	Klee, 1993
L. esculentum	Ailsa Craig	Leaf disk	AT LBA 4404	Kanamycin	Carbenicillin	Fillati *et al.*, 1987	PG and PE cDNAs	Seymour *et al.*, 1993
L. esculentum	Ailsa Craig	Cotyledons	AT LBA4404	Kanamycin	Carbenicillin	Bird *et al.*, 1988	TOM5 cDNA	Fray and Grierson, 1993
L. esculentum	Ailsa Craig	Cotyledon and hypocotyl	AT C58C1	Kanamycin	Cefotaxime	Protocol optimization	*nptII*	Lipp João and Brown, 1993

Table 1. Continued

Species	Genotype	Explant Source	Transformation Method	Selective Antibiotic	Killing *Agrobacterium* Antibiotic	Protocol	Main Gene (s) Involved	Reference
L. esculentum	Seokwang	Leaves	AT LBA4404	Kanamycin	Cefotaxime	Cho *et al.*, 1992	Modified B.t. sbsp. tenebrionis	Cho *et al.*, 1993
L. esculentum	Moneymaker	Cotyledons	AT EHA105, AT MOG101	Kanamycin	Cefotaxime, vancomycin and carbenicillin	Protocol optimization	*nptII*	van Roekel *et al.*, 1993
L. esculentum	Better Boy hybrid VFN	Cotyledons	LBA 4404	Kanamycin	Carbenicillin and cefotaxime	Own protocol	Prosys antisense	Orozco-Cardenas *et al.*, 1993
L. esculentum	Hybrids OH7814 and OH1498	Stem segments	AT LBA 4404	Kanamycin	Carbenicillin	Lagrimi *et al.*, 1992	TobAn POD, *nptII* or PML 507 (POD over-producer) directed by CaMV35S	Lagrimi *et al.*, 1993

Table 1. Continued

Species	Genotype	Explant Source	Transformation Method	Selective Antibiotic	Killing *Agrobacterium* Antibiotic	Protocol	Main Gene (s) Involved	Reference
L. esculentum	Better Boy	Cotyledons	AT LBA4404	Kanamycin	Carbenicillin	Fillati *et al.*, 1987	CaMV 35S, anti-(*tap1*/*tap2*) and nos	Scherf *et al.*, 1993
L. esculentum and *L. peruvianum*	UC82B, Castone, Flerline, Monalbo and CMV INRA	Cotyledons	AT p35GUSINT	Kanamycin	Cefotaxime	Protocol optimization	*nptII* and *gus*	Hamza and Chupeau, 1993
L. esculentum	Money-maker	Cotyledons	AT LBA4404	Kanamycin	Timentin	Filati *et al.*, 1987; Frary, 1995	*Pto* gene	Martin *et al.*, 1993
L. esculentum	V36	Cotyledons	AT LBA4404	Kanamycin	Carbenicillin	Yoder *et al.*, 1988	Chimeric Ds/*gus*, *nptII*, and an Ac-derived transposase element	Goldsbrough *et al.*, 1993

Table 1. Continued

Species	Genotype	Explant Source	Transformation Method	Selective Antibiotic	Killing *Agrobacterium* Antibiotic	Protocol	Main Gene (s) Involved	Reference
L. esculentum	Geneva 80	Cotyledons	AT C58Z707	Kanamycin	Carbenicillin	McCormick *et al.*, 1986	CP of CMV-WL	Xue *et al.*, 1994
Lycopersicon spp.	San Marzano	Cotyledons	AT LBA4404	Kanamycin	Cefotaxime	Own protocol	CMV satellite RNA (Ra)	Valanzuolo *et al.*, 1994
Lycopersicon	MsK93	Leaf discs	AT GV3101	Kanamycin	Cefotaxime	Koornneef *et al.*, 1986	*cryIA(b)* and *cryIC*	van der Salm *et al.*, 1994
L. esculentum	UC82b and CL143	Cotyledons	AT (NA)	Kanamycin	Carbenicillin	Fillati *et al.*, 1987; McCormick *et al.*, 1986	CMV satellite RNA sequence	McGarvey *et al.*, 1994
L. esculentum	Better Boy	Cotyledons	AT LBA4404	Kanamycin	Carbenicillin and cefotaxime	Narvaez-Vasquez *et al.*, 1991	CaMV 35-S prosystemin	McGurl *et al.*, 1994

Table 1. Continued

Species	Genotype	Explant Source	Transformation Method	Selective Antibiotic	Killing *Agrobacterium* Antibiotic	Protocol	Main Gene (s) Involved	Reference
L. esculentum	Moneymaker	Cotyledons and hypocotyls	AT LBA4404	Kanamycin	Carbenicillin	McCormick *et al.*, 1986	*Fen* and *Pto* tomato genes	Martin *et al.*, 1994
L. esculentum × *L. pennellii*	F_1 Hybrid	Leaf disks from greenhouse-grown leaves	AT LBA4404	Kanamycin	Cefotaxime and vancomycin	McCormick *et al.*, 1986	TYLCV capsid protein V1	Kunik *et al.*, 1994
L. esculentum	VF36	Cotyledons	AT LBA4404	Kanamycin	Carbenicillin or ampicillin	Filatti *et al.*, 1987	TSWV N gene	Kim *et al.*, 1994
L. esculentum	Ailsa Craig	Cotyledons	ATLBA 4404	Kanamycin	Carbenicillin	Filatti *et al.*, 1987	PG and β-subunit antisense gene	Watson *et al.*, 1994
L. esculentum	UC82B	Leaves	AT ABI	Kanamycin	Carbenicillin	Klee *et al.*, 1993	ACCd gene	Reed *et al.*, 1995

Table 1. Continued

Species	Genotype	Explant Source	Transformation Method	Selective Antibiotic	Killing *Agrobacterium* Antibiotic	Protocol	Main Gene (s) Involved	Reference
L.esculentum	ATV847	Cotyledons	AT LBA4404	Kanamycin	Carbenicillin	Shahin *et al.*, 1986; Filatti *et al.*, 1987; Yoder *et al.*, 1988	TSWV N gene	Ultzen *et al.*, 1995
L. chilense	LA2747 and LA1930	Leaves	AT LBA4404	Kanamycin	Carbenicillin	Horsch *et al.*, 1985 modified	*nptII*, *CAT*	Agharbaoui *et al.*, 1995
L. esculentum	Money-maker	Cotyledons	AT MOG101	Kanamycin	Carbenicillin	van Roekel *et al.*, 1993	Endochitinase and β-1,3-glucanase genes	Jongedijk *et al.*, 1995
L. esculentum	House-Odoriko	Cotyledons	AT LBA4401	Kanamycin	Carbenicillin	NA	Tomato antisense acid invertase cDNA (*Aiv*-1)	Ohyama *et al.*, 1995

Table 1. Continued

Species	Genotype	Explant Source	Transformation Method	Selective Antibiotic	Killing *Agrobacterium* Antibiotic	Protocol	Main Gene (s) Involved	Reference
L. esculentum	Seokwang	Leaf-disk	AT LBA4404	Kanamycin	Cefotaxime	McCormick *et al.*, 1986	B.t. δ-endotoxin	Rhim *et al.*, 1995
L. esculentum	Moneymaker and Motelle	Cotyledons and hypocotyls	AT LBA4404	Kanamycin	Timentin	Fillati *et al.*, 1987	Tomato *Mi* cDNA and genomic constructs	Frary, 1995
L. esculentum	Ailsa Craig	Cotyledons	AR LBA9402	Kanamycin	NA	Tepfer, 1984	Oryzacy-sstatin I and derivative genes	Urwin *et al.*, 1995
L. esculentum	T5 and H100	Cotyledons	AT LBA 4404	Kanamycin	Carbenicillin	McCormick *et al.*, 1986	Antisense acid invertase (TIV1)	Klann *et al.*, 1996
L. esculentum	Rutgers	Cotyledons	AT EHA101	Kanamycin	Carbenicillin	McCormick *et al.*, 1986	BAC promoter – gus constructs	Koehler *et al.*, 1996

Table 1. Continued

Species	Genotype	Explant Source	Transformation Method	Selective Antibiotic	Killing *Agrobacterium* Antibiotic	Protocol	Main Gene (s) Involved	Reference
L. esculentum	VF36	Cotyledons	AT (NA)	Kanamycin	Carbenicillin	McCormick, 1991	Tobacco N gene	Whithan *et al.*, 1996
L. esculentum	Money-maker	Cotyledons and hypocotyls	AT LBA 4404	Kanamycin	Timentin	Protocol optimization	*nptII*	Frary and Earle, 1996
L. esculentum	ATV847	Cotyledons	AT LBA4404	Kanamycin	Carbenicillin	Ultzen *et al.*, 1995	CMV-CP	Gielen *et al.*, 1996
L. esculentum	Haubners Vollendung	Leaves	AT GV3101 AT C58C1	Kanamycin	Cefotaxime	McCormick *et al.*, 1986	Stilbene synthase	Thomzik *et al.*, 1997
L. esculentum	H722	Cotyledons	AT pCIB542/ A136	Kanamycin	Carbenicillin	Fillati *et al.*, 1987	Tomlox promoter β-glucuronidase gene	Beaudoin and Rothstein, 1997
L. esculentum	Better Boy hybrid VFN	Cotyledons	AT LBA4404	Kanamycin	Carbenicillin and cefotaxime	Orozco-Cardenas *et al.*, 1993	Prosys-GUS promoter	Jacinto *et al.*, 1997

Table 1. Continued

Species	Genotype	Explant Source	Transformation Method	Selective Antibiotic	Killing *Agrobacterium* Antibiotic	Protocol	Main Gene (s) Involved	Reference
L. esculentum	Ailsa Craig	Stem internodes	AT LBA4404	Kanamycin	Carbenicillin	Bird *et al.*, 1988	C4 gene from TLCV	Krake *et al.*, 1998
L. esculentum	Ailsa Craig	Cotyledons	AT LBA4404	Kanamycin	Carbenicillin	NA	35S-ACC-oxidase sense gene (ACO1)	Hamilton *et al.*, 1998
L. esculentum	UC82B	Cotyledons	AT LBA4404	Kanamycin	Cefotaxime	Fillati *et al.*, 1987; Hamza and Chupeau, 1993	Yeast *HAL2* gene	Arrilaga *et al.*, 1998
L. esculentum	NC8276	Cotyledons	AT LBA4404	Kanamycin	Carbenicillin	Fillati *et al.*, 1987	Homeobox gene *LeT6*	Janssen *et al.*, 1998
L. esculentum	Moneymaker	Cotyledons	AT EHA105	Kanamycin	NA	Van Roekel *et al.*, 1993	*Avr9* gene	Honée *et al.*, 1998

Table 1. Continued

Species	Genotype	Explant Source	Transformation Method	Selective Antibiotic	Killing *Agrobacterium* Antibiotic	Protocol	Main Gene (s) Involved	Reference
L. esculentum	Ailsa Craig	Stem internodes	AT LBA4404	Kanamycin	Carbenicillin	Bird *et al.*, 1988	Pepper *ccs* and *fib* genes	Kuntz *et al.*, 1998
L. esculentum	Seokwang	Leaves	AT LBA4404	Kanamycin and Hygromycin	Carbenicillin	Own protocol	Antisense RNA of *AL1*-gene from TGMV	Rhim, 1998a
L. esculentum	Moneymaker	Cotyledons	AT LBA4404	Kanamycin	Carbenicillin	Fillatti *et al.*, 1987	*Mi*	Milligan *et al.*, 1998
L. esculentum	Moneymaker	Cotyledons	AT LBA4404	Kanamycin	Ticarcillin/ potassium or cefotaxime	Protocol optimization	*npt*	Ling *et al.*, 1998
L. esculentum	AC	Stem segments	AT LBA4404	Kanamycin	Carbenicillin	Bird *et al.*, 1988	Alcohol dehydrogenase cDNA (ADH 2)	Speirs *et al.*, 1998

Table 1. Continued

Species	Genotype	Explant Source	Transformation Method	Selective Antibiotic	Killing *Agrobacterium* Antibiotic	Protocol	Main Gene (s) Involved	Reference
L. esculentum and L. *pennellii*	Mogeor LA716	Cotyledons	AT LBA4404	Kanamycin or Hygromycin	Timentin	Frary and Earle, 1996	BIBAC genomic DNA library	Hamilton *et al.*, 1999
L. esculentum	Pera	Cotyledons and hypocotyls	AT LBA4404	Kanamycin	Carbenicillin	Protocol development	Tomato basic peroxidase (tpx1) gene	Manssouri *et al.*, 1999
L. esculentum	Starfire	Greenhouse-derived leaves, *in vitro*-grown hypocotyls and cotyledons	ATL BA4404	Kanamycin	Carbenicillin	Horsch *et al.*, 1985; Agharbaoui *et al.*, 1995	*Endochitinase* (pcht28)	Tabaei-zadeh *et al.*, 1999
L. esculentum	Ceraciforme	Stems	AR 8196/ pRYH1	Kanamycin	Cefotaxime	NA	Cytosine deaminase (cd) and *nptII*	Hashimoto *et al.*, 1999

Table 1. Continued

Species	Genotype	Explant Source	Transformation Method	Selective Antibiotic	Killing *Agrobacterium* Antibiotic	Protocol	Main Gene (s) Involved	Reference
L. esculentum	Abunda	Greenhouse-derived leaf explants	AT C58pMP90	Kanamycin	Carbenicillin	Filatti *et al.*, 1987	*E. coli* glutathione reductase (*gor*) gene	Brüggemann *et al.*, 1999
L. esculentum	Summerset	Cotyledons	AT (NA)	Kanamycin	Carbenicillin	McCormick, 1991	Maize *Sh1* promoter fused to *gus*	D'Aoust *et al.*, 1999
L. esculentum	Chloronerva mutant X LA716	Cotyledons	AT LBA4404	Kanamycin	Ticarcillin	Ling *et al.*, 1998	*Chloronerva* gene	Ling *et al.*, 1999
L. esculentum	Peak and Meitin 11	Leaf-disc	AT LBA4404	Kanamycin	NA	NA	Trichosanthin (TCS)	Jiang *et al.*, 1999
L. pennellii	PE-47	Leaf	AT LBA4404	Kanamycin	Carbenicillin	Shahin *et al.*, 1996	*nptII*, *gus*	Gisbert *et al.*, 1999

Table 1. Continued

Species	Genotype	Explant Source	Transformation Method	Selective Antibiotic	Killing *Agrobacterium* Antibiotic	Protocol	Main Gene (s) Involved	Reference
L. esculentum	"Santa Clara" and "Firme" mutant	Cotyledons	AT LBA4404	Kanamycin	Timentin	Frary and Earle, 1996	Antisense ACCS (pTACC2) cDNA	Nogueira *et al.*, 1999
Lycopersicon	VF36	Cotyledons	AT (NA)	Kanamycin	Carbenicillin	NA	TMV *N* gene	Erickson *et al.*, 1999
L. esculentum	Pusa ruby	Cotyledons	AT LBA4404	Kanamycin	Cefotaxime	NA	PhMV CP	Sree Vidya *et al.*, 2000
L. esculentum	P-73	Cotyledons	AT LBA4404	Kanamycin	Carbenicillin	Gisbert, 1997	Yeast *HAL1* gene	Gisbert *et al.*, 2000
L. esculentum	Potentat, mutant *nor*	Cotyledons	AT LBA4404 AT GV3101 AT EHA105	Kanamycin	Carbenicillin	McCormick *et al.*, 1991	TSWV N gene	Fedorowicz *et al.*, 2000

Table 1. Continued

Species	Genotype	Explant Source	Transformation Method	Selective Antibiotic	Killing *Agrobacterium* Antibiotic	Protocol	Main Gene (s) Involved	Reference
L. esculentum	NC8276	Cotyledons	AT LBA4404	Kanamycin	Carbenicillin	Fillati *et al.*, 1987	Wild type or mutated BV1 or BC1 MP from BDMV	Hou *et al.*, 2000
L. esculentum	IPA-5 and IPA-6 processing varieties	Cotyledons	AT LBA4404	Kanamycin	Timentin	Frary and Earle, 1996	Tomato *Sw-5* gene	Costa *et al.*, 2000b
L. esculentum	Pusa Ruby	Cotyledons	AT LBA4404	Kanamycin	Cefotaxime	McCormick *et al.*, 1996	Synthetic cry1Ac gene for ICP	Mandaokar *et al.*, 2000
L. esculentum	Money-maker	Cotyledons	AT LBA4404	Kanamycin	Timentin	Frary and Earle, 1996	Tomato Sw-5 gene	Brommonschenkel *et al.*, 2000

Table 1. Continued

Species	Genotype	Explant Source	Transformation Method	Selective Antibiotic	Killing *Agrobacterium* Antibiotic	Protocol	Main Gene (s) Involved	Reference
L. esculentum	T5	Cotyledons	AT (NA)	Kanamycin	Carbenicillin	Fillati *et al.*, 1987	PG inhibitor protein (pGPIP)	Powell *et al.*, 2000
L. esculentum	Ailsa Craig	Stem	AT LBA4404	Kanamycin	Carbenicillin	Bird *et al.*, 1988	Bacterial phytoene desaturase (*crtI*)	Römer *et al.*, 2000
L. esculentum	Ailsa Craig	Cotyledons	AT LBA4404	Kanamycin	Carbenicillin	Fillati *et al.*, 1987	Recombinant expansin (*CsExp1*) from cucumber	Rochange and McQueen-Mason, 2000
L. esculentum	Ailsa Craig	Cotyledons	AT EHA105	Kanamycin	Carbenicillin	McCormick *et al.*, 1986	Chimeric PG TAP G1:GUS or TAP G4:GUS	Hong *et al.*, 2000

Table 1. Continued

Species	Genotype	Explant Source	Transformation Method	Selective Antibiotic	Killing *Agrobacterium* Antibiotic	Protocol	Main Gene (s) Involved	Reference
L. esculentum	Ailsa Craig *NR ripe* mutant	Cotyledons	AT LBA4404	Kanamycin	Carbenicillin	Bird *et al.*, 1988	*pNR6.1 AS*	Hackett *et al.*, 2000
L. esculentum	Heinz FND902	Cotyledons	AT C58	Kanamycin	Carbenicillin	McCormick *et al.*, 1986 modified	ACCd directed by 35S, *rolD* or *prb-ib* promoter	Robison *et al.*, (Submitted)
L esculentum	UC82	Cotyledons	AT C58C1	Kanamycin	Vancomycin, ticarcilin or cefotaxime	Own protocol	*nptII*	Hu and Phillips, 2001
L. esculentum	Heinz FND902	Cotyledons	AT C58	Kanamycin	Carbenicillin	McCormick *et al.*, 1986 modified	ACCd directed by 35S, *rolD* or *prb-ib* promoter	Robison *et al.*, (Submitted)

Table 1. Continued

ACCd = 1-aminocyclopropane-1-carboxylic acid deaminase gene; *ACCS* = 1-aminocyclopropane-1-carboxylic acid synthase gene; AS = antisense; AT = *Agrobacterium tumefaciens;* AR = *Agrobacterium rhizogenes*; BDMV= *Bean dwarf mosaic virus*; *Fen* = gene that controls sensitivity to insecticide fenthion; *BAC* = bean abscission cellulase; BIBAC = binary bacterial artificial chromosome; B.t = *Bacillus thuringiensis*; *CAT* = chloroacetil transferase gene; *ccs* = capsanthin/capsorubin synthase gene; fib = fibrillin; IT, indirect transfer; ICP = insecticidal crystallized protein; MP = movement protein; NA = not available; *NR* = *Never-ripe*; NU = not used; N = nucleocapsid; *PG* = *polygalacturonase*; POD = peroxidase; *p*-prosystem-*GUS* = chimeric prosystemin β-glucuronidase promoter; *Pto* = gene for resistance to *Pseudomonas syringae* pv. tomato; TYLCV = *Tomato yellow leaf curl virus*; TGMV = *Tomato golden mosaic virus*; TMV = *Tobacco mosaic virus*; TLCV = *Tomato leaf curl geminivirus*; ToMV = *Tomato mosaic virus.*

*Unless otherwise stated, leaves and stems, hypocotyls and cotyledons are derived from *in vitro*-grown plant material.

Brown, 1993) were some of the variables in transformation protocols. And last but not least, the bacterial strain (Davis *et al.*, 1991a; Van Roekel *et al.*, 1993), concentration (Fillatti *et al.*, 1987) and pre and co-culture periods (Fillatti *et al.*, 1987; Davis *et al.*, 1991b; Hamza and Chupeau, 1993) affected significantly tomato transformation efficiency, being worthy of attention when looking for improvements or aiming at new transformation protocols.

Chyi and Phyllips (1987) pointed out that basically three factors were the key steps of transformation: 1) the use of stem segments to enhance adventitious bud regeneration; 2) the incubation of the inoculated stem explant in a medium containing kinetin to promote cell division activity at the inoculation site, and 3) the selection for kanamycin resistance at the bud development stage to favor the competition of transformed cells during regeneration.

Frary and Earle (1996) developed an improved protocol for *Agrobacterium*-mediated transformation of the tomato cultivar Moneymaker by examining the effects of six different factors on transformation efficiency. Explant size, explant orientation, gelling agent and plate sealant were found to affect transformation efficiency. Two other factors, type of explant (hypocotyl or cotyledon) and frequency of transfer to fresh selective regeneration medium did not have an effect on transformation efficiency. The *in vitro* morphogenic responses were benefited by the use of micropore tape as a plate sealant (micropore and parafilm). Micropore tape gave a higher rate of transformation (11.7%) than parafilm (7.6%); however, this difference was not significant (Frary, 1995). By combining the best treatments for each factor, an average transformation efficiency of 10.6% was obtained for Moneymaker (Fray and Earle, 1996).

Likewise, based on Frary and Earle's protocol, Costa *et al.* (2000b) and Nogueira (2000) evaluated factors influencing genetic transformation of fresh market and processing Brazilian tomato cultivars. Concurrently with previous findings, the importance of genotype was confirmed. In addition, the authors detected significant differences regarding temperature during the co-cultivation phase, obtaining significant differences in transformation efficiencies when co-cultivation was carried at 22°C.

Kanamycin (25-500 mg l^{-1}) is the most widely used selectable marker in all surveyed publications (See Table 1). Exceptions are the works by Hamilton *et al.* (1998) and Rhim (1998b), which used hygromycin.

However, in the latter a double selection using hygromycin in addition to kanamycin was used. It should be noted that the wide range of antibiotic concentrations used may reflect genotypic variation and culture requirements among the various materials used in the surveyed literature. For instance, two Brazilian processing varieties, namely 'IPA-5' and 'IPA-6', showed different sensitivities to kanamycin incorporated to the medium. 'IPA-6' required a higher concentration of the selective antibiotic, otherwise a lot of escapes were obtained out of transformation experiments (Costa *et al.*, 2000b). A follow up of a transformation experiment with 'IPA-6' is presented in Figure 2, starting from regeneration until the fruit setting stage (R_1 generation), where tomato morphogenesis and its suppression in non-transformed explants in the presence of kanamycin is observed. In addition to the use of selective markers, reporter genes are also used to confirm T-DNA presence in the regenerant genome. Figure 3 displays an antisense ACC synthase transformed tomato where stable T-DNA insertion and expression was confirmed by specific PCR and Southern Blot as well as GUS histochemical assay.

For successful transformation using *Agrobacterium*, effective elimination of bacteria from the tissue is necessary as soon as their presence is no longer required. This is accomplished by the addition of antibiotics into the culture medium (Ling *et al.*, 1998). For tomato, the antibiotics that have been used to control bacteria are listed on Table 1. The use of Timentin was first proposed for tomato by Frary (1995) and Frary and Earle (1996). The first report on the stimulatory effect of timentin on *in vitro* organogenesis of tobacco (*Nicotiana tabacum* Petit Havana SR1) was done by Nauerby *et al.* (1997). Since then few reports on the use of Timentin for the improvement of *in vitro* morphogenesis have been published (Cheng *et al.*, 1998; Ling *et al.*, 1998). To our knowledge, positive effects of Timentin on *in vitro* morphogenesis of cultured tomato tissues was reported only by Ling *et al.* (1998) and Costa *et al.* (2000a). Ling *et al.* (1998), in experiments involving transformation of the tomato cultivar Moneymaker, were the first to set up comparisons on the effect of the antibiotics cefotaxime and ticarcillin/potassium clavulanate upon transformation efficiency. Positive effects of ticarcillin/potassium clavulanate (150 mg L^{-1}) on callus growth and shoot regeneration were observed. Interestingly, the authors analyzed the combined effect of these antibiotics under different selection conditions (0, 50 or 100 mg L^{-1} kanamycin). Significant differences were observed and transformation frequencies on the ticarcillin/potassium

Figure 2 : Steps involved in *Agrobacterium*-mediated transformation of a processing tomato cultivar ('IPA-6:: TC 134' lineage with resistance to *Tospovirus*). **A**. Cotyledon explants in shoot induction selective medium (MS plus 1.0 mg dm^{-3} zeatin, 0.01 mg dm^{-3} NAA, 150 mg dm^{-3} kanamycin and 300 mg dm^{-3} Timentin) four weeks after transformation. **B.** Non-co-cultivated tomato cotyledon explants in shoot induction selective medium (MS plus 1.0 mg dm^{-3} zeatin, 0.01 mg dm^{-3} NAA, 150 mg dm^{-3} kanamycin and 300 mg dm^{-3} Timentin) four weeks after transformation. **C**. Non-transformed (left) and transformed shoot (right). Note that the non-transformed shoot lacks roots and bleaches due to the presence of kanamycin in the rooting medium, whereas the transformed counterpart develops normal roots and remains green during culture in rooting medium. **D**. Acclimatized transgenic plants growing under greenhouse conditions. **E.** Detail of a transgenic 'IPA6::TC134'tomato plant with ripened fruits.

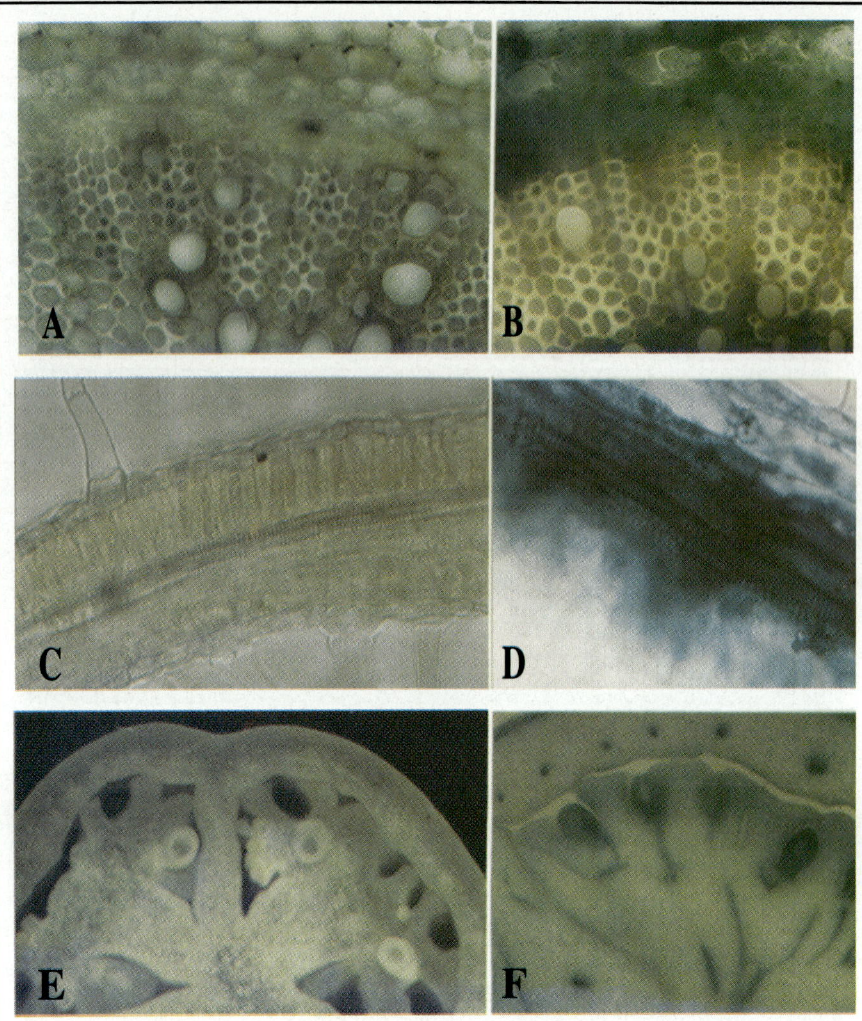

Figure 3 : Histochemical *in situ* detection of *uidA* gene expression in tomato plants (cv. Santa Clara and its natural mutant 'Firme' transformed with *Agrobacterium tumefaciens* [LBA4404 - pCAMBIACCS2- antisense]. **A** – Detail of the region surrounding the major vascular bundle of a non-transformed tomato regenerant; **B** – Micrograph showing a transverse section of the region surrounding the major vascular bundle of leaves from 'Firme' transgenic plants, evidencing the *in situ* activity of the *uidA* gene product in the main vein; **C** – Longitudinal leaf section of a non-transformed 'Santa Clara' plant, evidencing the vascular region. **D** - Longitudinal leaf section evidencing the activity of the *uidA* gene product in the region surrounding a leaf vascular bundle of 'Santa Clara' transformant; **E** - Micrograph showing a transverse section of the ovary of a non-transformed plant. **F** -Micrograph showing a transverse section of the ovary of transgenic tomato plant evidencing *gus* activity.

clavulanate-containing medium were about 40-50% higher than on the cefotaxime-containing medium, with four to five-fold more transformed plants obtained using ticarcillin/potassium clavulanate. Further on, it was observed that Timentin (300 mg L^{-1}) caused an increase in the morphogenesis of in vitro cotyledonary explants of Brazilian tomato fresh market and processing cultivars (*Lycopersicon esculentum* Mill. 'Santa Clara', 'IPA –5' and 'IPA-6'), and the 'Firme' natural mutant. In two out of three cultivars tested, rooting of shoots was positively influenced, both in the presence and absence of Timentin in the rooting medium, among shoots regenerated from explants derived from Timentin-supplemented medium (Costa *et al.*, 2000a).

4. TOMATO BIOTECHNOLOGY : FROM ACTUAL TO POTENTIAL TRAITS

The application of genetic manipulation to the improvement of food crops, including tomato, has enormous basic and applied aspects. In this section we will highlight some of the potential traits that have been or could be manipulated in tomato, as illustrated in Figure 4. A number of important phenotypes have been obtained in transgenic tomato. Many of these phenotypes improve agronomic characteristics of the plant by confering insect and disease resistance, herbicide tolerance, and improvement of fruit quality traits in the harvested product (Houck *et al.*, 1993). A listing potential transgenic crops for commercialization from 2000 onwards includes tomato with virus resistance (improved control of CMV and TMV virus diseases), insect resistance (reduced losses to looper, hornworm, and fruit worm) and herbicide tolerance for improved weed control (James, 2000).

4.1. Fruit quality

Studies involving fruit ripening are considered important landmarks in tomato biotechnology that generated applied and consistent knowledge at both the scientific and practical points of view. The development and maturation of fruits has received considerable scientific scrutiny due to both the uniqueness of such processes to the biology of plants and the importance of the fruit as a significant component of the human diet. Moreover, molecular and genetic analysis of fruit development, and especially ripening of fleshy fruits, has resulted in significant gains in knowledge over recent years (Giovannoni, 2001). The ripening process is a highly regulated developmental process requiring the expression of a large number of gene products. Enzymes involved in the degradation

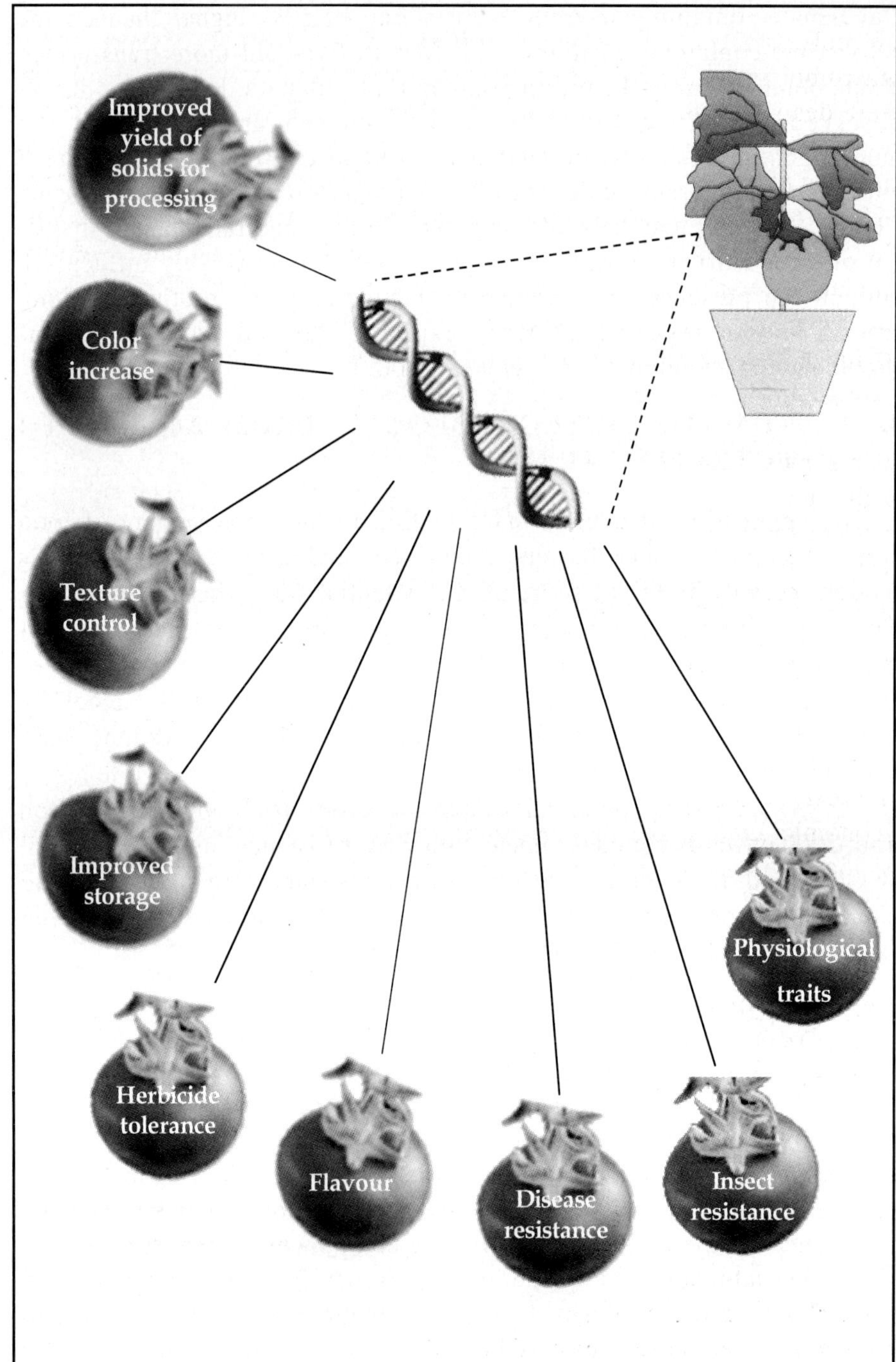

Figure 4 : Attributes and traits of tomato which could be manipulated by genetic engineering (Adapted from Tucker, 1993).

of cell walls, complex carbohydrates, chlorophyll and other macromolecules must be coordinately expressed with enzymes that make fruit desirable nutritionally and aesthetically (Klee, 1993). Excellent reviews have been presented on the molecular biology of tomato fruit ripening (Theologis *et al*., 1992; Hobson and Grierson, 1993; Gray *et al*., 1994; Theologis, 1994; Klee and Lanaham, 1995; Picton *et al*., 1995a; Picton *et al*., 1995b; Tucker and Zhang, 1996; Giovannoni, 2001). Allied to that, genetic mapping of loci controlling fruit ripening (Giovannoni *et al*., 1999; Doganlar *et al*., 2000) will bring a fundamental scientific basis to the already well-studied ripening process in tomato. This scientific background has benefited several other groups that deal on fruit postharvest physiology. One major area in which genetically engineered plants may become commercially important is in the improvement of fruit quality (Tucker, 1993). Coincidentally, the first transgenic plant product marketed commercially by Calgene was the well known 'Flavr Savr™' tomato which had been modified to contain reduced levels of cell wall softening enzyme polygalacturonase (Kramer and Redenbaugh, 1994). Tomato purée with a similar type of modification has been on the market in the UK since February 1996 (Dunwell, 2000).

The role of ethylene in promoting the ripening of tomato fruits has received particular attention, partly because of its intrinsic scientific interest, but also for reasons of experimental convenience and the economic importance of tomato as a major food crop (Picton *et al*., 1995a). Synthesis of the phytohormone ethylene is believed to be essential for many plant developmental processes (Klee and Lanahan, 1995). Ethylene synthesis has been examined in transgenic plants using several independent approaches to reduce its synthesis by means of antisense gene constructs against ACC oxidase or ACC synthase (Hamilton *et al*., 1990; Oeller *et al*., 1991; Theologis *et al*., 1992) and by expressing ACC deaminase (Klee *et al*., 1991; Klee, 1993).

In the last 12 years, notable progress has been obtained on the molecular biology of tomato ripening, mainly by using the antisense RNA approach (Smith *et al*., 1988; Sheehy *et al*., 1988; Hamilton *et al*., 1990; Smith *et al*., 1990; Bird *et al*., 1991; Oeller *et al*., 1991; Gray *et al*., 1992; Peñarrubia *et al*., 1992; Hall *et al*., 1993; Picton *et al*., 1993; Watson *et al*., 1994; Ohyama *et al*., 1995; Picton *et al*., 1995b; Klann *et al*., 1996; Hackett *et al*., 2000) or cosuppression (Seymour *et al*., 1993; Hamilton *et al*., 1998; Simons and Tucker, 2000). So far, plant antisense technology has been the most widely form of gene silencing because of its simplicity in obtaining reduced expression (Senior, 1998).

The RNA antisense technique has an advantage over the use of ripening resistant mutants, where a 15 days ethylene treatment reverted the transgenic fruit phenotype allowing ripening (Oeller *et al*., 1991). In mutant tomatoes, especially "ripening inhibitor" (*rin*) and "Never-ripe" (*Nr*) (Lanahan *et al*., 1994; Wilkinson *et al*., 1995), the ethylene treatment did not restore normal ripening, possibly because the mutation affected a gene which product participates in the signal transduction pathway, blocking the ethylene effect on ripening (Gray *et al*., 1992).

4.1.1. ACC oxidase

Hamilton *et al*. (1990) showed, by using the antisense RNA approach, that a cDNA (pTOM13) inhibited ethylene synthesis by 97% in transgenic plants. As a result it was found out that the cDNA encoded a polypeptide involved in the conversion of ACC to ethylene, called ACC oxidase. Hamilton *et al*. (1998) reported a reduced ACC-oxidase activity in transgenic tomato plants transformed wiih a 35S-ACC-oxidase sense gene. Fifteen per cent of the transformed plants had reduced ACC-oxidase activity and 96% of the plants transformed with a ACC-oxidase sense gene, containing two additional inverted copies of its 5' untranslated region, exhibited reduced ACC-oxidase activity compared to wild type plants. It was suggested that the introduction of a small repeat into a transgene would greatly increase sense suppression frequency of homologous, endogenous genes. The use of this homology dependent gene silencing (HDGS) may have practical applications in situations where reliable silencing of endogenous genes is required (Hamilton *et al*., 1998).

4.1.2. ACC synthase

An antisense approach was also used to turb ethylene biosynthesis in tomato transgenic plants expressing the cDNA ACS2 (encoding ACC synthase); down-regulation of ACC synthase led to a marked inhibition of both isozymes - LE-ACS2 and LE-ACS4 - leading to an overall inhibition on fruit ripening (Oeller *et al*., 1991). Therefore, adding ethylene could restore ripening. A difficulty with antisense gene shutoff is linked to the fact that ACC synthase is a divergent gene family. As a consequence, since different gene family members can be quite divergent in DNA sequence, the degree of ethylene inhibition can vary from tissue to tissue (Klee and Lanahan, 1995).

4.1.3. ACC deaminase

Tomato plants that are delayed in fruit ripening have also been developed by *Agrobacterium tumefaciens*-mediated transfer of a gene encoding a bacterial gene for ACC deaminase, which metabolizes the ethylene precursor, ACC, to an inactive product (Gray *et al.*, 1994; Klee and Lanahan, 1995). This approach was first envisaged by Klee *et al.* (1991). Further characterization was carried out, and fruit ethylene synthesis was significantly reduced, and the time for fruit to ripen was extended for in both delayed ripening lines relative to the parental control line UC82B (Reed *et al.*, 1995).

4.1.4. Cell wall metabolism genes : PG and PE

The first gene to be manipulated in transgenic tomato fruit was that encoding the polygalacturonase (PG) enzyme (Gray *et al.*, 1994; Senior, 1998). Antisense technology has been used to silence the PG gene and has resulted in genetically modified fruit with less than 1% of normal PG activity (Smith *et al.*, 1988). This reduction was reflected by a reduced depolymerisation, but not solubilisation of pectin during ripening (Smith *et al.*, 1990). Transgenic tomatoes were more resistant to mechanical damage and cracking (Schuch *et al.*, 1991) and the low level PG phenotype surprisingly increased pathogen resistance to *Rhizopus stolonifer* and *Geothrichum candidum* (Kramer *et al.*, 1992). Also, Powell *et al.* (2000) showed that transgenic expression of a pear PG inhibitor protein (pPGIP) reduced the growth of *Botrytis cinerea* on ripe tomato fruit.

The involvement of PG and PE genes in fruit ripening has been reviewed elsewhere (Hobson and Grierson, 1993; Tucker, 1993; Picton *et al.*, 1995a; Tucker and Zhang, 1996). To elucidate the role of polygalacturonase during tomato fruit ripening a pleiotropic genetic mutation, *rin,* was utilized. The results indicated that PG is the primary determinant of cell wall polyuronide degradation, but suggest that this degradation is not enough for the induction of softening, of elevated levels of ethylene, and the biosynthesis of lycopene in the *rin* fruit (Giovannoni *et al.*, 1989). Collectively, the results obtained with transgenic tomato having altered PG levels are consistent with the hypothesis that PG-mediated pectin disassembly does not contribute to early fruit softening but contributes significantly to tissue deterioration in the late stages of fruit ripening.

Transgenic tomato fruit in which both PG and PE have been down regulated have also been produced (Seymour *et al.*, 1993). Later on, a PG/PE primary transformant, which displayed the strongest down-regulation of PG and PE, was selected for further evaluations. Simultaneous down-regulation has been achieved for the two cell wall hydrolases (PE and PG) in tomato fruit, and the stable inheritance of this coordinated gene silencing over two generations was reported (Simons and Tucker, 1999). Only two of the three isoforms of PE in fruit were silenced by this chimeric construction, thus providing indication of the relative homology between the gene sequences for these isoforms. Several works allowed the evaluation of quality traits in transgenic tomatoes from the food science perspective (Errington *et al.*, 1997, 1998; Tucker, 1993).

4.1.5. Carotenoid biosynthesis genes

A tomato phytoene synthase gene (*Psy-1*), correspondent to the cDNA clone TOM5, which product is involved in carotenoid biosynthesis during tomato fruit ripening, and which enzyme is required for the production of phytoene, an essential intermediate in the production of carotenoids, was characterized by Gray *et al.* (1994). Transgenic tomato plants expressing an antisense RNA to pTOM5 were found to display disrupted carotenoid biosynthesis (more than 97% reduction), resulting in yellow fruits and pale flowers (Bird *et al.*, 1991). Bramley *et al.* (1992) proceeded with further biochemical characterization of the transgenic plants in which carotenoid synthesis was inhibited through the expression of antisense RNA to pTOM5. Analysis of different tissues of transformed and control plants indicated that although ripe fruit and flower carotenoid levels were reduced in modified plants, leaf carotenoid levels were not decreased (Bramley *et al.*, 1992), suggesting that the pTOM5 gene product was not involved in carotenoid synthesis in the leaf. Later on, Fray and Grierson (1993) performed genetic analysis and constitutive overexpression of TOM5 cDNA in a mutant *yellow flesh* mutant, similar to antisense TOM5 plants. Interestingly, synthesis of the carotenoid lycopene was restored in *yellow flesh* fruits, confirming its deficiency in phytoene synthase, whilst in some mutant transgenic plants, inhibition of carotenoid production synthesis in immature green fruit, leaves and flowers was observed, due to cosuppression of a functionally distinct phytoene synthase expressed in green tissues (Fray and Grierson, 1993; Gray *et al.*, 1994).

A phytoene synthase gene present in tomato chloroplasts (*Psy-2*) has been recently characterized, and differences in biochemical properties between this and other higher plant and chromoplast-derived phytoene synthases have been observed (Fraser *et al.*, 2000). These findings will benefit the design and evaluation of transgenic plants with altered carotenoid contents in both leaf and fruit tissues. Aiming to enhance the carotenoid content and profile of tomato fruit, transgenic lines were produced containing a bacterial carotenoid gene (*crtl*) encoding the enzyme phytoene desaturase, which converts phytoene to lycopene (Römmer *et al.*, 2000). It was verified that expression of this gene in transgenic tomatoes did not increase total carotenoid levels, and levels of noncarotenoid isoprenoids remained unchanged in transformants. However, the β-carotene content increased about threefold, up to 45% of the total carotenoid content. Endogenous carotenoid genes were concurrently upregulated, except for phytoene synthase, which was repressed.

4.1.6. Acid invertases

The acid invertases belong to a class of enzymes that play an important role on the metabolism of sucrose in ripening tomato fruits, hydrolyzing sucrose to glucose and fructose (Klann *et al.*, 1996). Suppression of acid invertase activity by antisense RNA was reported by Ohyama *et al.* (1995) and Klann *et al.* (1996). A cDNA for acid invertase, *Aiv-1,* was introduced into tomato in an antisense orientation with respect to the heterologous promoter (Ohyama *et al.*, 1995). The antisense gene effectively suppressed the invertase acid in soluble and cell wall fractions from ripening fruits. The sucrose content of transformed fruits was markedly increased, while the hexose content decreased, suggesting that acid invertase is one of the main determinants on sugar composition of tomato fruit. Further on, Klann *et al.* (1996) reported that an antisense acid invertase (*Tlv1*) gene had a marked effect in transgenic tomato fruits, increasing sucrose composition, decreasing hexose content, reducing overall size and highly disturbing growth habits. Interestingly, sucrose-accumulating fruits had elevated ethylene evolution relative to control counterparts.

4.1.7. Lypoxigenases (LOX)

The investigation of developmental and environmental signal regulation over the expression of two tomato lipoxygenese (*tomlox*) genes led Beaudoin and Rothstein (1997) to isolate their 5'-untranslated regions and use them to drive the expression of a β-glucuronidase reporter gene in transgenic tomato and tobacco. In tomato fruit, expression of *tomloxA-gus* was observed throughout fruit ripening, with the highest expression at the orange stage, while *tomloxB-gus* tomato fruits GUS activity was detected somewhat earlier than expected, even thought in tobacco it was regulated as predicted. In both cases activity was found mostly in the outer pericarp, although *tomloxA-gus* fruits also displayed intense staining in the collumela and placental tissues. More recently, Griffiths *et al.* (1999) reported the generation of transgenic antisense LOX tomato fruit and their analysis in terms of mRNA expression studies, LOX enzyme activity and volatiles analysis. To determine the importance of the enzyme lipoxygenase in generation of volatile C6 aldehyde and alcohol flavour-related compounds, five antisense LOX genes were constructed and transferred to tomato plants (Griffiths *et al.*, 1999). Reduced levels of endogenous *tomloxA* and *tomloxB* were detected in transgenic fruit containing the p2ALX construct compared to non-transformed plants, whereas the levels of mRNA for a distinct isoform, *tomloxC,* were either unaffected or even increased. In the case of p2ALX plants, LOX enzyme activity was also greatly reduced compared to wild-type plants. Analysis of flavour-related volatiles, however, indicated that there were no significant changes. These findings suggest that either very low levels of LOX are sufficient for the generation of C6 aldehydes and alcohols, or that another isoform, such as *tomloxC*, in the absence of *tomloxA* and *tomloxB,* is responsible for the production of these compounds (Griffiths *et al.*, 1999).

4.2. Disease and insect resistance

Tomato crops are affected by several pathogens and disease resistance genetic control is very used in practice (Laterrot, 2000). Biotechnology approaches using recombinant DNA are enhancing tomato breeding programs with advantages like not being dependent on sexual compatible crosses, less expenditure of resources and time, availability of new genes and traits, and the circumvention of the transfer of linked, undesired genes. Many tomato resistance genes of various functional classes will be cloned in the next few years and some of them are already accessible (Table 2) to be employed in transformation experiments. The speed and ease in which this will be achieved will depend on the genetic definition of the target gene, the complexity of the class of genes, and the pace of technology development (Michelmore, 1995). Technology developments will continue to decrease the gap between the ability to identify molecular markers that are physically close to resistance genes and the size of genomic fragments that can be cloned and used in complementation experiments. One of the features that facilitate cloning and identification of R genes are their conserved domains, such as leucine rich repeats (LRRs), leucine zipers, and nucleotide binding site (NBS). Such is the

Table 2. Cloned tomato resistance genes (adapted from Hammond-Kosack and Jones, 1997).

Gene	Pathogen	Infection type/organ attacked	Predicted features of R protein	Reference
Pto	*Pseudomonas syringae* p.v. tomato (avrPto)	Extracellular bacteria/leaf	Intracellular serine/ threonine protein kinase	Martin *et al.*, 1993
Prf	*Pseudomonas syringae* p.v. tomato (*avrPto*)	Extracellular bacteria/leaf	Leucine rich repeat, nucleotide binding and leucine zipper motifs	Salmeron *et al.*, 1996
Cf-9, Cf-2, Cf-4, Cf-5,	*Cladosporium fulvum (Avr9, Avr2, Avr4, Avr5)*	Biotrophic extracellular fungus without haustoria/ leaf	Extracellular LRR protein with single membrane spanning region and short cytoplasmic carboxyl terminus	Jones *et al.*, 1994; Dixon *et al.*, 1997
Sw-5	*Tospovirus*	Concentric rings lesions, Mosaic, yellowing, chlorosis/ leaf	Cytoplasmic protein with a potential nucleotide binding site (NBS) domain and a C-terminal end consisting of a leucine rich repeats (LRRs)	Brommonschenkel *et al.*, 1997, 2000
Mi	*Meloydogine*	Root knots or galls/roots	Leucine zipper	Milligan *et al.*, 1998
Mi-3	*Meloydogine*	Root knots or galls/roots		Yaghoobi *et al.*, 1995

case of the genetic and physical analysis and identification of loci associated to disease resistance genes in tomato, such as resistance to *Clavibacter michiganensis* (van Heusden *et al.*, 1999), *Alternaria alternata* f. sp. *lycopersici* (Mesbah *et al.*, 1999), *Cucumber mosaic virus* (CMV) (Stamova and Chetelat, 2000) and *Tomato spotted wilt virus* (TSWV) (Brommonschenkel *et al.*, 2000). The final steps in cloning often require complementation, either using *Agrobacterium* (respecting the maximum size of genomic fragments), BAC clones, *in planta Agrobacterium*-mediated transformation, or alternatively, transient assays through co-bombardment in stable complementation experiments (Michelmore, 1995).

The availability of cloned resistance (R) genes now opens the possibility of the addition of new R genes to a plant line by genetic transformation (Bent, 1996). According to this author, the extensive use of the transgenesis approach will occur if it matches or exceeds traditional plant breeding in reliability, versatility and cost effectiveness. Some resistance genes cloned from tomato, which might be used in homologous and heterologous transformation systems, are seen in Table 3. Already, R genes were successfully introduced into commercial varieties (Table 4), and will continue to be. However, it must be kept in mind that these biotechnology techniques are complementary tools to conventional breeding and do not aim to replace other breeding practices.

4.2.1. Disease resistance

4.2.1.1. Bacterial diseases : *Pseudomonas syringae* pv. *tomato* is the causal agent of an important disease, bacterial speck, mainly because of the absence of resistant varieties or hybrids (Martin *et al.*, 1993). *Pto* and *Prf* genes confer to tomato plants resistance to *Pseudomonas syringae* pv. *tomato*. Several studies demonstrated the existance of a signaling pathway leading to resistance (Salmeron *et al.*, 1996). An YAC spanning the *Pto* locus region was identified and a cDNA clone that represents a gene family, at least six members of which genetically co-segregate with *Pto,* was obtained (Martin *et al.*, 1993). *Agrobacterium tumefaciens* harboring a cDNA insert of a member of this family was used to transform susceptible *L. esculentum* cv. Moneymaker. Two plants, containing the transgene complementing Pto in the susceptible plants, were resistant to *P. syringae* pv. *tomato*, displaying inhibited bacterial growth and symptom suppression. Analysis of the *Pto* amino acid sequence revealed similarities with catalytic domains of plant, mammalian and lower eukaryote serine-threonine

Table 3. Summary of requested patents involving tomato transgenesis from January 2000 to December 2000. Data adapted from Delphion Intellectual Property Network (http://www.delphion.com/).

Patent and title	Applicant(s)	Application Number	Issued
US6124528 : Modification of soluble solids in fruit using sucrose phosphate synthase encoding sequence	Calgene LLC, Davis, CA	US1998000051341	Sept. 26, 2000
EP577252B1 : Transgenic tomato plants with altered polygalacturonase isoforms	Arizona Board of Regents on behalf of the University of Arizona	EP1993000303533	Sept. 20, 2000
US6100449 : Transgenic tomato plants containing a fusarium resistance gene	Yeda Research and Development Co. Ltd., Rehovot, Israel Yissum Research and Development Company, Jerusalem, Israel	US1997000930996	Aug. 8, 2000
US6087162 : Transgenic plants resistant to geminivirus infection	Seminis Vegetable Seeds, Inc., Saticoy, CA	US1996000643779	July 11, 2000
US6069299 : Fungus and insect control with chitinolytic enzymes	Cornell Research Foundation, Inc., Ithaca, NY	US1998000025691	May 30, 2000
US6022739 : Systemin	Washington State University Research Foundation, Inc., Pullman, WA	US1997000881094	Feb. 8, 2000

Table 4. Pathogens controlled by genetic resistance in commercial varieties (according to Laterrot, 2000).

Pathogens	Genes	Frequency of resistant varieties (1)
Fungi		
Verticillium dahliae	*Ve*	★★★★★
Fusarium oxysporum f. sp. *lycopersici*		
Pathotype 0 (ex1)	*I*	★★★★★
Pathotype 1 (ex2)	*I-2*	★★★
Fusarium oxysporum f. sp. *radicis lycopersici*	*Frl*	★★
Pyrenochaeta lycopersici	*Py1*	★
Alternaria alternata f. sp. *lycopersici*	*Asc*	★★★★★
Fulvia fulva = *Cladosporium fulvum*	*Cf* (series)	★★★
Stemphylium spp.	*Sm*	★★★
Phytophthora infestans	*Ph-2*	★
Leveillula taurica	*Lv*	★
Oidium lycopersicum	Oligogenic	★

Table 4. Continued

Pathogens	Genes	Frequency of resistant varieties (1)
Bacteria		
Pseudomonas syringae pv tomato	*Pto*	★
Raustonia solanacearum	Oligogenic	★
Viruses		
Tobacco mosaic virus (TMV)	*Tm-2*^{2}	★★★
Tomato spotted wilt virus (TSWV)	*Sw-5*	★
Tomato yellow leaf curl virus (TYLCV)	Oligogenic	★
Nematodes		
Meloidogyne ssp.	*Mi*	★★

(1) ★ (= weak) to ★★★★★(= very frequent)

protein kinases, suggesting its possible role in a signal transduction pathway (Martin *et al.*, 1993).

Prf is located within the *Pto* gene cluster (Salmeron *et al.*, 1996) and besides resistance phenotype has a pleiotropic in the tomato response to the organoposphate insecticide Fenthion (Salmeron *et al.*, 1994). Nucleotide sequence of this gene demonstrated that it encodes a protein with leucine rich repeats, nucleotide binding sites and leucine zipper motifs, characteristic of a plant disease resistance genes (Salmeron *et al.*, 1996). *Pto*-closely linked markers allowed *Prf* gene localization within cosmid contigs and the obtention of *Agrobacterium*-mediated transgenic tomatoes with a *Prf* containing cosmid. Inoculation of transformed plants with P. *syringae* pv. *tomato* strains, test-crosses with a mutant plant (*prf-3*) and Fenthion sensitivity test were conducted, providing evidence for the presence of the *Prf* gene in the introduced cosmid DNA (Salmeron *et al.*, 1996).

Heterologous expression of tomato *Pto* in transgenic tobacco (Thilmony *et al.*, 1995 and Rommens *et al.*, 1995) indicated conservation of the signal transduction pathway, demonstrating the functionality and importance of host-specific resistance genes being transferred between species within related taxonomic groups.

4.2.1.2. Fungal diseases : The tomato–*Cladosporium fulvum* pathosystem is particulary amenable to the determination of the molecular mechanisms driving the evolution of novel specificities at complex *R* gene clusters. The *Cf* genes in tomato confer gene-for-gene resistance to strains of *Cladosporium fulvum* and reside in complex loci carrying multiple genes (Parniske *et al.*, 1997). Measurements of fungal growth revealed that the relative efficiency of the *Cf* genes decreased in the order *Cf-2, Cf-5, Cf-9, Cf-4, Cf-11, Cf-3,* and that the resistant phenotype controlled by *Cf-2* is epistatic over *Cf-5* or *Cf-9,* and *Cf-5* and *Cf-9* genes are epistatic in action to *Cf-4* (Hammond–Kosack *et al.*, 1994). Positional cloning allowed *Cf-2* locus isolation, with two cosmid clones conferring resistance to *C. fulvum* in complementation tests (Dixon *et al.*, 1996). Experiments using specific probes from a tomato genomic library made possible to determine that this locus comprised two independently functional genes. *Cf-2.1* and *Cf-2.2* did not confer phenotypic differences to transformed plants.

The tomato *Cf-9* gene, which confers resistance to infection by races of *C. fulvum* carrying the avirulence gene *Avr9*, was isolated by transposon targeting using a maize *Activator-Dissociation* element (*Ac-Ds*)-based system to target a specific gene from tomato (Jones *et al.* 1994). *Cf-9* heterologous expression led to hypersensitive cell death in transformed potato and tobacco subjected to AVR9 peptide injection (Hammond–Kosack *et al.*, 1998), which enforces the hypothesis that plant mechanisms used to combat microbes may be conserved in different plant families. In order to verify if the resistance phenotype was introduced with the *Cf-9* genomic sequences, transformed tomatoes cv. Moneymaker with the same cosmids used for potato and tobacco transformation were inoculated with *C. fulvum*. It was observed that the susceptible tomato cultivar was resistant to a *C. fulvum* race that expressed the corresponding functional avirulence gene *Avr9*. Honée *et al.* (1998), using cell suspension cultures of tomatoes carrying the *Cf-9* resistance gene, showed that treatment of *Cf-9*-expressing cells with various elicitors, except the AVR9 peptide from *C. fulvum,* induced the expression of defense-related genes. The regeneration capacity of transgenic *Cf-9* leaf discs was not influenced by the expression of *Avr9* transgene, but transgenic plants expressing both the *Cf-9* and *Avr-9* genes were never obtained. Transgenic F_1 seedlings generated from crosses between *Avr-9* expressing plants and wild-type Cf-9 plants died within a few weeks. However, cotyledon-derived callus from these seedlings developed similarly to those that contained only *Cf-9* or the *Avr-9* gene. The induction of defense responses in *Cf-9* tomato cells by the *AVR-9* elicitor was developmentally regulated and is absent in callus tissue and cell-suspension cultures, which consist of undifferentiated cells (Honée *et al.*, 1998).

Another approach is the use of chitinases, which are members of a group of pathogenesis-related (PR) proteins that are up regulated in response to pathogen infection and certain abiotic stresses. In transgenic tomato plants, GUS expression was induced around necrotic lesions caused by *Alternaria solani* and *Phytophthora infenstans* (Samac and Shah, 1991). Tabaeizadeh *et al.* (1999) reported the transformation of the tomato cv. Starfire with a *L. chilense* acidic endochitinase gene (*pcht28*), the first work reporting the *in vivo* effectiveness of an acidic endochitinase. Northern blot analysis revealed a high level of chitinase transcripts in most of the transgenic tomatoes, and inoculation with two *V. dahaliae* races showed that *pcht28*-transformed tomatoes had significantly better resistance than non-transformed plants. This is an entirely distinct resistance process compared to the *Cf* genes. Chitinases are believed to promote chitin hydrolysis, the major fungal cell wall structural polysaccharide, thus limiting fungal growth. On the other hand, *Cf* genes encode receptors that recognize fungi avirulence (*Avr*) genes, which activate plant defense mechanisms leading to hypersensitive cell death, thus limiting colonization of the host by the pathogen.

The use of polygalacturonase inhibitor proteins has been claimed to be another potential defense strategy (Powell *et al.*, 2000). This enzyme inhibits fungal enzymes (endopolygalacturonases) that promote plant cell wall degradation, hence allowing the pathogen entrance in the plant tissue, which will lead to the disease. The coding sequence of a pear polygalacturonase inhibitor protein was transferred to tomato plants by Stotz *et al.* (1993). Transformed plants were morphologically indistinguishable from untransformed plants; yet transgenic pPGIP protein was abundantly produced in all tissues of transgenic ones (Powell *et al.*, 2000). Fruits of *pPGIP* plants were inoculated with *Botrytis cinerea* conidia and up to 15% reduction in symptom severity were observed. Leaves from transformed plants also displayed smaller lesions when inoculated with two strains of *B. cinerea*.

4.2.1.3. Viral diseases : Virus diseases are particularly difficult to manage since they cannot be controlled by chemical treatment (Roselló *et al.*, 1996). Several viruses are known to infect tomato, including ToMV, TSWV, TYLCV, ToMoV, TMV and CMV, among others. Some may lead to complete crop losses, such as TYLCV in the Middle Esat and Mediterranean countries (Czosnek *et al.*, 1990). Because genetic resistance offers the best means of protecting crop plants against viral infections, considerable efforts have been devoted to the development of resistant varieties. The concept of pathogen-derived resistance (PDR) arose with the work of Sanford and Johnston (1985), and created an alternative to introduce virus resistance in agronomic crops, including the tomato. Coat protein-mediated protection (CPMP), movement protein-mediated protection (MPMP) and RNA or DNA-mediated resistance are PDR approaches applied to virus resistance that do not depend on gene silencing (Baucombe, 1996). Conversely, resistance mechanisms dependent on gene silencing, such as antisense RNA or homology–dependent resistance (based on PTGS) are also available (Lindbo *et al.*, 1993; de Carvalho Niebel *et al.*, 1995). This latter mechanism has been intensely studied in recent years, as it seems to be part of a general defense response of plants against viral infection (Anandalakshmi *et al.*, 2000; Covey *et al.*, 1997; Ratcliff *et al.*, 1999; Voinnet *et al.*, 1999). Effective PTGS requires a high degree of sequence specificity, and therefore resistance is highly strain-specific (Baulcombe, 1994, 1996).

Xue *et al.* (1994) obtained transgenic tomatoes expressing a high level of resistance to CMV using CPMP. The practical impact of these plants for controlling CMV was emphasized, since it was the first report of transgenic tomato expressing the CP gene of CMV.

Transgenic tomato plants expressing full-length, translatable forms of the CMV coat protein (CP) were challenged with CMV strains of distinct virulence, and demonstrated tolerance or resistance not only upon mechanical inoculation but also upon natural inoculation by aphid vectors in open field (Gielen *et al.*, 1996). Hybrids carrying hemizygous copies of the CMV CP had diminished levels of protection, although some of the transformant lines produced hybrids that remained fully protected. In their work, Gielen *et al.* (1996) also demonstrated the feasibility of a CMV subgroup I coat protein gene mediating protection to CMV from subgroups I and II, which may support the durability of this resistance, which does not involve gene silencing. Kaniewski *et al.* (1999) noticed that the use of a native CMV-D CP gene that generated highly CMV-resistant tobacco plants was ineffective to produce CMV-resistant tomato lines. It was reasoned that lower gene expression levels led to poor resistance in tomato. Later, additional trials with a CMV CP gene construct under control of the CaMV 35S promoter resulted in a high frequency of tomato plants with extreme resistance to CMV. Although transgenic lines containing two copies of the CP gene could not be distinguished, in terms of CMV resistance, from lines containing a single copy, CP expression levels in the former were higher. Furthermore, plants did not become infected despite the high inoculum concentration used, and it was observed that the best transgenic lines containing a single copy of the CP gene displayed resistance to both CMV subgroups under mechanical inoculation and aphid vector transmission (Kaniewski *et al.*, 1999). More recently, Sree Vidya *et al.* (2000) reported on the successful development of transgenic tomato plants harboring and expressing the CP gene of PhMV that showed at least partial resistance to PhMV.

Several studies evaluating virus resistance of transgenic tomato under field conditions were performed in the past twelve years (Nelson *et al.*, 1988; Sanders *et al.*, 1992; Fuchs and Provvidenti, 1996; Tomassoli *et al.*, 1999). Nelson *et al.* (1988) generated the progeny of two plants containing the TMV coat protein. CP-expressing tomato plants showed a delay or absence of systemic infection under field conditions regardless of the TMV concentration used for challenge. The plants displayed resistance to the highly virulent U1 strain of TMV and also, albeit to a lesser extent, against a ToMV strain. CP expression did not alter fruit yield, growth and development of the transgenic plants. Sanders *et al.* (1992) also studied field resistance of transgenic tomatoes expressing TMV and ToMV coat proteins, though with somewhat distinct results from those obtained by Nelson *et al.* (1988). TMV CP-expressing plants displayed no reduction of fruit production when inoculated with TMV U1 and PV230 strains, whereas yields for control plants decreased 20 and 69% respectively. However, TMV CP-expressing tomato plants were susceptible to ToMV inoculation while ToMV CP-expressing plants were resistance to ToMV.

Fuchs and Provvidenti (1996) evaluated CMV resistance of transgenic tomatoes transformed with the CMV-WL coat protein. Progeny of two hemizygous inbred lines (one homozygous and one hybrid) displayed high levels of resistance and good yield performance, with no recover of CMV from transgenic leaves. Field trials conducted from 1993 to 1997 demonstrated that homozygous lines derived from these plants displayed delayed and milder symptoms compared to control (non-transformed) plants (Tomassoli *et al.*, 1999). These tests also demonstrated the durability of resistance in advanced generations, being the authors convinced that the best line tested (UC82b) is ready for commercialization as a new CMV resistant variety.

The pathogen specificity associated with PDR might be a disadvantage, as observed by Prins *et al.* (1996), Van Dun *et al.* (1988), Sanders *et al.* (1992), Van Dun and Bol (1988) and Anderson *et al.* (1989). Even though, there are examples of broad-spectrum resistance mediated by PDR (Xue *et al.*, 1994; Namba *et al.*, 1991). Besides the level of homology between the transgene and the challenge virus, other factors that might affect the spectrum of PDR include the expression level of the transgene (Quemada *et al.*, 1991; Sanders *et al.*, 1992; Kaniewski *et al.*, 1999) and inoculum concentration (Abel *et al.*, 1986; Nelson *et al.*, 1988). It is also possible that additional unknown factors might play a role on this on this resistance system.

Tomato plants expressing an attenuating CMV satellite RNA sequence were tolerant to CMV when challenged with CMV virion or RNA preparations, or with TAV (McGarvey *et al.*, 1994). Although the progeny of R_0 transgenic plants became infected with CMV, systemic spread seemed blocked and disease symptoms were dramatically attenuated in transformed plants by the third week post-inoculation. Reduction in virus accumulation was also detected compared to control plants. Transformed plants were essentially identical in morphology and fruiting behavior when compared to non-transformed ones. Similar results were obtained by Saito *et al.* (1992). Tomato plants with stable integration and expression of a cDNA for the CMV satRNA T73, exhibited very mild symptoms after challenge with CMV-O. Overall plant growth and fruit set was normal.

Provvidenti and Gonsalves (1995) studied the inheritance of CMV-WL coat protein mediated resistance in a homozygous tomato line. The resistance gene, designated *Cmv*, behaved as a single dominant gene when transformed tomatoes were challenged with CMV-China, and conferred resistance to strains belonging to both CMV subgroups (I and II). Although *npt-II* and *gus* were useful as selectable markers during tissue culture regeneration of the transgenic plants, the authors stressed that they were not completely reliable to predict resistance since they did not follow a mendelian segregation pattern, with 5 to 15% of progeny plants being negative for one or both markers.

Gonsalves *et al.* (1996) reported the development of virus resistant transgenic plants, including tomato, based on PDR. The same strategy was employed by Federowicz *et al.* (2000), which incorporated the TSWV nucleoprotein (N) gene into *L. esculentum*. These authors obtained a population of eighteen plants from three cultivated genotypes, which will be used for further analysis of gene expression and level of resistance to TSWV.

Whithan *et al.* (1996) transformed tomato with the tobacco *N* gene, which confers natural resistance to TMV, and observed that it mediated a hypersensitive response to TMV U1 in transformed plants. Tomato plants carrying the *N* gene also displayed resistance to TMV Cg and ToMV. As observed by Erickson *et al.* (1999), this nicely demonstrated that naturally-ocurring *R* genes can be transferred across species that cannot be crossed or breached by any conventional breeding methods, thus introducing "natural" pathogen resistance into heterologous plant species.

Natural resistance to TSWV has been described in *L. esculentum, L. peruvianum, L. chilense* and *L. hirsutum* (Finlay, 1953; Kumar *et al.*, 1993; Stevens *et al.*, 1994; Barineau *et al.*, 2001). Unfortunately, incorporating this trait into commercial tomato cultivars has been a difficult task as it is associated to undesirable characteristics and because of the *L. esculentum* incompatibility in crosses with *L. peruvianum* and *L.*

chilense. Even though, the dominant *Sw-5* gene was introgressed into cultivated tomato from the wild species *L. peruvianum* and provides resistance against TSWV isolates from different locations (Stevens *et al*., 1992). In addition to TSWV, *Sw-5* or closely linked gene(s) also provide resistance against two other tospovirus species (TSCV and GRSV), thus confering a broad-spectrum resistance to *Tospovirus* (Boiteux and Giordano, 1993; Giordano *et al*., 1994). The resistance phenotype conferred by *Sw-5* varies from immunity to hypersensitive response. The *Sw-5* gene has been recently cloned by chromosome walking (Brommonschenkel, 1996; Brommonschenkel and Tanksley, 1997; Brommonschenkel *et al*., 1997), allowing the incorporation of this trait into commercial tomato cultivars more rapidly by using genetic transformation techniques. Through *Agrobacterium*-mediated transformation experiments it was possible to refine the position of the *Sw-5* locus within the cosmid contig. The resistance phenotype in transgenic plants cosegregated with the introduced TC134 cosmid-derived sequence in several progenies, indicating that resistance was due to the introduced sequence and not to genetic changes due to tissue culture manipulations (Brommonschenkel *et al*., 1998). This association was also observed when the progenies were evaluated for resistance to TCSV and GRSV, indicating that the *Sw-5* tospovirus resistance specificity resides into the TC134 genomic clone. Assuming that all plant-pathogen recognition events fit into the same basic signal transduction pathway, it should be possible to engineer resistance to a pathogen in any plant species merely through the insertion of the genes that mediated the initial phases of plant-pathogen interaction (Brommonschenkel *et al*., 1998). A true test of whether or not resistance genes can provide this kind of protection should utilize an R gene for a pathogen that attacks a variety of plant species. TSWV/*Sw-5* is a good model system for investigating this issue due to the wide host range of TSWV and the broad-spectrum resistance provided by *Sw-5*. Attempting to evaluate the heterologous expression of *Sw-5* gene, Brommonschenkel *et al*. (1998) addressed introduced *Sw-5* into the TSWV susceptible tobacco cv. Havana 425 via *Agrobacterium*-mediated transformation. Disease screening of progenies from *Sw-5* transgenic tobacco plants revealed that *Sw-5* was capable of providing resistance in transgenic tobacco. However, none of the TSWV resistant progenies showed resistance to TCSV or GRSV, indicating that in contrast to the broad-spectrum resistance observed in tomato, the *Sw-5* resistance in transgenic tobacco is species-specific.

In our laboratory, based on protocol suggested by Frary (1995) and Frary and Earle (1996), several tomato cultivars have been transformed. A general procedure is described in Figure 1, as used in transformation experiments for processing and fresh market tomato cultivars (Costa, 1999; Costa *et al.*, 2000b; Nogueira, 2000). By means of *A. tumefaciens*-mediated transformation of the Brazilian processing tomato cultivar 'IPA-6', a transgenic tomato lineage named 'IPA-6::TC134', containing the *Sw-5* gene, was obtained (Costa, 1999) (Figure 2). Analysis of *Sw-5* segregation in 'IPA-6::TC134' using PCR and the chi-square test showed that the proportion of plants carrying and not carrying the transgene fitted a 15:1 ratio, indicating the incorporation of at least two copies of the gene into the recipient's genome. The 'IPA-6::TC134' line was resistant to infection by TSWV, GRSV and TCSV. The resistance phenotype varied from a hypersensitivity response (HR) to apparent immunity, similar to that conferred by *Sw-5* in 'Stevens', the transgene donor. Therefore, the use of the *Sw-5* gene in transformation experiments can be a fast and efficient alternative to incorporate resistance against *Tospovirus* into susceptible tomato cultivars. The resistance provided by *Sw-5* was also studied in heterologous systems such as tobacco, eggplant and lettuce (Picoli, 2000; Lau, 2000; Picoli *et al.*, 2000; Karasawa *et al.*, 2000), resulting in TSWV resistant plants in the first two cases, but not in the latter. Such results may contribute to the understanding of the *Sw-5*-derived resistance and its relation to the signal transduction machinery of the host plant.

Geminiviruses are also important viruses causing severe disease in many crops, including the tomato. Several strategies have been adopted to engineer resistance against geminiviruses. The gene encoding the capsid protein of TYLCV was cloned into a Ti-derived plasmid and used to transform *L. esculentum* × *L. pennellii* hybrids (Kunik *et al.*, 1994). Two groups were observed, one that was undistinguishable from untransformed plants and another where delayed symptoms and recovery from disease were observed. This work demonstrated the successful use of the TYLCV capsid protein in a PDR approach where transformed tomato was resistant to the geminivirus upon whitefly-mediated inoculation. Krake *et al.* (1998) used wild type and frameshift forms of the *C4* gene from TLCV to produce transgenic tomato and tobacco. Similar levels of the transcript were observed for all combinations of plant and construct, although plants transformed with the wild type *C4* gene displayed virus-like symptoms while plants expressing the modified gene were resistant.

Recently, transgenic tomato plants expressing wild-type or mutated BV1 (NS) or BC1 (MP) movement proteins from BDMV were generated and examined for phenotypic effects and resistance to ToMoV (Hou *et al.*, 2000). The authors used this strategy to determine whether nonfunctional geminiviral MPs can act as dominant negative mutants providing broad-spectrum geminivirus resistance. The reduced regeneration and abnormal phenotypes of plants transformed with wild type MP suggested its deleterious effect on tomato development and probably a viral pathogenicity function for this protein. Plants transformed with wild type NS did not display an abnormal phenotype. Transformed plants with wild type MP and NS had the lowest rates of ToMoV infection fourteen days after inoculation, followed by the mutated MP transformants. In all cases (wild type and mutated MP and NS) infection rates were lower than in non-transformed plants. An overall delay in infection was also observed in transformed tomatoes compared to non-transformed plants. It was concluded that expression of wild type viral movement proteins in transgenic plants might have deleterious effects on various aspects of plant development, but the use of mutated movement proteins might be an attractive approach to generate plants with broad-spectrum resistance to geminiviruses.

4.2.1.4. Nematode resistance : Nowadays, the use of modified cultural practices, chemicals and resistant varieties are the main strategies to control nematodes. Nevertheless, efforts are being made to diminish or even cease with the use of nematicides due to the toxicological and environmental problems caused by them. This leads to a demand for alternative nematode management techniques in tomato crops (Rich and Olson, 1999), and the increased use of host plant resistance (HPR) to solve this problem (Roberts, 1992). In tomato, resistance to root-knot nematodes (*Meloidogyne* spp.) is conferred by a single dominant gene, *Mi,* which was introgressed into cultivated tomato from its wild relative species *L. peruvianum* (Smith, 1944). It has proven useful for the management of *M. arenaria, M. incognita* and *M. javanica* and is the basis of all the root-knot resistance in commercial use today (Roberts, 1992). Rich and Olson (1999) observed that cultivars PSR 8991994 and Sanibel, which hold the *Mi* gene, exhibited good resistance to *M. javanica* in three consecutive field trials and that root gall severity did not increase in the resistant cultivars from one test to the next. Even considering the occurrence of resistance-breaking biotypes (Roberts, 1992) and decreased resistance at high temperatures (Dropkin, 1969; Haroon *et al.*, 1993), a more detailed study of *Mi* gene was performed.

In different tomato lines several RFLP markers were found and used to construct a high-resolution map around the *Mi* gene. The closest distance between two known markers containing the *Mi* gene is calculated to be about 2.9 cM, equivalent to about 1.5 Mb. This distance made it possible to apply chromosome-walking techniques with the aid of YAC libraries. The *Mi* gene was recently cloned by Milligan *et al.* (1998). Two genes, *Mi-1.1* and *Mi-1.2,* were found, although only the latter conferred resistance to a root nematode susceptible cultivar. This gene is most similar to another tomato gene that confers resistance to *Pseudomonas syringae* (*Prf*) and its protein has structural motifs found in a protein family that protects plants from viral, bacterial and fungal pathogens (Milligan *et al.*, 1998).

Heat stable resistance genes, named *Mi-2* and *Mi-3,* were found in *L. peruvianum* conferring nematode resistance at higher temperatures in which *Mi* is not effective (Cap *et al.*, 1993; Yaghoobi *et al.*, 1995). Segregation analysis suggested that single dominant genes control the traits, and segregate independently from *Mi*. Yaghoobi *et al.* (1995) applied bulk segregant analysis to identify RAPD molecular markers associated to *Mi-3* and found two markers that are strongly correlated with this gene. These and RFLP data allowed genetic mapping of *Mi-3*, which should facilitate its incorporation into commercial cultivars and provide a tool for map-based cloning of *Mi-3*. Considering the occurrence of *Mi* resistance-breaking biotypes (Roberts, 1992) and decreased resistance at high temperatures (Dropkin, 1969; Haroon *et al.*, 1993), these genes should be candidates to incorporation into commercial genotypes by means of genetic transformation. This approach might also provide less time expenditure in breeding programs aiming to recover desirable agronomic traits. Atkinson *et al.* (1995) pointed out the main advantages of the biotechnological approach, which are: (1) no other changes in agronomic practices are required; (2) a reduction in toxicological and environmental risks associated with chemical control; and (3) the provision of effective, appropriate and inexpensive crop protection.

Engineered resistance to nematodes may be based on strategies such as induced resistance at the sites in which the nematode invades the plant or at the feeding cells, developing plant immunogenic responses to the nematode cuticle or to secretions, and production of anti-nutritional factors (Atkinson *et al.*, 1995; Grundler, 1996; Vrain, 1999). The search and use of root-specific promoters that respond precisely to nematode invasion would facilitate the acceptability of these approaches (Atkinson *et al.*, 1995; Vrain, 1999). One example is the promoter of a defense-related gene (hydroxymethylglutaryl CoA reductase), triggered by fungal and bacterial pathogens, that drove strong GUS expression in root-knot nematode giant cells in tomato plants (Cramer *et al.*, 1993).

Due to the unpredictable durability of natural resistance genes, as already forewarned by Roberts (1992), the search and introduction of new nematode resistance genes must be a priority in breeding programs. A screening for new genes conferring resistance to root-knot nematodes, performed by Ammati *et al.* (1985), suggested the existence of new resistance gene(s) for nematode resistance in *Lycopersicon* ssp. Some of these genes control the expression of protein inhibitors (PI), which are defense-related proteins induced by wounding and herbivory. There are some examples of the use of transgenic plants expressing PIs aiming at insect resistance (Hilder *et al.*, 1987). Nematode control based on PIs such as cowpea trypsin inhibitor and cystatins (Hilder *et al.*, 1987; Atkinson *et al.*, 1995) may soon become a reality for some crops. *Agrobacterium rhizogenes*-transformed tomato lines expressing cystatins, orizacystatin I (Oc-I) and one variant form (Oc-IΔ86), induced a slower-than-normal growth of *Globodera pallida*. A decrease in female size and the prevention of egg production were also observed (Urwin *et al.*, 1995; Atkinson *et al.*, 1995).

Engineered resistance may provide protection against nematodes when conventional means are not effective, too expensive or not available, besides avoiding environmental hazards. However, the molecular interactions between the host and the parasite and the possible effects of transgenic plants in the environment must be elucidated (Grundler, 1996). To date, in tomato the only report on that was by Frary (1995) *Agrobacterium*-mediated transformation of tomato was used in an attempt to identify *Mi* gene for resistance to root-knot nematodes. To test for complementation of the susceptible phenotype, antisense of the resistant phenotype, or cosuppression of the resistant phenotype, experiments were performed to introduce each of the 5 different *Mi* cDNA sense and antisense constructs into both root-knot nematode susceptible (Moneymaker) and resistant (Motelle) tomato cultivars, and each of 15 genomic DNA sense constructs into Moneymaker (Frary, 1995). Root-knot nematode screening revealed that none of the clones isolated and used for transformation corresponded to the root-knot nematode gene, *Mi*. As reported before (section 4.2.1.3) only recently Brommonschenkel *et al.* (2000) reported on the homology between tomato *Sw-5* gene and *Mi*.

4.2.2. Insect resistance

This is an important agronomical trait that has been approached using a variety of methods. By far the most successful strategy used to date is the expression of the *Bacillus thuringiensis* δ-endotoxin (or a truncated version of it) in transgenic plants (Houck *et al.*, 1993). A slightly different version of the *B. thuringiensis* toxin confers resistance to *Manduca sexta, Helliothis zea,* and *Keiferia lycopersicella* in transgenic tomato under greenhouse conditions (Fischoff *et al.*, 1987) or in the field (Delannay *et al.*, 1989). Tolerance to coleoptera attack was achieved by transforming tomato with a modified *B. thurigiensis* sbsp. *tenebrionis* δ-endotoxin gene (Cho *et al.*, 1993; Rhim *et al.*, 1995; Rhim, 1998b). Rhim *et al.*, 1995 obtained three transgenic tomato plants with tolerance to coleoptera larvae, and the same transgene conferred tomato plants insecticidal activity against coleoptera larvae bred to the 4th generation. The expression caused a significant insecticidal activity of tomato plants against larvae of the potato colorado beetle (Rhim, 1998b).

Proteinase inhibitors are produced locally and systemically in response to mechanical wounding (Ryan, 1990) such as insect wound. McGurl *et al.* (1992) suggest that systemin is the polypeptide associated with signal transduction and wound induction of defense proteins in tomato given the systemic movement at the same rate of the endogenous wound signal and activation of proteinase expression. Transgenic tomato expressing anti-sense prosystemin, the systemin precursor, displaying reduced proteinase inhibitors synthesis, and constitutive expression of prosystemin transgene in tomato leading to high levels of proteinase inhibitors I and II, corroborate this hypothesis (McGurl *et al.*, 1992; Orozco-Cardenas *et al.*, 1993; McGurl *et al.*, 1994).

Employing a tomato prosystemin-β-glucuronidase (*Prosys*-Gus) fusion reporter gene in tomato plants and tissue printing with specific tomato prosystemin antibodies, Jacinto *et al.* (1997) provided evidence for prosystemin synthesis in the leaf vascular bundles, petioles and stems. GUS activity increased in response to wound and methyl jasmonate (MJ) application, which also induces defense genes, and correlated with increased prosystemin mRNA levels in the main veins of tomato leaves. Later on, Jacinto *et al.* (1999) recovered tobacco plants transformed with the same construct that exhibited wound- and MJ-inducible GUS activity and with the same localization pattern observed for tomato, indicating the correct recognition by tobacco transcription factors. Although the isolation of prosystemin cDNA in tobacco was not successful, increased trypsin inhibitor activity in MJ treated leaves suggests that wound inducible defense signals are operational in tobacco regulating a prosystemin-like promoter. Hence, novel insect resistance genes from homologous or heterologous systems, regulated by a prosystemin or a constitutive promoter, are amenable to be used for tomato given the suggested similarity between the two systems.

Tomato plants expressing *Bacillus thuringiensis cry* genes, partially or fully modified, were resistant to a variety of insects (Fischhcoff *et al.*, 1987; Perlak *et al.*, 1991; van der Salm *et al.*, 1994; Mandaokar *et al.*, 2000). In general, transformation with wild-type genes yields low expression levels and confers resistance to only some insect species, modifications of the coding sequences may led to marked increases in expression levels and broad-spectrum resistance (Perlak *et al.*, 1991; van der Salm *et al.*, 1994).

Transgenic tomatoes displayed resistance to *Manduca sexta, Heliothis virescens* and *Heliothis zea, Leptinotarsa decemlineata* and *Helicoverpa armigera* (Fischhcoff *et al.*, 1987; van der Salm *et al.*, 1994; Rhim, 1998; Mandaokar *et al.*, 2000), although this resistance ranged from lethal to a stunting effect on larval growth. Differences in protein expression suggested that protection was related to higher levels of transgene expression (Mandaokar *et al.*, 2000). Usually modified genes exhibited superior expression levels (Fischhcoff *et al.*, 1987; van der Salm *et al.*, 1994), as these altered genes had potential instability motifs and putative poliadenilation signals removed (Perlak *et al.*, 1991; van der Salm *et al.*, 1994).

Relevant arguments and data are raised about the development of insects with resistance to *Bt* crystal proteins (Rie *et al.*, 1990). However, expression of *cryIC-cryIA(b)* fused gene construction opens up an alternative to broad-spectrum insect resistance and the use of different gene classes in resistance management strategies to control pest insects applying transgene technology as described by van der Salm *et al.* (1994).

4.3. Other traits

In addition to disease resistance, insect resistance and fruit quality, other tomato characteristics have been modified by *Agrobacterium*-mediated transformation. Physiological aspects such as photosynthetic enhancement, yield increase, altered carbohydrate metabolism, and improvement to stress responses have been studied. The area of metabolism has been a focus for fundamental and applied research recently (Dunwell, 2000). Of all the specific enzymes which activity has been modified, possibly the most well studied is sucrose phosphate synthase (SPS), a key enzyme involved in the regulation of sucrose metabolism (Dunwell, 2000).

Transformed tomato plants expressing both the native form of SPS and the SPS gene from maize have been produced (Worrell *et al.*, 1991). When expressed from a cDNA sequence for the SPS gene from maize under control of a ribulose bisphosphate carboxylase small subunit promoter in transgenic tomatoes, total SPS activity was boosted up to sixfold in leaves and appeared to be physiologically uncoupled from the tomato regulatory machinery (Worrell *et al.*, 1991). Also, the elevated SPS activity caused a reduction in starch and increase in sucrose levels in tomato leaves, indicating that SPS is involved in the regulation of carbon partitioning in the leaves. Several studies on the physiology and metabolism of transformed tomato plants overexpressing the maize SPS and untransformed control tomatoes were further carried out (Galtier *et al.*, 1993; Galtier *et al.*, 1995; Micallef *et al.*, 1995; Laporte *et al.*, 1997; Murchie *et al.*, 1999). SPS overexpression led to increased foliar sucrose and decreased starch accumulation in the leaves such that sucrose/starch ratio was positively correlated with total extractable SPS (Murchie *et al.*, 1999). In general, a positive correlation was observed between the ratio of sucrose to starch and SPS activity, and the photosynthetic capacity was increased in plants with high SPS activity. For the food industry the increase of SPS activity is of great importance since it improves the yield of solids for processing (Tucker, 1993). D'Aoust *et al.* (1999) obtained transgenic plants expressing two versions (*Sh35* and and *Sh1*) of the maize promoter *Shrunken-1* (*Sh1*), which in addition to *Sus1* controls the expression of sucrose synthase (*Susy*), fused with *gus*. For both promoters very faint gus expression was detected in vegetative tissues, and no expression was detected in fruit pericarp tissues. However, in seeds *Sh1* promoted low gus expression, but *Sh35* directed higher *gus* expression. (D'Aoust *et al.*, 1999).

At present, our laboratory is involved in a collaborative work with Prof. E.P.B. Fontes developing transgenic tomato plants expressing sense or antisense copies of a sucrose binding protein (SBP) homologue, S64, which has been recently reported by Pedra *et al.* (2000), and also an ER-resident protein, BiP (Alvim *et al.*, 2001). In tobacco, constitutive expression of S64 or BiP positively affected photosynthetic rates, carbohydrate partitioning and increased the tolerance to water deficit.

These genes appear then as relevant candidates for further physiological studies in tomato.

Tomato plants were transformed with gene constructs containing a tomato alcohol dehydrogenase (ADH) cDNA coupled with a sense-oriented CaMV 35S promoter or the fruit specific PG promoter (Speirs *et al.*, 1998). Ripening fruits from plants transformed with the constitutively expressed transgene(s) or with fruit-specific genes displayed a range of ADH activities, but supression was observed only in the former. The phenotypes of modified fruit were transmitted to second generation plants in accordance to the inheritance pattern of the transgenes. Modified ADH levels in ripening fruit positively influenced the balance between some of the aldehydes and the corresponding alcohols associated with flavour production. In a preliminary taste trial, fruit with elevated ADH activity and higher levels of alcohol were identified as having a more intense "ripe fruit" flavour (Speirs *et al.*, 1998).

Manipulation of intracellular phytochrome levels and analysis of the subsequent effects on plant phenotype has been approached by overexpressing a phytochrome gene in transgenic tomato plants. Boylan and Quail (1989) have introduced an oat phytochrome cDNA into tomato plants under transcriptional control of the CaMV 35S promoter. Overexpression of the gene resulted in the production and assembly of the active oat phytochrome and a dramatic and pleiotropic effect on tomato phenotype: seedlings developed short hypocotyls with a high anthocyanin content and adult plants were short and bushy, and dark green with more deeply pigmented and firmer fruit than untransformed counterparts.

Klee and Lanahan (1995) pointed out the reliability of using transgenic plants in basic studies, among them hormone biology. Many studies have demonstrated the influence of phytohormones on plant gene expression (Martineau *et al.*, 1994; Storti *et al.*, 1994; Bettini *et al.*, 1998). Martineau *et al.* (1994) created a chimeric *ipt* gene consisting of the *A. tumefaciens* coding region and a promoter from a gene that was highly expressed in tomato ovaries. When this construct was introduced into cultivar UC82B, the resulting transformants had significantly higher levels of total soluble fruit solids than non-transformed plants. Elevated levels of PR-1 and chitinase mRNAs were also detected in transformed plants, but once fruits and flowers were removed the mRNA levels were similar to those from non-transformed plants. This supports the hypothesis that cytokinin produced in fruits can be transported to leaves where it influences expression of defense-related genes (Martineau *et al.*, 1994). Bettini *et al.* (1998) and Storti *et al.* (1994) scrutinized phytormone balance in transgenic plants associated to the competence of tomato cells to defend themselves against *Fusarium oxysporum* infection. The expression of *Agrobacterium* phytohormone-related genes led a susceptible cultivar to inhibit mycelial growth and germ–tube elongation of *Fusarium* conidia. Higher callose content, peroxidase activation after pathogen challenge and resistance to fusaric acid might have contributed to the acquisition of this defense capacity. Bettini *et al.* (1998) observed extracellular PR-proteins (acidic chitinase and PR-1) being constitutively expressed in transformed plants besides a higher ethylene evolution in these plants in response to treatment with fungal cell-wall components, which was not observed in the control. As already suggested by these authors, and supported by the above-mentioned results, the altered hormone balance in such plants indicates a possible strategy for obtaining pathogen resistance through a general or target-organ specific modification.

Aloni *et al.* (1998) confirmed the hypothesis that in addition to the well-defined roles of auxin and cytokinin, there is a critical role for ethylene in determining morphogenesis. The authors compared *Agrobacterium tumefaciens*-induced crown-gall in an ethylene-insensitive mutant, *Never ripe (Nr)*, and its isogenic wild-type parent. It was shown that infection by *A. tumefaciens* in the latter resulted in higher rates of ethylene evolution (up to 50-fold greater than control), decreased vessel differentiation, and a typical unorganized callus shape of the gall. Conversely, *Nr* stems displayed galls with a smooth surface and very limited effect on vessel differentiation.

Considering that salt stress greatly affects and sometimes impairs the productivity in several areas, and given the multigenic nature of this trait, genetic transformation of crops with single transgenes may offer interesting alternatives to conventional breeding. It could lead to a slight improvement in tolerance levels, which could be sufficient from a breeding point of view (Gisbert *et al.*, 2000). The regeneration of transgenic plants overexpressing these genes may clarify if their expression is part of a salt tolerance machinery, or alternatively it is a consequence of the salt stress situation (Arrilaga *et al.*, 1998). In this way, several halotolerance (HAL) genes have been isolated and expressed successfully in heterologous systems such as tomato (Arrilaga *et al.*, 1998; Gisbert *et al.*, 2000). *Agrobacterium*-mediated transformation was used to overexpress two genes from *Saccharomyces cerevisae, HAL2* (Arrilaga *et al*., 1998) and *HAL1* (Gisbert *et al*., 2000), which are involved in the regulation of K^+ and Na^+ transport, respectively. Under salt stress, callus formation from hypocotyl explants was higher on the *HAL2*-expressing transgenic line than in control plants. In addition, the former also showed a higher level of shoot production on NaCl-supplemented medium. *HAL2* overexpression increased tolerance to high lithium and sodium concentrations because it encodes a cation-sensitive nucleotidase required for sulfate assimilation (Arrilaga *et al*., 1998). Results from different tests indicated a high level of salt tolerance in the two different transgenic plants bearing four copies or one copy of the *HAL1* gene (Gisbert *et al*., 2000). In addition, measurement of the intracellular K^+ to Na^+ ratios showed that transgenic lines were able to retain more K^+ than the control under salt stress. Although plants and yeast cannot be compared in an absolute sense, these results indicate that the mechanism controlling the positive effect of the *HAL1* gene on salt tolerance may be similar in transgenic plants and yeast (Gisbert *et al*., 2000).

Recently, the role of ethylene in the susceptibility of tomato to race 2 of *Verticillium dahlie* was examined by creating transgenic plants in which ethylene exposure synthesis was prevented or reduced by the expression of the bacterial ACC deaminase transgene under the control of one of three promoters: 35S CaMV, *rol*D and the promoter for the pathogenesis-related protein *prb*-1 of tobacco (Robinson *et al*., submitted). Significant reductions in wilt symptoms were obtained for *rol*-D and *prb*-1b, but not for 35S-transformants. The pathogen was detected in stem sections of plants with reduced symptoms suggesting that reduced ethylene synthesis resulted in increased disease tolerance. Such ACC deaminase-transformed plants, when compared to normal plants, displayed an increased ability to grow in the presence of Cd, Co, Cu, Mg, Ni, Pb or Zn and to accumulate these metals (Grichko *et al*., 2000). In general, transgenic tomato plants expressing ACC deaminase, especially those controlled by *prb*-1b promoter, acquired greater amount of metals within the plant tissues, and were less subject to the deleterious effects of these metals on plant growth than non-transgenic counterparts (Grichko *et al*., 2000). Further on, using the same tomato lines generated by Robison *et al*., Grichko and Glick (2001) evaluated flooding tolerance of homozygous transgenic tomato plants. All of the transgenic tomato plants expressing ACC deaminase showed some increased tolerance to flooding stress and were less subject to the deleterious effects of root hypoxia on plant growth than were non-transformed plants. Plants that included an ACC deaminase gene under the control of the *rol*D promoter were protected to the greatest extent.

In an attempt to create tomato fruit with improved freezing tolerance, Hightower *et al.* (1991) transformed a fusion construct containing staphylococcal protein A and a synthetic antifreeze gene into cultivated tomato. The researchers found that the gene fusion was expressed and inhibited ice recrystallization in protein extracts from the transgenic plants. Bruggemann *et al.* (1999) studied antioxidants and antioxidative enzymes and the possible role of glutathione reductase (GR) in tomato genotypes differing in chilling tolerance and in transgenic tomato plants overexpressing GR. Three lines with significantly increased GR activity were identified. However, these plants presented identical chilling sensitivity of the photosynthetic apparatus when compared with wild-type plants, as measured after a photoinhibition treatment and by the effect of long-term chilling on rubisco activity. It was concluded that GR was not the limiting factor for avoidance of oxidative stress.

Nowadays, considerable attention has been directed towards the possibility of using plants to remove heavy metals from the environment, a process called phytodetoxification or phytoremediation (Bisizyly *et al.*, 2000; Salt *et al.*, 1998). As defined by Grinchko *et al.* (2000), it is the use of plants to remove, destroy or sequester hazardous substances from the environment. The authors came to the conclusion that plants with lowered levels of stressed ethylene may be useful components to phytoremediation strategies and, based on transgenic tomato responses, good results may be obtained in engineering other plants to be more effective in heavy metal phytoremediation.

Transgenic tomato plants overexpressing peroxidase have been obtained (Lagrimi *et al.*, 1992; Sherf *et al.*, 1993; Mansouri *et al.*, 1999). These enzymes are involved in several physiological and biochemical processes, including cell growth and expansion, differentiation and development, auxin catabolism, lignification and abiotic and biotic stresses (see Mansouri *et al.*, 1999). Based on the expression pattern of a basic tomato peroxidase gene, *tpx1*, a role for this gene in the stress adaptation process was suggested, being probably involved in the biosynthesis of lignin (Mansouri *et al.*, 1999). Tomato plants overexpressing the gene *tpx1* under control of the CaMV 35S promoter were obtained. Transgenic plants showed a 2-5 fold increase in the activity of peroxidase ionically bound to the cell wall, whereas soluble peroxidase activity remained similar or even lower than wild-type plants

(Mansouri *et al.*, 1999). The effects of the gene on leaf lignin levels, rootability of shoots and *in vitro* morphogenesis of different tomato tissues were evaluated. Interestingly, the root system was underdeveloped in transgenic plants, but rootability of the stem was not affected by overexpression of peroxidase. There was no difference in the morphogenetic response of cotyledon explants, however when hypocotyl explants were tested one of the transgenic lines showed a 30% reduction in the frequency of shoot organogenesis. Also, a 40-220% increment of leaf lignin content was found in transgenic plants. The results supported the involvement of the basic *tpx1* isoperoxidase in the lignification process.

Agrabacterium-mediated transformation has also been a useful tool in studies of plant morphogenesis. Jenssen *et al.* (1998) demonstrated that overexpression of a tomato homeobox class I *knox* gene, *LeT6,* led to dramatic changes in leaf morphology. Otherwise, overexpression of a heterologous homeobox gene, *kn1,* did not produce such variability. The authors suggest a fundamental role for *LeT6* in tomato leaf morphogenesis and that variability in homeobox gene expression may account for some of the diversity in leaf forms seen in nature. Rochange and McQueen-Mason (2000) reported successful expression of a recombinant expansin [*CsExp1,* isolated from cucumber hypocotyls] in transgenic tomato plants, under the control of a constitutive promoter. In some transformants, *CsExpe1* transcript and protein accumulated to high levels, and expansin activity from the cell wall was increased up to about 20-fold the activity measured in wild-type plants. Expansins, which comprise multigene families, have been suggested to play an active role in growth and cell wall disassembly during fruit ripening (Rochange and McQueen-Mason, 2000).

Minor importance has been given to herbicide tolerance studies in tomato, although they are also available. Greenhouse experiments were carried out with *aroA* gene- transformed tomatoes, which displayed glyphosate tolerance (Fillatti *et al.*, 1987). Resistance to phosphonothricin (an inhibitor of glutamine synthase) has also been achieved in transgenic tomato by expressing the *bar* gene from *Streptomyces hygroscopicus* (De Block *et al.*, 1987). Incidentally, Fillatis' work was fundamental since it provided a basic protocol for tomato transformation used by many other groups (Table 1).

5. PUBLIC ACCEPTANCE OF TRANSGENIC PLANTS

The scenery and context of tomato transgenesis are similar to those in other crops, where economics, crop yield and the particular interests of the target commodity market led to the use of this technology. The broad utilization of pest resistance traits, especially virus resistance, in tomato transformation efforts is noticeable in the literature. Conversely, herbicide resistance is not detected, at least not to the level observed for soybean or cotton. Nevertheless, increasing issues on fruit quality, storage and more refined applications such as vaccines, hormones and human protein production are now being unraveled. Patents involving varieties with these novel traits are already being requested as seen in Table 3.

Solanaceous crops were the first to be tested in transgenic field trials because they were the first to be transformed experimentally (Houck *et al.*, 1993). Tomato is self-fertile and fruit production is dependent on the completion of the reproductive cycle, like other fertile species in the family. Its heavy pollen is not wind-borne to any significant degree. Current regulations require the use of borders and pollen traps to limit pollen dispersal to within field trials. Thus pollen spread is dependent on insect pollination (Houck *et al.*, 1993).

Debate about the risks connected with plant genome organization, gene flow to wild species, horizontal transfer, impact of the products on non-target species, allergenic potential and even social-economic impacts that transgenics may cause are in vogue. The subject has been extensively reviewed elsewhere (Sasson, 1998; Dunwell, 2000). Hails (2000) discussed some of these wide open questions involving transgenic 'fitness' and ecological effects. Concerns about the safety and eventual hazards associated with the use of antibiotic resistance as selectable markers is a current issue connected with transgenics, especially those bound for human nourishment.

Risk assessment of selectable marker genes has focused on neomycin phosphotransferase (*nptII*), used for instance in the Flvr Savr™ tomatoes (Flavell *et al.*, 1992; Fuchs *et al.*, 1993). The data confirmed that the NPTII protein is rapidly degraded under simulated mammalian digestion conditions and caused no deleterious effects in mice fed with exaggerated doses (up to 5000 mg kg^{-1} body weight) (Fuchs *et al.*, 1993). Selectable markers substitutes, which do not involve antibiotic resistance, such as the oncogenes *IPT, ROL* A, B and C (Ebinuma *et al.*, submitted) and the phosphomannose-isomerase (*pmi*) gene, have recently been developed and used in other crops (Negrotto *et al.*, 2000; Wang *et al.*, 2000). Hashimoto *et al.* (1999) investigated the potential use of the cytosine deaminase (*cd*) gene, which deaminates cytosine to uracil, as a conditional negative marker in tomato hairy roots. Negative markers, also called suicidal genes, are disadvantageous for cells expressing them. For instance, the *cd* gene converts a non-toxic compound, 5-fluorocytosine (5FC), into a lethal one, 5-fluorouracil (Hashimoto *et al.*, 1999). It was used in a co-transformation strategy were the wild *A. rhizogenes* strain 8196 harbored two plasmids with transferable T-DNA, one located in the pRi8196 and other on the pRHY1 binary vector. The latter contained the *cd* and *nptII* genes. The results supported the hypothesis of negative selection being a good selection strategy, since every tomato root clone that did display kanamycin resistance did also exhibit a lethal phenotype when 5FC was present in the media. The authors did point out the possible use of this method in mutagenesis and transposable element studies since this negative marker confers a dominant lethal characteristic. Although some adaptations to tomato will probably be required, these types pf markers may increase public acceptance of transgenic tomatoes, since the putative hazardous effects associated with the use of antibiotics and antibiotic resistance genes will be excluded.

In the co-transformation system the marker gene and the gene(s) of interest are placed in separate DNA molecules introduced in plant genomes as unlinked fragments (Komari *et al.*, 1996). Hence, it allows segregation of these genes in the progeny (Komari *et al.*, 1996), besides providing an easier way to pyramid new transgenes into one cultivar. It might be an interesting alternative for forthcoming breeding programs or research involving tomato transformation.

Goldsbrough *et al.* (1993) described a tomato transformation system that utilizes the maize Ac/Ds transposable element system to reposition or eliminate transgenes subsequent to a primary transformation event. Tomato plants were transformed with T-DNA vectors containing an Ac transposase gene, *nptII* and a chimeric Ds element bearing the GUS reporter gene. GUS activity in the progeny plants with Ds/GUS elements moved to different and some cases multiple locations via transposition, showing a range of expression levels. In a number of the cases, these elements were transposed to distant sites, which allow differential elimination of the transgenes by genetic segregation and generated selective marker-free transgenic tomato lines.

6. CONCLUSIONS AND FUTURE PROSPECTS

Efficient tomato transformation protocols are available in the literature, yet small adjustments are sometimes necessary. The development of new cultivars with resistance to old and emerging pests and pathogens will certainly benefit from the cloning and characterization of R genes from related species, since evidence for signal transduction pathways conservation is suggested from studies on heterologous systems. The same may be stated regarding cloned genes from tomato being used in other species.

Novel genes, especially those related to product quality and medical applications, are now being employed in genetically modified plants. This is a corolary of the success of transgenics, which is however linked to public acceptance and general belief concerning its safety. The production of genetically modified tomato will certainly follow the use of marker-free OGMs (Carr, 2000), which seem to be better accepted commercially. Transgene manipulation and engineering, no matter what is the characteristic involved, has a molecular, ecological, economical and ethical context to be evaluated. The possibilities are almost endless, including crop resistance and quality improvement, management of environmental problems, using of plants as bioreactors yielding medicines, hormones or other willed substances, and undoubtfully granting benefits to humans and generating a differentiated organism or products that may have increased aggregated value.

ACKNOWLEDGMENTS

Authors thank to Dr. JJ Giovannoni, Dr. GA Tuker, Dr. BR Glick, Dr. MM Robison, Dr. SL Rhim and other colleagues who kindly made available their publications upon our request, easing a lot our lack of updated literature facilities. The authors are indebt with Fundação de Amparo à Pesquisa do Estado de Minas Gerais (FAPEMIG) for financial support to tomato project in our laboratory.

REFERENCES

Abel PP, Nelson RS, De B, Hoffmann N, Rogers SR, Fraley RT and Beachy RN (1996) Delay of disease development in transgenic plants that express the tobacco mosaic virus coat protein gene. *Science,* **232** : 738-743.

Adams TL and Quiros CF (1985) Somatic hybridization between *Lycopersicon peruvianum* and *L. pennellii*: regenerating ability and antibiotic resistance as selection systems. *Plant Sci.,* **40** : 209-212.

Agharbaoui Z, Greer AF and Tabaeizadeh Z (1995) Transformation of the wild tomato *Lycopersicon chilense* Dun. by *Agrobacterium tumefaciens. Plant Cell Rep.,* **15** : 102-105.

Aloni R, Wolf A, Feigenbaum P, Avni A and Klee HJ (1998) The *Never ripe* mutant provides evidence that tumor-induced ethylene controls the morphogenesis of *Agrobacterium tumefaciens*-induced crown galls on tomato stems. *Plant Physiol.*, **117** : 841-849.

Alvim FC, Carolino SMB, Cascardo JCM, Martinez CA, Otoni WC and Fontes EPB. Enhanced accumulation of BiP in transgenic plants confers tolerance to abiotic stresses. *Plant Physiol. (Submitted).*

Ammati M, Thomason IJ and Roberts PA (1985) Screening *Lycopersicon* ssp. for new genes imparting resistance to root-knot nematodes (*Meloidogyne* spp.). *Plant Dis.*, **69** : 112-115.

Anderson EJ, Stark DM, Nelson RS, Powell PA, Tumer NE and Beachy RN (1989) Transgenic plants that express the coat protein genes of tobacco mosaic virus or alfalfa mosaic virus interfere with disease development of some nonrelated viruses. *Phytopathology,* **79** : 1284-1290.

Anandalakshmi R, Marathe R, Ge X, Herr Jr JM, Mau C, Mallory A, Pruss G, Bowman L and Vance VB (2000) A calmodulin-related protein that suppresses post transcriptional gene silencing in plants. *Science,* **290** : 142-144.

Arrilaga I, Gill-Mascarell R, Gisbert C, Sales E, Montesinos C, Serrano R and Moreno V (1998) Expression of the yeast *HAL2* gene in tomato increases the *in vitro* salt tolerance of transgenic progenies. *Plant Sci.*, **136** : 219-226.

Atkinson HJ, Urwin PE, Hansen E and McPherson MJ (1995) Designs for engineered resistance to root-parasitic nematodes. *TIBTECH.*, **13** : 369-374.

Barineau MS, Scott JW, Canady MA and Stevens MR (2001) Tomato spotted wilt virus (TSWV) resistance in tomato derived from *Lycopersicon chilense* Dun. LA 1938. *Euphytica,* **117** : 19-25.

Baulcombe DC (1994) Novel strategies for engineering virus resistance in plants. *Curr. Op. Biotech.*, **5** : 117-124.

Baulcombe DC (1996) Mechanisms of plant-derived resistance to viruses in transgenic plants. *Plant Cell,* **8** : 1833-1844.

Baum K, Gröning B and Meier I (1997) Improved ballistic transient transformation conditions for tomato fruit allow identification of organ-specific contributions of I-box and G-box to the *RBCS2* promoter activity. *Plant J.*, **12** : 463-469.

Beaudoin N and Rothstein SJ (1997) Developmental regulation of two tomato lipoxygenase promoters in transgenic tobacco and tomato. *Plant Mol. Biol.*, **33** : 835-846.

Bec S, Chen L, Ferrière NM, Legravre T, Fauquet C and Guiderdoni E (1998) Comparative histology of microprojectile-mediated gene transfer to embryogenic calli in japonica rice (*Oryza sativa* L.): influence of the structural organization of target tissues on genotype transformation ability. *Plant Sci.*, **138** : 177-190.

Bellini C, Chupeau MC, Guerche P, Vastra G and Chupeau Y (1989) Transformation of *Lycopersicon peruvianum* and *Lycopersicon esculentum* mesophyll protoplasts by electroporation. *Plant Sci.*, **65** : 63-75.

Bent AF (1996) Plant disease resistance genes: function meets structure. *Plant Cell,* **8** : 1757-1771.

Bettini P, Cosi E, Pellegrini MG, Turbanti L, Vendramin GG and Buiatti M (1998) Modification of competence for *in vitro* response to *Fusarium oxysporum* in tomato cells. III. PR-protein gene expression and ethylene evolution in tomato cell lines transgenic for phytohormone-related genes. *Theor. Appl. Genet.*, **97** : 575-583.

Bird CR, Ray JÁ, Fletcher JD, Boniwell JM, Bird AS, Teulieres C, Blain I, Bramley PM and Schuch W (1991) Using antisense RNA to study gene function: inhibition of carotenoid biosynthesis in transgenic tomatoes. *Bio/Technology,* **9** : 635-639.

Bird CR, Smith CJ, Ray JA, Moureau P, Bevan MW, Bird AS, Hughes S, Morris PC, Grierson D and Schuch W (1988) The tomato polygalacturonase gene and ripening-specific expression in transgenic plants. *Plant Mol. Biol.,* **11** : 651-662.

Bizily SP, Rugh CL and Meagher RB (2000) Phytodetoxification of hazardous organomercurials by genetically engineered plants. *Nature Biotech.,* **18** : 213-217.

Boiteux LS and Giordano L de B (1993) Genetic basis of resistance against two Tospovirus species in tomato (*Lycopersicon esculentum*). *Euphytica,* **71** : 151-154.

Boylan MT and Quail PH (1989) Oat phytochrome is biologically active in transgenic tomatoes. *Plant Cell,* **1** : 765-773.

Branca C, Bucci G, Domiano O, Ricci A, Torelli A and Bassi M (1991) Auxin structure and activity on tomato morphogenesis *in vitro* and pea stem elongation. *Plant Cell. Tiss. Org. Cult.,* **24** : 105-114.

Bramley PM, Teulieres C, Blain I, Bird CR and Schuch W (1992) Biochemical characterization of transgenic plants in which carotenoid synthesis has been inhibited through the expression of antisense RNA to pTOM5. *Plant J.,* **2** : 343-349.

Brommonschenkel SH (1996) Genetic mapping and progress towards the positional cloning of the *Sw-5* tomato spotted wilt virus (TSWV) resistance gene in tomato (*Lycopersicon esculentum* Mill.) *Ph.D. Thesis,* Cornell University, Cornell.

Brommonschenkel SH and Otoni WC (1998) Santa Clara-TSW, a new fresh market tomato cultivar resistant to tospoviruses obtained by genetic transformation. *Genet. Mol. Biol.,* **21 [Suppl.]** : 391.

Brommonschenkel SH and Tanksley SD (1997) Map-based cloning of the tomato genomic region that spans the *Sw-5* tospovirus resistance gene in tomato. *Mol. Gen. Genet.,* **256** : 121-126.

Brommonschenkel SH, Tanksley SD, Frary A, Frary A, Otoni WC and Cheavegatti A (1998) Positional cloning, molecular characterization and heterologous expression of the *Sw-5* of tomato Tospovirus resistance gene. In: *7th International Congress of Plant Pathology,* Vol. 3. International Society for Plant Pathology, Edinburgh, Scotland (Abstract 5.4.7).

Brommonschenkel SH, Frary A, Frary A and Tanksley SD (2000) The broad-spectrum Tospovirus resistance gene *Sw-5* of tomato is a homolog of the root-knot nematode resistance gene *Mi. Mol. Plant-Microbe Interact.,* **10** : 1130-1138.

Brüggemann W, Beyel V, Brodka M, Poth H, Weil M and Stockhaus J (1999) Antioxidants and antioxidative enzymes in wild-type and transgenic *Lycopersicon* genotypes of different chilling tolerance. *Plant Sci.,* **140** : 145-154.

Brunetti A, Tavazza M, Noris E, Tavazza R, Caciagli P, Ancora G, Crespi S and Accotto GP (1997) High expression of truncated viral rep protein confers resistance to tomato yellow leaf curl virus in transgenic tomato plants. *Mol. Plant-Microbe Interact.,* **10** : 571-579.

Cap GB, Roberts PA and Thomason IJ (1993) Inheritance of heat-stable resistance to *Meloidogyne incognita* in *Lycopersicon peruvianum* and its relationship to the *Mi* gene. *Theor. Appl. Genet.*, **85** : 777-783.

Carr S (2000) EU safety regulation of genetically-modified crops - Summary of a tem country study funded by DGXII under its Biotechnology program. The Open University, 15p.

de Carvalho Niebel F, Frendo P, van Mantagu M and Cornelissen (1995) Post-transcriptional cosuppression of β-1,3-glucanase genes does not affect accumulation of transgene nuclear mRNA. *Plant Cell,* **7** : 347-358.

Chen LZ and Adachi T (1994) Plant regeneration via somatic embryogenesis from cotyledon protoplasts of tomato (*Lycopersicon esculentum* Mill.) *Breed. Sci,* **44** : 257-262.

Chen LZ and Adachi T (1998) Protoplast fusion between *Lycopersicon esculentum* and *L. peruvianum*-complex: somatic embryogenesis, plant regeneration and morphology. *Plant Cell Rep.,* **17** : 508-514.

Cheng ZM, Schnurr JÁ and Kapaun JÁ (1998) Timentin as an alternative antibiotic for suppression of *Agrobacterium tumefaciens* in genetic transformation. *Plant Cell Rep.,* **17** : 646-649.

Chlyah A, Taarji H and Chlyah H (1990) Tomato (*Lycopersicon esculentum* L.): anther cuture and induction of androgenesis. In: *Biotechnology in Agriculture and Forestry, Vol. 12, Haploids in Crop Improvement I* (Ed. Bajaj YPS), Springer-Verlag, Berlin, pp. 443-457

Cho HJ, Kim SJ, Kang JH, Rhim SL and Kim BD (1993) Transgenic tomato and potato plants expressing insecticidal activity against Coleopteran larvae. *Advances in Developmental Biology and Biotechnology of Higher Plants,* Korean Society Plant Tissue Culture, pp. 311-324.

Chyi YS and Phillips GC (1987) High efficiency *Agrobacterium*-mediated transformation of *Lycopersicon* based on conditions favorable for regeneration. *Plant Cell Rep.,* **6** : 105-108.

Compton ME and Veilleux RE (1991) Shoot, root and flower morphogenesis on tomato inflorescence explants. *Plant Cell Tis. Org. Cult.,* **24** : 223-231.

Costa MGC (1999) Transformação genética de dois cultivares de tomateiro industrial mediada por *Agrobacterium tumefaciens* e incorporação do gene *Sw-5. MS Thesis*, UFV, Viçosa, Minas Gerais, Brazil.

Costa MGC, Brommonschenkel SH and Otoni WC (1998) Incorporation of resistance to tospovirus (TSWV) in processing tomato cultivars by means of *Agrobacterium*-mendiated transformation. *Genet. Mol. Biol.,* **21 [Suppl.]** : 391.

Costa MGC, Nogueira FTS, Figueira ML, Otoni WC, Brommonschenkel SH and Cecon PR (2000a) Influence of the antibiotic timentin on plant regeneration of tomato (*Lycopersicon esculentum* Mill.) cultivars. *Plant Cell Rep.,* **19** : 327-332.

Costa MGC, Nogueira FTS, Otoni, WC and Brommonschenkel SH (2000b) Transformação genética de cultivares de tomateiro industrial mediada por *Agrobacterium tumefaciens. R. Bras. Fisiol. Veg.,* **12** : 107-118

Costa MGC, Nogueira FTS, Otoni, WC and Brommonschenkel SH (2000c) Regeneração *in vitro* de cultivares de tomateiro (*Lycopersicon esculentum* Mill.) industrial IPA-5 e IPA-6. Ciênc. *Agrotec.,* **24** : 671-678

Covey SN, Al-Kaff NS, Langara A and Turner DS (1997) Plants combat infection by gene silencing. *Nature,* **385** : 781-782.

Cramer CL, Weissenborn D, Cottinghan CK, Denbow CJ, Eisenback JD, Radin DN and Yu X (1993) Regulation of defense-related gene expression during plant-pathogen interactions. *J. Nematol.,* **25** : 507-518.

Czosnek H, Navot N and Laterrot H (1990) Geographical distribution of the tomato yellow leaf curl virus. A first survey using a specific DNA probe. *Phytopathol. Medit.* **29** : 1-6.

D'Aoust MA, Nguyen-Quoc B, Le VQ and Yelle S (1999) Upstream regulatory regions from the maize Sh1 promoter confer tissue-specific expression of the β-glucuronidase gene in tomato. *Plant Cell Rep.,* **18** : 803-808.

Davis ME, Lineberger RD and Miller AR (1991a) Effects of tomato cultivar, leaf age, and bacterial strain on transformation by *Agrobacterium tumefaciens. Plant Cell Tiss. Org. Cult.,* **24** : 115-121.

Davis ME, Miller AR and Lineberger RD (1991b) Temporal competence for transformation of *Lycopersicon esculentum* (L.) Mill. cotyledons by *Agrobacterium tumefaciens*: relation to wound-healing and soluble plant factors. *J. Exp. Bot.,* **42** : 359-364.

Deblaere R, Reynaerts A, Hofte H, Hernalsteens JP, Leemans J and Van Montagu M (1987) *Meth. Enzymol.,* **153** : 205-208.

De Block M, Botterman J, Vanderwiele M, Dockx J, Thoen C, Grossele V, Rao Movva N, Thompson C, Van Montagu M and Leemans J (1987) Engineering herbicide resistance in plants by expression of a detoxifying enzyme. *EMBO J.,* **6** : 2513-2518.

Delannay X, La Vallee BJ, Proksch RK, Fuchs RL, Sims SR, Greenplate JT, Marrone PF, Dodson RB, Augustine JJ, Layton JG and Fischhoff DA (1989) Field performance of transgenic tomato plants expressing the *Bacillus thuringiensis* var. *kurstaki* insect control protein. *Bio/Technology,* **7** : 1265-1269.

Dixon MS, Jones DA, Keddie JS, Thomas CM and Harrison K (1996) The tomato *Cf-2* disease resistance locus comprises two functional genes encoding leucine-rich repeat proteins. *Cell,* **84** : 451-459.

Dobigny A, Tizroutine S, Gaisne C, Haicour R, Ducreux G and Sihachkr D (1996) Direct regeneration of transformed plants from stem fragments of potato inoculated with *Agrobacterium rhizogenes. Plant Cell Tiss. Org. Cult.,* **45** : 115-121.

Doganlar S, Tanksley SD and Mutschler MA (2000) Identification and molecular mapping of loci controlling fruit ripening time in tomato. *Theor. Appl. Genet.,* **100** : 249-255.

Dropkin VH (1969) The necrotic reaction of tomatoes and other hosts resistant to *Meloidogyne*: reversal by temperature. *Phytopathology,* **59** : 163-1637.

Dunwell JM (2000) Transgenic approaches to crop improvement. *J. Exp. Bot.,* **51** : 487:496.

Düzyaman E, Tanrisever A and Günver G (1994) Comparative studies on regeneration of different tissues of tomato *in vitro. Acta Hortic.,* **366** : 235-242.

Ebinuma H, Sugita K, Matunaga E, Endo S and Kasahara T. Selection of marker-free transgenic plants using the oncogenes (*ipt, rol A, B, C*) of *Agrobacterium* as selectable markers. In: *Molecular Biology of Woody Plants, Vol. 2,* (Eds. Jain MS and Minocha SC). Kluwer Academic Press, The Netherlands (In press).

Erickson FL, Dinesh-Kumar SP, Holzberg S, Ustach CV, Dutton M, Handley V, Corr C and Baker BJ (1999) Interactions between tobacco mosaic virus and the tobacco *N* gene. *Phil. Trans. R. Soc. Lond.* B, **354** : 653-658.

Errington N, Mitchell JR and Tucker GA (1997) Changes in the force relaxation and compression responses of tomatoes during ripening: the effect of continual testing and polygalacturonase activity. *Postharvest Biol. Technol.*, **11** : 141-147.

Errington N, Tucker GA and Mitchell JR (1998) Effect of genetic down-regulation of polygalacturonase and pectin esterase activity on rheology and composition of tomato juice. *J. Sci. Food Agric.*, **76** : 515-519.

Evans DA (1987) Somaclonal variation. In: *Tomato Biotechnology.* Alan R, Liss Inc., New York. pp. 59-69.

Fári M, Resende GM and Melo NF (2000) Avaliação da capacidade de regeneração *in vitro* em tomateiro industrial. *Pesq. Agrop. Bras.*, **35** : 1523-1529.

Faria RT and Illg RD (1996) Inheritance of *in vitro* plant regeneration ability in the tomato. *Braz. J. Genet.*, **19** : 113-116.

Federowicz O, Bartoszewski G, Stoeva P and Niemirowicz-Szczytt K (2000) *Agrobacterium*-mediated transformation of cultivated tomato with construct carrying the nucleoprotein (N) gene from tomato spotted wilt virus (TSWV). *Acta Physiol. Plant.*, **22** : 277-281.

Fillatti JJ, Kiser J, Rose R and Comai L (1987) Efficient transfer of a glyphosate tolerance gene into tomato using a binary *Agrobacterium tumefaciens* vector. *Bio/Technology,* **7** : 726-730.

Finlay KW (1953) Inheritance of spotted wilt resistance in tomato. II. Five genes controlling spotted wilt resistance if four tomato types. *Aust. J. Biol. Sci.* **6** : 153-163.

Fischhoff DA, Bowdish KS, Perlack FJ, Marrone FJ, McCormick SM, Niedermayer JG, Dean DA, Kusano-Kretzmer K, Mayer EJ, Rochester DE, Rogers SG and Fraley RT (1987) Insect tolerant transgenic tomato plants. *Bio/Technology,* **5** : 807-813.

Flavell RB, Dart E, Fuchs RL and Fraley RT (1992) Selectable marker genes: safe for plants? *Bio/Technology,* **10** : 141-144.

Frankenberger EA, Hasegawa PM and Tighelaar EC (1981) Diallel analysis of shoot forming capacity among selected tomato genotypes. *Z. Pflanzenphysiol,* **102** : 233-242.

Frary A (1995) The use of *Agrobacterium tumefaciens*-mediated transformation in the map-based cloning of tomato genes and an analysis of factors affecting transformation efficiency. *Ph.D. Thesis,* Cornell University, Ithaca, NY.

Frary A and Earle ED (1996) An examination of factors affecting the efficiency of *Agrobacterium*-mediated transformation of tomato. *Plant Cell Rep.,* **16** : 235-240.

Fraser PD, Schuch W and Bramley PM (2000) Phytoene synthase from tomato (*Lycopersicon esculentum*) chloroplasts partial purification and biochemical properties. *Planta,* **211** : 361-369.

Fray RG and Grierson D (1993) Identification and genetic analysis of normal and mutant phytoene synthase genes of tomato by sequencing, complementation and co-suppression. *Plant Mol. Biol.,* **22** : 589-602.

Fuchs M and Provvidenti R (1996) Evaluation of transgenic tomato plants expressing the coat protein gene of cucumber mosaic virus strain WL under field conditions. *Plant Dis.,* **80** : 270-275.

Fuchs RL, Ream JE, Hammond BG, Naylor MW, Leimgrumber RM and Berberich SA (1993) Safety assessment of the neomycin phosphotransferase II (NPTII) protein. *Bio/Technology,* **11** : 1543-1547.

Galtier N, Foyer CH, Huber J, Voelker TA and Huber SC (1993) Effects of elevated sucrose phosphate synthase activity on photosynthesis, assimilate partitioning and growth in tomato (*Lycopersicon esculentum* var. UC82B). *Plant Physiol.,* **101** : 535-543.

Galtier N, Foyer CH, Murchie E, Alred R, Quick P, Voelker TA, Thépenier C, Lascève G and Betsche T (1995) Effects of light and atmospheric carbon dioxide enrichment on photosynthesis and carbon partitioning in the leaves of tomato (*Lycopersicon esculentum* L.) plants over-expressing sucrose phosphate synthase. *J. Exp. Bot.,* **46** : 1335-1344.

Gavazzi G, Tonelli C, Todesco G, Arreghini E, Raffaldi F, Vecchio F, Barbuzzi G, Biasini MG and Sala F (1987) Somaclonal variation versus chemically induced mutagenesis in tomato (*Lycopersicon esculentum* L.). *Theor. Appl. Genet.,* **74** : 733-738.

Gielen JJL, de Haan P, Kool AJ, Peters D, van Grinsven MQJM and Goldbach RW (1991) Engineered resistance to spotted wilt virus, a negative-strand RNA virus. *Bio/Technology,* **9** : 1363-1367.

Gielen J, Ultzen T, Bontems S, Loots W, van Schepen A, Westbroek A, Haan P and van Grinsven M (1996) Coat protein-mediated protection to cucumber mosaic virus infections in cultivated tomato. *Euphytica,* **88** : 139-149.

Gill R, Malik KA, Sanago MHM and Saxena PK (1995) Somatic embryogenesis and plant regeneration from seedling cultures of tomato (*Lycopersicon esculentum* Mill.) *J. Plant Physiol.,* **47** : 273-276.

Giordano LB, Boiteux LS and Horino Y (1994) Avaliação em condições de campo de genótipos de tomate para resistência a tospoviroses. *Hort. Bras,* **12** : 176-178.

Giovannoni J, Yen H, Shelton B, Miller S, Vrebalov J, Kannan P, Tieman D, Hackett R, Grierson D and Klee H (1999) Genetic mapping of ripening and ethylene-related loci in tomato. *Theor. Appl. Genet.,* **98** : 1005-1013.

Giovannoni J (2001) Molecular biology of fruit maturation and ripening. *Annu. Rev. Plant Physiol. Plant Mol. Biol.,* **52** : 725-749.

Giovannoni JJ, DellaPenna D, Bennet AB and Fischer RL (1989) Expression of a chimeric polygalacturonase gene in transgenic *rin* (*ripening inhibitor*) tomato fruits results in polyuronide degradation but not in fruit softening. *Plant Cell,* **1** : 53-63.

Gisbert C (1997) Transformación genética en *Lycopersicon*: introducción de genes relacionados con la tolerancia a la salinidad en *L. esculentum* Mill. cv. P-73 y de genes marcadores en *L. pennellii* (Corr.) D'Arcy entrada PE 47. *Ph.D. Thesis,* Universidad Politécnica, Madrid.

Gisbert C, Arrilaga I, Roig LA and Moreno V (1999) Acquisition of a collection of *Lycopersicon pennellii* (Corr. D'Arcy) transgenic plants with *uidA* and *nptII* marker genes. *J. Hort. Sci. Biotech.,* **74** : 105-109.

Gisbert C, Rus AM, Bolarín MC, Coronado-López JM, Arrilaga I, Montesinos C, Caro M, Serrano R and Moreno V (2000) The yeast *HAL1* gene improves salt tolerance of transgenic tomato. *Plant Physiol.,* **123** : 393-402.

Goldsbrough AP, Lastrella CN and Yoder JI (1993) Transposition mediated re-positioning and subsequent elimination of marker genes from transgenic tomato. *Bio/Technology,* **11** : 1286-1292.

Gonsalves D, Pang SZ, Gonsalves C, Xue B, Yepes M and Jan FJ (1996) Developing transgenic crops that are resistant to tospoviruses. *Acta Hortic.,* **431** : 427-431.

Gray JE, Picton S, Shabber J, Schuch W and Grierson D (1992) Molecular biology of fruit ripening and its manipulation with antisense genes. *Plant Mol. Biol.,* **19** : 69-87.

Gray JE, Picton S, Giovannoni JJ and Grierson D (1994) The use of transgenic and naturally occurring mutants to understand and manipulate tomato fruit ripening. *Plant Cell Environ.,* **17** : 557-571.

Gresshoff PM and Doy CH (1972) Development and differentiation of haploid *Lycopersicon esculentum* (tomato). *Planta,* **107** : 161-170.

Grichko VP, Filby B and Glick BR (2000) Increased ability of transgenic plants expressing the bacterial enzyme ACC deaminase to accumulate Cd, Co, Cu, Ni, Pb and Zn. *J. Biotech.,* **81** : 45-53

Grichko VP and Glick BR (2001) Flooding tolerance of transgenic tomato plants expressing the bacterial enzyme ACC deaminase controlled by the *35S, rolD* or PRB-1*b* promoter. *Plant Physiol. Biochem.,* **39** : 19-25.

Griffiths A, Prestage S, Linforth R, Zhang J, Taylor A and Grierson D (1999) Fruit-specific lipoxygenase suppression in antisense-transgenic tomatoes. *Postharvest Biol. Technol.,* **17** : 163-173.

Groot SP, Bouwer R, Busscher M, Lindhout P and Dons HJ (1995) Increase of endogenous zeatin riboside by introduction of the *ipt* gene in wild type and the *lateral suppressor* mutant of tomato. *Plant Growth Regul,* **16** : 27-36.

Grundler FMW (1996) Engineering resistance against plant-parasitic nematodes. *Field Crops Res.,* **45** : 99-109.

Hackett RM, Ho CW, Lin Z, Foote CC, Fray RG and Grierson D (2000) Antisense inhibition of the *Nr* gene restores normal ripening to the tomato *Never-Ripe* mutant, consistent with the ethylene receptor-inhibition model. *Plant Physiol.,* **124** : 1079-1085.

Hails RS (2000) Genetically modified plants – the debate continues. *Tree,* **15** : 14-18.

Hall LN, Tucker GA, Smith CJS, Watson CF, Seymour GB, Bundick Y, Boniwell JM, Fletcher JD, Ray JA, Schuch W, Bird CR and Grierson D (1993) Antisense inhibition pectin esterase gene expression in transgenic tomatoes. *Plant J.,* **3** : 121-129.

Hamilton AJ, Lycett GW and Grierson D (1990) Antisense gene that inhibits synthesis of the hormone ethylene in transgenic plants. *Nature,* **346** : 284-287.

Hamilton AJ, Brown S, Yuanhai H, Ishizula M, Lowe A, Solis AGA, Grierson D (1998) A transgene with repeated DNA causes high frequency, post-transcriptional suppression of ACC-oxidase gene expression in tomato. *Plant J.,* **15** : 737-746.

Hamilton CM, Frary A, Xu Y, Tanksley SD and Zhang HB (1999) Construction of tomato genomic DNA libraries in a binary-BAC (BIBAC) vector. *Plant J.,* **18** : 223-229.

Hammond-Kosack KE and Jones JDG (1994) Incomplete dominance of tomato *Cf* genes for resistance to *Cladosporium fulvum. Mol. Plant-Microbe Interact.,* **7** : 58-70.

Hammond-Kosack KE and Jones JDG (1997) Plant disease resistance genes. *Annu. Rev. Plant Physiol. Plant Mol. Biol.,* **48** : 575-607.

Hamond-Kosack KE, Tang S, Harrison K and Jones JDG (1998) The tomato *Cf-9* disease resistance gene functions in tobacco and potato to confer responsiveness to the fungal avirulence gene product *Avr9. Plant Cell,* **10** : 1251-1266.

Hamza S and Chupeau Y (1993) Re-evaluation of conditions for plant regeneration and *Agrobacterium*-mediated transformation from tomato (*Lycopersicon esculentum*). *J. Exp. Bot.,* **44** : 1837-1845.

Hanus-Fajerska E, Lech M, Pindel A and Miczynski K (2000) Selection for virus resistance in tomato exposed to tissue culture procedures. *Acta Physiol. Plant.* **22** : 317-324.

Haroon SA, Baki AA, Huettel RN (1993) An *in vitro* test for temperature sensitivity and resistance to *Meloidogyne incognita* in tomato. *J. Nemat.,* **25** : 83-88.

Hashimoto RY, Menck CF and Sluys MAV (1999) Negative selection driven by cytosine deaminase gene in *Lycopersicon esculentum* hairy roots. *Plant Sci.,* **141** : 175-181.

Hightower R, Baden C, Penzes E, Lund P and Dunsmuir P (1991) Expression of antifreeze proteins in transgenic plants. *Plant Mol. Biol.,* **17** : 1013-1021.

Hilder VA, Gatehouse AMR, Sheerman SE, Baker F and Bouter D (1987) A novel mechanism of insect resistance engineered into tobacco. *Nature,* **330** : 160-163.

Hobson G and Grierson D (1993) Tomato. In: *Biochemistry of Fruit Ripening* (Eds. Seymour GB, Taylor J E and Tucker GA). Chapman & Hall, London, pp. 405-442.

Honée G, Buitink J, Jabs T, De Kloe J, Sijbolts F, Apotheker M, Weide R, Sijen T, Stuiver M and De Wit PJGM (1998) Induction of defense-related responses in *Cf 9* tomato cells by the *AVR9* elicitor peptide of *Cladosporium fulvum* is developmentally regulated. *Plant Physiol.,* **117** : 809-820.

Hong SB, Sexton R and Tucker ML (2000) Analysis of gene promoters for two tomato polygalacturonases expressed in abscission zones and the stigma. *Plant Physiol.,* **123** : 869-881.

Horsch RB, Fry JE, Hoffmenn NL, Eichholtz D, Rogers SG and Fraley RT (1985) A simple and general method for transferring genes into plants. *Science,* **227** : 1229-1231.

Hou YM, Sanders R, Ursin VM and Gilbertson RL (2000) Transgenic plants expressing geminivirus movement proteins: abnormal phenotypes and delayed infection by Tomato mottle virus in transgenic tomatoes expressing the Bean dwarf mosaic virus BV1 or BC1 proteins. *Mol. Plant-Microbe Interact.,* **13** : 297-308.

Houck CM, Facciotti D and Goodman (1993) Transgenic plants from Solanaceae. *In: Transgenic Plants, Vol. 2, Present Status and Social and Economic Impacts.* (Eds. Kung SD and Wu R) Academic Press, Inc., London, pp. 49-78.

Hu W and Phillips GC (2001) A combination of overgrowth-control antibiotics improves *Agrobacterium tumefaciens*-mediated transformation efficiency for cultivated tomato (*L. esculentum*). *In Vitro Cell. Dev. Biol. Plant,* **37** : 12-18.

Ichimura K, Uchiumi T, Tsuji K, Oda M and Nagaoka M (1995) Shoot regeneration of tomato (*Lycopersicon esculentum* Mill.) in tissue culture using several kinds of supporting materials. *Plant Sci.,* **108** : 93-100.

Jacinto T, McGurl B, Franceschi V, Delano-Freier J and Ryan CA (1997) Tomato prosystemin promoter confers wound-inducible, vascular bundle-specific expression of the β-glucuronidase gene in transgenic tomato plants. *Planta,* **203** : 406-412.

Jacinto T, McGurl B and Ryan CA (1999) Wound-regulation and tissue specificity of the tomato prosystemin promoter in transgenic tobacco plants. *Plant Sci.,* **140** : 155-159.

James C (2000) Global status of commercialized transgenic crops:1999. ISAAA Briefs, **17** : 1-65.

Janssen BJ, Lund L and Sinha N (1998) Overexpression of a homeobox gene, *LeT6*, reveals indeterminate features in the tomato compound leaf. *Plant Physiol.*, **117** : 771-786.

Jiang GY, Jin DM, Weng ML, Guo BT and Wang B (1999) Transformation and expression of trichosanthin gene in tomato. *Acta Bot. Sin.*, **41** : 334-336.

Jones CG, Seymour GB, Bird CR, Schuch W, Boniwell J, Lycett GW and Tucker GA (1994) Down regulation of two non-homologous genes with a single chimeric gene construct. *In: Mechanisms and Applications of Gene Silencing*, (Eds Grierson D, Lycett GW and Tucker GA) Nottingham University Press, Nottingham, pp. 85-95.

Jones DA, Thomas CM, Hammond-Kosack KE, Balint-Kurti PJ and Jones JDG (1994) Isolation of the Tomato *Cf-9* gene for resistance to *Cladosporium fulvum* by transposon tagging. *Science,* **266** : 789-793.

Jongedijk E, Tigelaar H, van Roekel JSC, Bres-Vloemans SA, Dekker I, van den Elzen PJM, Cornelissen BJC and Melchers LS (1995) Synergistic activity of chitinases and β-1,3-glucanases enhances fungal resistance in transgenic tomato plants. *Euphytica,* **85** : 173-180.

Jongsma M, Koornneef, Zabel P and Hille J (1987) Tomato protoplast DNA transformation: physical linkage and recombination of exogenous DNA sequences. *Plant Mol. Biol.,* **8** : 383-394.

Kaniewski W, Ilardi V, Tomassoli L, Mitsky T, Layton J and Barba M (1999) Extreme resistance to cucumber mosaic virus (CMV) in transgenic tomato expressing one or two viral coat proteins. *Mol. Breed.,* **5** : 111-119.

Karasawa M, Brommonschenkel SH, Yamazaki E, Otoni WC, Lau D and Lima GSA (2000) Transformação de alface (*Lactuca sativa* L.) com o gene *Sw-5* que confere resistência a tospovírus em tomateiro. *Gen. Mol. Biol.,* **23** : 239-240 (abstracts).

Kartha KK, Gamborg OL, Shyluk JP and Constabel F (1976) Morphogenic investigations on in vitro leaf culture of tomato (*Lycopersicon esculentum* Mill. cv. Starfire) and high frequency plant regeneration. *Z. Pflanzenphysiol.,* **77** : 292-301.

Kim JW, Sun SSM and German TL (1994) Disease resistance in tobacco plants transformed with the tomato spotted wilt virus nucleocapsid gene. *Plant Dis.,* **78** : 615-621.

Kinsara A, Patnaik SN, Cocking EC and Power JB (1986) Somatic hybrid plants of *L. esculentum* Mill. with *L. peruvianum. J. Plant Physiol.,* **125** : 225-234.

Klann EM, Hall B and Bennett AB (1996) Antisense acid invertase (TLV1) gene alters soluble sugar composition and size in transgenic tomato fruit. *Plant Physiol.,* **112** : 1321-1330.

Klee HJ (1993) Ripening physiology of fruit from transgenic tomato (*Lycopersicon esculentum*) plants with reduced ethylene synthesis. *Plant Physiol.,* **102** : 911-916.

Klee HJ, Hayford MB, Kretzmer KA, Barry GF and Kishore GM (1991) Control of ethylene synthesis by expression of a bacterial enzyme in transgenic tomato plants. *Plant Cell,* **3** : 1187-1193.

Klee HJ and Lanahan MB (1995) Transgenic plants in hormone biology. In: *Plant Hormones: Physiology, Biochemistry and Molecular Biology* (Ed. Davies PJ). Kluwer Academic Publishers, Netherlands. p. 340-353.

Koblitz H (1991) Cell, tissue and organ culture of *Lycopersicon.* In: *Genetic Improvement of Tomato* (Ed. Kaloo G). Springer-Verlag, New York, pp. 231-245.

Kochevenko A, Ratushnyak Y, Korneyeyev D, Stasik O, Porublyova L, Kochubey S, Suprunova T and Gleba Y (2000) Functional cybrid plants of *Lycopersicon peruvianum* var 'dentatum' with chloroplasts of *Lycopersicon esculentum. Plant Cell Rep.,* **19** : 588-597.

Koehler SM, Matters GL, Nath P, Kemmerer EC and Tucker ML (1996) The gene promoter for a bean cellulase is ethylene-induced in transgenic tomato and shows high sequence conservation with a soybean abscission cellulase. *Plant Mol. Biol.,* **31** : 595-606.

Komari T, Hiei Y, Saito Y, Murai N and Kumashiro (1996) Vectors carrying separate T-DNAs for co-transformation of higher plants mediated by *Agrobacterium tumefaciens* and segregation of transformants free from selection markers. *Plant J.,* **10** : 165-174.

Koorneef M, Bade J, Hanhart C, Horsman K, Schel J, Soppe W, Verkerk R and Zabel P (1993) Characterization and mapping of a gene controlling shoot regeneration in tomato. *Plant J.,* **3** : 131-141.

Koornneef M, Hanhart C, Jongsma M, Toma I, Weide R, Zabel P and Hille J (1986) Breeding of a tomato genotype readily accessible to genetic manipulation. *Plant Sci.,* **45** : 201-208.

Koornneef M, Hanhart C and Martinelli LA (1987a) A genetic analysis of cell culture traits in tomato. *Theor. Appl. Genet.,* **74** : 633-641.

Koornneef M, Jongsma M, Weide R, Zabel P and Hille J (1987b) Transformation in tomato. In: *Tomato Biotechnology* (Eds Nevis DJ and Jones RA), Alan R. Liss, Inc., New York, pp. 169-178.

Koornneef M, van Diepen JAM, Hanhart CJ, Kieboom-de Waart AC, Martinelli L, Schoenmakers HCH and Wijbrandi J (1989) Chromosomal instability in cell and tissue cultures of tomato haploids and diploids. *Euphytica,* **43** : 179-186.

Krake LR, Rezaian MA and Dry IB (1998) Expression of the tomato leaf curl geminivirus *C4* gene produces virus-like symptoms in transgenic plants. *Mol. Plant-Micr. Interact.,* **11** : 413-417.

Kramer M, Sanders R, Bolkan H, Waters C, Sheehy RE and Hiatt RW (1992) Post harvest evaluation of transgenic tomatoes with reduced levels of polygalacturonase: processing, firmness and disease. *Postharvest Biol. Technol.,* **1** : 241-255.

Kramer MG and Redenbaugh K (1994) Commercialization of a tomato with an antisense polygalacturonase gene: The FLAVR SAVR™ tomato story. *Euphytica,* **79** : 293-297.

Kumar NK, Ullman DE and Cho JJ (1993) Evaluation of *Lycopersicon* germplasm for tomato spotted wilt tospovírus resistance by mechanical and thrips transmission. *Plant Dis.,* **77** : 938-941.

Kunik T, Salomon R, Zamir D, Navot N, Zeidan M, Michelson I, Gafni Y and Czosnek H (1994) Transgenic tomato plants expressing the tomato yellow leaf curl virus capsid protein are resistant to the virus. *Bio/Technology,* **12** : 500-504.

Kuntz M, Chen HC, Simkin AJ, Römer S, Shipton CA, Drake R, Schuch W and Bramley P (1998) Upregulation of two ripening-related genes from a non-climacteric plant (pepper) in a transgenic climacteric plant (tomato). *Plant J.,* **13** : 351-361.

Kurtz SM and Lineberger RD (1983) Genotypic differences in morphogenic capacity of cultured leaf explants of tomato. *J. Am. Soc. Hort. Sci.,* **108** : 710-714.

Kut SA, Bravo JE and Evans DA (1984) Tomato. In: *Handbook of Plant Cell Culture* (Eds Evans DA, Sharp R and Yamada Y). New York, McMillan. pp. 247-89.

Lagrimi LM, Vaughn J, Finer J, Klotz K and Rubaihayo P (1992) Expression of a chimeric tobacco peroxidase gene in transgenic tomato plants. *J. Am. Soc. Hortic. Sci.,* **117** : 1012-1016.

Lagrimi LM, Vaughn J, Erb WA and Miller SA (1993) Peroxidase overproduction in tomato: wound-induced polyphenol deposition and disease resistance. *HortSci.,* **28** : 218-221.

Lamproye A, Hofinger M, Berthon JY and Gaspar T (1990) [Benzo(*b*) selenienyl-3] acetic acid: a potent synthetic auxin in somatic embryogenesis. *C.R. Acad. Sci. Paris,* **311** : 127-132.

Lanahan MB, Yen HC, Giovannoni JJ and Klee HJ (1994) The *never ripe* mutation blocks ethylene perception in tomato. *Plant Cell,* **6** : 521-530.

Laporte MM, Galagan JA, Shapiro JA, Boersig MR, Shewmaker CK and Sharkey TD (1997) Sucrose-phosphate synthase activity and yield analysis of tomato plants with maize sucrose-phosphate synthase. *Planta,* **203** : 253-259.

Laterrot H (2000) Disease resistance in tomato: practical situation. *Acta Physiol. Plant.,* **22** : 328-331.

Latif M (1993) Genetic improvement in *L. esculentum* Mill. through somatic hybridization and transformation. *Ph.D. Thesis,* University of Nottingham, Nottingham.

Latif M, Numtaz N, Davey MR and Power JB (1993) Plant regeneration from protoplasts isolated from cell suspension cultures of wild tomato, L. chilense Dun. *Plant Cell Tiss. Org. Cult.,* **32** : 311-317.

Lau D, Brommonschenkel SH, Carvalho ECS, Lima GSA and Zerbini FM (1999) Analysis of the *Sw-5* mediated resistance response to *Tospovirus* species. *Virus Rev. and Res.,* **4** : 154-155.

Leemans J (1993) *Ti* to tomato, tomato to market: a decade of plant biotechnology. *Bio/Technology,* **11** : 22-26.

Lindbo JA, Silva-Rosales L, Proebsting WM and Dougherty WG (1993) Induction of a higly specific antiviral state in transgenic plants: implications for regulation gene expression and viral resistance. *Plant Cell,* **5** : 1749-1759.

Ling HQ, Koch G, Baumlein H and Ganal MW (1999) Map-based cloning of *chloronerva,* a gene involved in iron uptake of higher plants encoding nicotianamine synthase. *Proc. Natl. Acad. Sci. USA,* **96** : 7098-7103.

Ling HQ, Kriseleit D and Ganal MW (1998) Effect of ticarcillin/potassium clavulanate on callus growth and shoot regeneration in *Agrobacterium*-mediated transformation of tomato (*Lycopersicon esculentum* Mill.). *Plant Cell Rep.,* **17** : 843-847.

Lipp João KH and Brown TA (1993) Enhanced transformation of tomato co-cultivated with *Agrobacterium tumefaciens* C58C1Rif^r::pGSFR1161 in the presence of acetosyringone. *Plant Cell Rep.,* **12** : 422-425.

Locy RD (1983) Callus formation and organogenesis by explants of six *Lycopersicon* species. *Can. J. Bot.,* **6** : 1072-1079.

Mandaokar AD, Goyal RK, Shukla A, Bisaria S, Bhalla R, Reddy VS, Chaurasia A, Sharma RP, Altosaar I and Kumar PA (2000) Transgenic tomato plants resistant to fruit borer (*Helicoverpa armigera* Hubner). *Crop Protection,* **19** : 307-312.

Mansouri IE, Mercado JA, Santiago-Dómenech N, Pliego-Alfaro F, Valpuesta V and Quesada MA (1999) Biochemical and phenotypical characterization of transgenic tomato plants overexpressing a basic peroxidase. *Physiol. Plant.,* **106** : 355-362.

Martin GB, Brommonschenkel SH, Chunwongse J, Frary A, Ganal MW, Spivey R, Wu T, Earle ED and Tanksley SD (1993) Map-based cloning of a protein kinase gene conferring disease resistance in tomato. *Science,* **262** : 1432-1436.

Martin GB, Frary A, Wu T, Brommonschenkel SH, Chunwongse J, Earle ED and Tanksley SD (1994) A member of the tomato *Pto* gene family confers sensitivity to fenthion resulting in rapid cell death. *Plant Cell,* **6** : 1543-1552.

Martineau B, Houck CM, Sheehy E and Hiatt WR (1994) Fruit-specific expression of the *A. tumefaciens* isopentenyl transferase gene in tomato: effects on fruit ripening and defense-related gene expression in leaves. *Plant J.,* **5** : 11-19.

Martineau B, Summerfelt KR, Adams DF and DeVerna JW (1995) Production of high solids tomatoes through molecular modification of levels of the plant growth regulator cytokinin. *Bio/Technology,* **13** : 250-254.

Masuta C, Komari T and Takanami Y (1989) Expression of a cucumber mosaic virus satellite RNA from cDNA copies in transgenic tobacco plants. *Annu. Phytopathol. Soc. Japan,* **55** : 49-55.

McCormick S, Niedermeyer JF, Barnason A, Horsch R and Fraley R (1986) Leaf disc transformation of cultivated tomato (*L. esculentum*) using *Agrobacterium tumefaciens. Plant Cell Rep.,* **5** : 81-84.

McCormick S (1991) Transformation of tomato with *Agrobacterium tumefaciens.* In: *Plant Tissue Culture Manual: Fundamentals and Applications* (Ed Lindsey K), Kluwer Academic Publishers, Dordrecht, pp. **B6** : 1-9.

McGarvey PB, Montasser MS and Kaper JM (1994) Transgenic tomato plants expressing satellite RNA are tolerant to some strains of cucumber mosaic virus. *J. Amer. Soc. Hort. Sci.,* **119** : 642-647.

McGurl B, Orozco-Cardenas M, Pearce G and Ryan CA (1994) Overexpression of the prosystemin gene in transgenic tomato plants generates a systemic signal that constitutively induces proteinase inhibitor synthesis. *Proc. Natl. Acad. Sci. USA,* **91** : 9799-9802.

McGurl B, Pearce G, Orozco-Cardenas M and Ryan CA (1992) Structure, expression, and antisense inhibition of the systemin precursor gene. *Science,* **255** : 1570-1573.

Meier I, Gröning B, Michalowski S and Spiker S (1997) The tomato *RBCS3A* promoter requires integration into the chromatin for correct organ-specific regulation. *FEBBS Lett.,* **415** : 91-95.

Mesbah LA, Kneppers TJA, Takken FLW, Laurent P, Hille J and Nijkamp HJJ (1999) Genetic and physical analysis of a YAC contig spanning the fungal disease resistance locus Asc of tomato (*Lycopersicon esculentum*). *Theor. Appl. Genet.,* **261** : 50-57.

Micallef BJ, Haskins KA, Vanderveer PJ, Roh KS, Shewmaker CK and Sharkey TD (1995) Altered photosynthesis, flowering, and fruiting in transgenic tomato plants that have an increased capacity for sucrose synthesis. *Planta,* **196** : 327-334.

Michelmore RW (1995) Isolation of disease resistance genes from crop plants. *Curr. Opin. Biotech.*, **6** : 145-152

Milligan SB, Bodeau J, Yaghoobi J, Kaloshian I, Zabel P and Williamson VM (1998) The root-knot resistance gene *Mi* from tomato is a member of the leucine zipper, nucleotide binding, leucine rich repeat family of plant genes. *Plant Cell* **10** : 1307-1321.

Morgan AJ, Cox PN, Turner DA, Peel E, Davey MR, Gartland KMA and Mulligan B (1987) Transformation of tomato using an *Ri* plasmid vector. *Plant Sci.*, **49** : 37-49.

Murchie EH, Sarrobert C, Contard P, Betsche T, Foyer CH and Galtier N (1999) Overexpression of sucrose-phosphate synthase in tomato plants grown with CO_2 enrichment leads to decreased foliar carbohydrate accumulation relative to untransformed controls. *Plant Physiol. Biochem.*, **37** : 251-260.

Nakata K, Tanaka H, Yano K and Takagi M (1992) An effective transformation system for *Lycopersicon peruvianum* by electroporation. *Jpn. J. Breed.*, **42** : 487-495.

Namba S, Ling KH, Gonsalves C, Gonsalves D, Slingtom JL (1991) Expression of the gene encoding the coat protein of cucumber mosaic virus (CMV) strain WL appears to provide protection to tobacco plants against infection by several different CMV strains. *Gene,* **107** : 181-188.

Nauerby B, Billing K and Wyndaele R (1997) Influence of the antibiotic timentin on plant regeneration compared to carbenicillin and cefotaxime in concentrations suitable for elimination of *Agrobacterium tumefaciens. Plant Sci.,* **123** : 169-177.

Negrotto D, Jolley M, Beer S and Wenck AR (2000) The use of phosphomannose-isomerase as a selectable marker to recover transgenic maize plants (*Zea mays* L.) via *Agrobacterium* transformation. *Plant Cell Rep.,* **19** : 798-803.

Nelson RS, McCormick SM, Delannay X, Dube P, Layton J, Anderson EJ, Kanieska M, Proksch RK, Horsch RB, Rogers SC, Fraley RT and Beachy RN (1988) Virus tolerance, plant growth and field performance of transgenic tomato plants expressing coat protein from tobacco mosaic virus. *Bio/Technology,* **6** : 403-409.

Newman PO, Krishnaraj S and Saxena PK (1996) Regeneration of tomato (*Lycopersicon esculentum* Mill.): somatic embryogenesis and shoot organogenesis from hypocotyl explants induced with 6-benzyladenine. *Int. J. Plant Sci.,* **157** : 554-560.

Niedz RP, Rutter SM, Handley LW and Sink KC (1985) Plant regeneration from leaf protoplasts of six tomato cultivars. *Plant Sci.,* **39** : 199-204.

Nogueira FTS (1999) Transformação genética e regeneração *in vitro* de tomateiros (*Lycopersicon esculentum* Mill.) 'Santa Clara' e mutante "Firme": inserção do cDNA ACCS na orientação anti-senso. MS Thesis, UFV, Viçosa, Minas Gerais, Brazil.

Nogueira FTS, Figueira ML, Sakyiama NS, Brommonschenkel SH and Otoni WC (1998) *Agrobacterium*-mediated transformation of tomato (*Lycopersicon esculentum* cv. Santa Clara). *Genet. Mol. Biol.,* **21 [Suppl.]** : 390 .

Oeller PW, Min-Wong L, Taylor LP, Pike DA and Theologis A (1991) Reversible inhibition of tomato fruit senescence by antisense RNA. *Science,* **254** : 437-439.

Ohki S, Bigot C and Mousseau J (1978) Analysis of shoot-forming capacity *in vitro* in two lines of tomato (*Lycopersicon esculentum* Mill.) and their hybrids. *Plant Cell Physiol.*, **19** : 27-42.

Ohyama A, Ito H, Sato T, Nishimura S, Imai T and Hirai M (1995) Suppression of acid invertase activity by antisense RNA modifies the sugar composition of tomato fruit. *Plant Cell Physiol.*, **36** : 369-376.

Orozco-Cardenas M, McGurl B and Ryan CA (1993) Expression of an antisense prosystemin gene in tomato plants reduces resistance toward *Manduca sexta* larvae. *Proc. Natl. Acad. Sci. USA,* **90** : 8273-8276.

Padmanabhan V, Paddock EF and Sharp WR (1974) Plantlet formation from *L. esculentum* leaf callus. *Can. J. Bot.*, **52** : 1429-1432.

Parniske M, Hammond-Kosack KE, Golstein C, Thomas CM, Jones DA, Harrison K, Wulff BBH and Jones JDG (1997) Novel disease resistance specificities result from sequence exchange between tandemly repeated genes at the *Cf-4/9* locus of tomato. *Cell,* **91** : 821-832.

Patil RS (1994) Genetic manipulation of tomato (*Lycopersicon esculentum* Mill.) for crop improvement. *Ph.D. Thesis,* University of Nottingham, Nottingham.

Patil RS, D'Utra Vaz FB, Davey MR and Power JB (1993) Hybridization through culture of embryos and immature seeds, of a wide range of tomato cultivars with a tomato somatic hybrid [*L. esculeuntum* (+) *L. peruvianum*]: emergence of a possible new marker gene for tomato breeding. *Plant Breed.,* **111** : 273-282.

Pedra JH, Delú-Filho N, Pirovani CP, Contim LAS, Dewey RE, Otoni WC and Fontes EPB (2000) Antisense and sense expression of a sucrose binding protein homologue gene from soybean in transgenic tobacco affects plant growth and carbohydrate partitioning in leaves. *Plant Sci.,* **152** : 87-98.

Peñarrubia L, Aguilar M, Margossian L and Fischer RL (1992) An antisense gene stimulates ethylene production during fruit ripening. *Plant Cell,* **4** : 681-687.

Perlak FJ, Deaton DW, Armstrong TA, Fuchs RL, Sims SR, Grrenplate JT and Fishchhoff DA (1990) Insect resistant cotton plants. *Bio/Technology,* **8** : 939-943.

Perlak FJ, Fuchs RL, Dean DA, McPheson SL and Fichhoff DA (1991) Modification of the coding sequence enhances plant expression of insect control protein genes. *Proc. Natl. Acad. Sci. USA,* **88** : 3324-3328.

Picoli EAT (2000) Morfogênese *in vitro* e transformação genética de berinjela (*Solanum melongena* cv. Embú L.) mediada por *Agrobacterium tumefaciens.* MS Thesis, UFV, Viçosa, Minas Gerais, Brazil.

Picoli EAT, Lima GSA, Lau D, Brommonschenkel SH, Otoni WC and Zerbini FM (1999) Heterologous expression of the *Sw-5* gene in eggplant (*Solanum melongena* L.) confers resistance to *Tospovirus* infection. *Virus Rev. and Research,* **4** : 155.

Picton S, Barton SL, Bouzayen M, Hamilton AJ and Grierson D (1993) Altered fruit ripening and leaf senescence in tomatoes expressing an antisense ethylene-forming enzyme transgene. *Plant J.,* **3** : 469-481.

Picton S, Gray JE and Grierson D (1995a) The manipulation and modification of tomato fruit ripening by expression of antisense RNA in transgenic plants. *Euphytica,* **85** : 193-202.

Picton S, Gray JE and Grierson D (1995b) Ethylene genes and fruit ripening. In: *Plant Hormones: Physiology, Biochemistry and Molecular Biology* (Ed Davies PJ) Kluwer Academic Publishers, Dordrecht, pp. 372-394.

Powell ALT, van Kan J, Have AT, Visser J, Greve LC, Bennett AB and Labavitch JM (2000) Transgenic expression of pear PGIP in tomato limits fungal colonization. *Mol. Plant-Microbe Interact.,* **113** : 942-950.

Pratta G, Zorzoli R and Picardi LA (1997) Intra and interespecific variability of *in vitro* culture response in *Lycopersicon* (tomatoes). *Braz. J. Genet.,* **20** : 75-78.

Prins M, Resende RO, Anker C, van Schepen A, Haan P and Goldbach R (1996) Engineered RNA-mediated resistance to Tomato Spotted Wilt Virus is sequence specific. *Mol. Plant-Microbe Interact.,* **9** : 416-418.

Provvidenti R and Gonsalves D (1995) Inheritannce of resistance to cucumber mosaic virus in a transgenic tomato line expressing the coat protein gene of the white leaf strain. *J. Hered.,* **86** : 85-88.

Quemada HD, Gonsalves D and Slighton JL (1991) Expression of coat protein gene from cucumber mosaic virus strain C in tobacco: protection against infections by CMV strains transmitted mechanically or by aphids. *Mol. Plant Pathol.,* **81** : 794-802.

Ratcliff FG; MacFarlane SA and Baulcombe DC (1999) Gene silencing without DNA: RNA-Mediated cross-protection between viruses. *Plant Cell,* **11** : 1207-1215.

Reed AJ, Magin KM, Anderson JS, Austin GD, Rangwala T, Linde DC, Love JN, Rogers SG and Fuchs RL (1995) Delayed ripening tomato plants expressing the enzyme 1-aminocyclopropane-1-carboxylic acid deaminase. 1. Molecular characterization, enzyme expression, and fruit ripening traits. *J. Agric. Food Chem.,* **43** : 1954-1962.

Rhim SL (1998a) Transgenic plants expressing an antisense RNA of *AL1*-gene from tomato golden mosaic virus (TGMV) *Kor. J. Plant Tiss. Cult.,* **25** : 147-152.

Rhim SL (1998b) Molecular breeding of transgenic tomato plants expressing the δ-endotoxin gene of *Bacillus thuringiensis* subsp. *tenebrionis. Agric. Chem. Biotech.,* **41** : 137-140.

Rhim SL, Cho HJ, Kim BD, Schnetter W and Geider K (1995) Development of insect resistance in tomato plants expressing the δ-endotoxin gene of *Bacillus thuringiensis* subsp. *tenebrionis. Mol. Breed.,* **1** : 229-236.

Rich JR and Olson SM (1999) Utility of *Mi* gene resistance in tomato to manage *Meloidogyne javanica* in North Florida., *J. Nemat.,* **31** : 715-718.

Rie JV, McGaughey WH, Johnson DE, Barnett BD and Mellaert HV (1990) Mechanism of insect resistance to the microbial insecticide *Bacillus thuringiensis. Science,* **247** : 72-74.

Roberts PA (1992) Current status of the availability, development, and use of host plant resistance to nematodes. *J. Nemat.,* **24** : 213-227.

Robison MM, Shah S, Tamot B, Pauls KP, Moffart BA and Glick BR. Reduced symptoms of *Verticillium* wilt in transgenic tomato expressing a bacterial ACC deaminase. *Plant Mol. Biol.* (Submitted).

Rochange SF and McQueen-Masin SJ (2000) Expression of a heterologous expansin in transgenic tomato plants. *Planta,* **211** : 583-586.

Römer S, Fraser PD, Kiano JW, Shipton CA, Misawa N, Schuch W and Bramley PM (2000) Elevation of the provitamin A content of transgenic tomato plants. *Nature Biotech.,* **18** : 666-669.

Rommens CMT, Salmeron JM, Oldroyd GED and Staskawicz BJ (1995) Intergeneric transfer and functional expression of the tomato disease resistance gene *Pto. Plant Cell,* **7** : 1537-1544.

Roselló S, Díez MJ and Nuez F (1996) Viral diseases causing the greatest economic losses to tomato crop. I. The tomato spotted wilt virus - a review. *Sci. Hortic.*, **67** : 117-150.

Ryan CA (1990) Protease inhibitors in plants: genes for improving defenses against insects and pathogens. *Annu. Rev. Phytopathol.*, **28** : 425-449.

Saito Y, Komari T, Masuta C, Hayashi Y, Kumashiro T and Takanami Y (1992) Cucumber mosaic virus-tolerant transgenic tomato plants expressing a satellite RNA. *Theor. Appl. Genet.*, **83** : 679-683.

Salmeron JM, Baker SJ, Carland FM, Mehta AY and Staskawicz BJ (1994) Tomato mutants altered bacterial disease resistance provide evidence for a new locus controlling pathogen recognition. *Plant Cell,* **6** : 511-520.

Salmeron JM, Oldroyd GED, Rommers CMT, Scofield SR, Kim HS, Lavelle DT, Dahalbeck D and Staskawicz BJ (1996) Tomato *Prf* is a member of the leucine-rich repeat class of plant disease resistance genes and lies embedded within the *Pto* kinase gene cluster. *Cell,* **86** : 123-133.

Salt DE, Smith RD and Raskin I (1998) Phytoremediation. *Annu. Rev. Plant Physiol. Plant Mol. Biol.*, **49** : 643-668.

Samac DA and Shah DM (1991) Developmental and pathogen-induced activation of the *Arabidopsis* acid chitinase promoter. *Plant Cell,* **3** : 1063-1072.

Sanders PR, Sammons B, Kaniewski W, Haley L, Layton J, LaVallee BJ, Delannay X and Tumer NE (1992) Field resistance of transgenic tomatoes expressing the tobacco mosaic virus or tomato mosaic virus coat protein genes. *Mol. Plant. Pathol.*, **82** : 683-690.

Sandford JC and Jonston SA (1985) The concept of parasite-derived resistance - Deriving resistance genes from the parasite's own genome. *J. Theor. Biol.*, **113** : 395-405.

Sasson A (1998) Plant biotechnology- derived products: market-value estimates and public acceptance. *In: IX Intenational Congress on Plant Tissue and Cell Culture,* Jerusalem, Israel. Kluwer Academic Press, Dordrecht. pp. 160.

Scherf BA, Bajar AD and Kolattukudy PE (1993) Abolition of an inducible highly anionic peroxidase activity in transgenic tomato. *Plant Physiol.*, **101** : 201-208.

Schuch W, Kanczler J, Robertson D, Hobson G, Tucker G, Grierson D, Bright S and Bird C (1991) Fruit quality characteristics of transgenic tomato fruit with altered polygalacturonase activity. *HortSci.*, **26** : 1517-1520.

Senior IJ (1998) Uses of plant gene silencing. *Biotech. Genet. Engineer Rev.*, **15** : 79-119

Seymour GB, Fray RG, Hill P and Tucker GA (1993) Down-regulation of two non-homologous endogenous tomato genes with a single chimaeric sense gene construct. *Plant Mol. Biol.*, **23** : 1-9.

Shahin EA, Sukhapinda K, Simpson RB and Spivey R (1986) Transformation of cultivated tomato by a binary vector in *Agrobacterium rhizogenes:* transgenic plants with normal phenotypes harbor binary vector T-DNA but no Ri plasmid DNA. *Theor. Appl. Genet.*, **72** : 770-777.

Sheehy RE, Kramer M and Hiatt WR (19880 Reduction of polygalacturonase activity in tomato fruit by antisense RNA. *Proc. Natl. Acad. Sci. USA,* **85** : 8805-8809.

Shtereva L, Atanassova B and Balacheva E (2000) *In vivo* and *in vitro* investigation of salt tolerance in three anthocyaninless mutants in tomato (*Lycopersicon esculentum* Mill.). *Acta Physiol. Plant.*, **22** : 254-256.

Shtereva LA, Zagorska NA, Dimitrov BD, Kruleva MM and Oanh HK (1998) Induced androgenesis in tomato (*Lycopersicon esculentum* Mill). II. Factors affecting induction of androgenesis. *Plant Cell Rep.,* **18** : 312–317.

Simons H and Tucker GA (1999) Simultaneous co-suppression of polygalacturonase and pectinesterase in tomato fruit: inheritance and effect on isoform profiles. *Phytochemistry,* **52** : 1017-1022.

Sink KC and Reynolds JF (1986) Tomato (*Lycopersicom esculentum* L.) In: *Biotechnology in Agriculture and Forestry, Vol. 2, Crops I,* (Ed Bajaj YPS) Springer-Verlag, Berlin, p. 319-344.

Smith PG (1944) Embryo culture of a tomato species hybrid. *Proc. Amer. Soc. Hort. Sci.,* **44** : 413-416.

Smith CJS, Watson CF, Ray J, Bird CR, Morris PC, Schuch W and Grierson D (1988) Antisense RNA inhibition of polygalacturonase gene expression in transgenic tomatoes. *Nature,* **334** : 724-726.

Smith CJS, Watson CF, Morris PC, Bird CR, Seymour GB, Gray JE, Arnold C, Tucker GA, Schuch W, Harding S and Grierson D (1990) Inheritance and effect on ripening of antisense polygalacturonase genes in transgenic tomatoes. *Plant Mol. Biol.,* **14** : 369-379.

Speirs J, Lee E, Holt K, Yong-Duk K, Scott NS, Loveys B and Schuch W (1998) Genetic manipulation of alcohol dehydrogenase levels in ripening tomato fruit affects the balance of some flavor aldehydes and alcohols. *Plant Physiol.,* **117** : 1047-1058.

Sree Vidya CS, Manoharan M, Ranjit Kumar CT, Savithri HS and Lakshmi Sita G (2000) *Agrobacterium*-mediated transformation of tomato (*Lycopersicon esculentum* var. Pusa Ruby) with coat-protein gene of *Physalis* mottle tymovirus. *J. Plant Physiol.,* **156** : 106-110.

Stamova BS and Chetelat RT (2000) Inheritance and genetic mapping of cucumber mosaic virus resistance introgressed from *Lycopersicon chilense* into tomato. *Theor. Appl. Genet.,* **101** : 527-537.

Stevens MR, Scott SJ and Gergerich RC (1992) Inheritance of a gene for resistance of tomato spotted wilt virus from *Lycopersicon peruvianum*). *Euphytica,* **59** : 9-17.

Stevens MR, Scott SJ and Gergerich RC (1994) Evaluation of seven *Lycopersicon* species for resistance to tomato spotted wilt virus (TSWV). *Euphytica,* **80** : 79-84.

Stommel JR, Tousignant ME, Wai T, Pasini R and Kaper JM (1998) Viral satellite RNA expression in transgenic tomato confers field tolerance to cucumber mosaic virus. *Plant Dis.,* **82** : 391-396.

Storti E, Bogani P, Bettini P, Bittini P, Guardiola ML, Pellegrini MG, Inzé D and Buiatti M (1994) Modification of competence for *in vitro* response to *Fusarium oxysporum* in tomato cells. II. Effect of the integration of *Agrobacterium tumefaciens* genes for auxin and cytokinin synthesis. *Theor. Appl. Genet.,* **88** : 89-96.

Stotz HU, Powell ALT, Damon SE, Greve LC, Bennett AB and Labavitch JM (1993) Molecular characterization of a polygalacturonase inhibitor from *Pyrus communis* L. cv. Bartlett. *Plant Physiol.,* **102** : 133-138.

Sugiyama M (2000) Genetic analysis of plant morphogenesis *in vitro*. *Int. Rev. Cytol.,* **196** : 67-84.

Sukhapinda K, Spivey R, Simpson RB and Shahin EA (1987) Transgenic tomato (*Lycopersicon esculentum* L.) transformed with a binary vector in *Agrobacterium rhizogenes*: Non-chimeric origin of callus clone and low copy numbers of integrated vector T-DNA. *Mol. Gen. Genet.,* **206** : 491-497.

Summers WL, Jaramillo J and Bailey T (1992) Microspore developmental stage and anther length influence the induction of tomato anther callus. *HortSci.,* **27** : 838-840.

Tabaeizadeh Z, Agharbaoui Z, Harrak H and Poysa V (1999) Transgenic tomato plants expressing a *Lycopersicon chilense* chitinase gene demonstrate improved resistance to *Verticillium dahliae* race 2. *Plant Cell Rep.,* **19** : 197-202.

Tal M, Dehan K and Heiken H (1977) Morphogenetic potential of cultured leaf sections of cultivated and wild species of tomato. *Ann. Bot.* **41** : 937-941.

Tan MMC, Colijn-Hooymans CM, Lindhout WH and Kool AJ (1987a) A comparison of shoot regeneration from protoplasts and leaf discs of different genotypes of the cultivated tomato. *Theor. Appl. Genet.,* **75** : 105-108.

Tan MMC, Rietveld EM, Van Merrewijk GAM and Kool AJ (1987b) Regeneration of leaf mesophyll protoplasts of tomato cultivars (*L. esculentum*): factors important for efficient protoplast culture and plant regeneration. *Plant Cell Rep.,* **6** : 172-175.

Tepfer D (1984) Transformation of several species of higher plants by *Agrobacterium rhizogenes*: sexual transmission of the transformed genotype and phenotype. *Cell,* **37** : 959-967.

Theologis A, Zarembinski TI, Oeller PW, Liang X and Abel S (1992) Modification of fruit ripening by suppressing gene expression. *Plant Physiol.,* **100** : 549-551.

Theologis A (1994) Control of ripening. *Curr. Opi. Biotech.,* **5** : 152-157.

Thilmony RL, Chen Z, Bressan RA and Martin GB (1995) Expression of the tomato *Pto* gene in tobacco enhances resistance to *Pseudomonas syringae* pv tabaci expressing *avrPto. Plant Cell,* **7** : 1529-1536.

Thomzik JE, Stenzel K, Stocker R, Schreier PH, Hain R and Sthal DJ (1997) Synthesis of a grapevine phytoalexin in transgenic tomatoes (*Lycopersicon esculentum* Mill.) conditions resistance against *Phytophythora infestans. Physiol. Mol. Plant Pathol.,* **51** : 265-278.

Tomassoli L, Ilardi M, Barba M and Kaniewski W (1999) Resistance of transgenic tomato to cucumber mosaic cucumovirus under field conditions. *Mol. Breed.,* **5** : 121-130.

Toyoda H, Matsuda Y, Utsumi R and Ouchi S (1988) Intranuclear microinjection for transformation of tomato callus cells. *Plant Cell Rep.,* **7** : 293-296.

Tucker GA (1993) Improvement of tomato fruit quality and processing characteristics by genetic engineering. *Food Sci. Technol. Today,* **7** : 103-108.

Tucker GA and Zhang J (1996) Expression of polygalacturonase and pectinesterase in normal and transgenic tomatoes. In: *Pectins and Pectinases - Progress in Biotechnology, Vol. 14.* (Eds. Visser J and Voragen AGJ) Elsevier Science, Amsterdam, pp. 347-354.

Uddin MR, Berry SZ and Bisges AD (1988) An improved shoot regeneration system for somaclone production in tomatoes. *HortSci.,* **23** : 1062-1064.

Ultzen T, Gielen J, Venema F, Westerbroek A, de Haan P, Tan ML, Schram A, van Grinsven M and Goldbach R (1995) Resistance to tomato spotted wilt virus in transgenic tomato hybrids. *Euphytica,* **85** : 159-168.

Urwin PE, Atkinson HJ, Waller DA and McPherson MJ (1995) Engineered oryzacystatin-I expressed in transgenic hairy roots confers resistance to *Globodera pallida. Plant J.,* **8** : 121-131.

van der Salm T, Bosh D, Honée G, Feng L, Munsterman E, Bakker P, Stiekema WJ and Visser (1994) Insect resistance of transgenic plants that express modified *Bacillus thuringiensis cryIA(b)* and *cryIC* genes: a resistance management strategy. *Plant Mol. Biol.,* **26** : 51-59.

Van Dun CMP and Bol JF (1988) Transgenic tobacco plants accumulating rattle virus coat protein resist infection with tobacco rattle virus and pea early browning virus. *Virology,* **167** : 649-652.

Van Dun CMP, Overduin B, Van Vloten-Doting L and Bol JF (1988) Transgenic tobacco expressing tobacco streak virus or mutated alfalfa mosaic virus cat protein does not cross-protect against alfalfa mosaic virus infection. *Virology,* **164** : 383-389.

Van Eck JM, Blowers AD and Earle ED (1995) Stable transformation of tomato cell cultures after bombardment with plasmid and YAC DNA. *Plant Cell Rep.,* **14** : 299-304.

Van Heusden AW, Koornneef M, Voorrips RE, Brüggemann W, Pet G, Vrielink-van Ginkel R, Chen X and Lindhout P (1999) Three QTLs from *Lycopersicon peruvianum* confer a high level of resistance to *Clavibacter michiganensis* spp. *michiganensis. Theor. Appl. Genet.,* **99** : 1068-1074.

Van Roekel JCS, Damm B, Melchers LS and Hoekema A (1993) Factors influencing transformation frequency of tomato (*Lycopersicon esculentum*). *Plant Cell Rep.,* **12** : 644-647.

Voinnet O, Pinto YM and Baulcombe DC (1999) Suppression of gene silencing: A general strategy used by diverse DNA and RNA viruses of plants. *Proc. Natl. Acad. Sci. USA,* **96** : 14147-14152.

Vrain CT (1999) Engineering natural and synthetic resistance for nematode management. *J. Nemat.,* **31** : 424-436.

Xue B, Gonsalves C, Provvidenti R, Slightom JL, Fuchs M and Gonsalves D (1994) Development of transgenic tomato expressing a high level of resistance to cucumber mosaic virus strains of subgroups I and II. *Plant Dis.,* **78** : 1038-1041.

Yaghoobi J, Kaloshian I, Wen Y and Williamson VM (1995) Mapping a new nematode resistance locus in *Lycopersicon peruvianum. Theor. Appl. Genet.,* **91** : 457-464.

Yoder JI, Palys L, Alpert K and Lassner M (1988) Ac transposition in transgenic tomato plants. *Mol. Gen. Genet.,* **213** : 291-296.

Young R, Kaul V and Williams (1987) Clonal propagation *in vitro* from immature embryos and flower buds of *Lycopersicon peruvianum* and *L. esculentum. Plant Sci.,* **52** : 237-242.

Wang AS, Evans RA, Alterdorf PR, Hanten JA, Doyle MC and Rosichan JL (2000) A manose selection system for production of fertile transgenic maize plants from protoplasts. *Plant Cell Rep.,* **19** : 654-660.

Watson CF, Zheng L and DellaPenna D (1994) Reduction of tomato polygalacturonase β subunit expression affects pectin solubilization and degradation during fruit ripening. *Plant Cell,* **6** : 1623-1634.

White PR (1934) Potentially unlimited growth of excised tomato root tips in a liquid medium. *Plant Physiol.,* **9** : 585-600.

Whithan S, McCormick S and Baker B (1996) The N gene of tobacco confers resistance to tobacco mosaic virus in transgenic tomato. *Proc. Natl. Acad. Sci. USA,* **93** : 8776-8781.

Wijbrandi J and De Both MTJ (1993) Temperate vegetable crops. *Sci. Hortic.,* **55** : 37-63.

Wijbrandi J, Vos JGM and Koornneef M (1988) Transfer of regeneration capacity from *Lycopersicon peruvianum* to *L. esculentum* by protoplast fusion. *Plant Cell Tiss. Org. Cult.,* **12** : 193-196.

Wijbrandi J, van Capelle WV, Hanhart CJ, van Leonen Matinet-Schuringa EP and Koornneef M (1990) Selection and characterization of somatic hybrids between *L. esculentum* and *L. peruvianum. Plant Sci.,* **70** : 197-208.

Wilkinson JQ, Lanahan MB, Yen HC, Giovannoni JJ and Klee HJ (1995) An ethylene-inducible component of signal transduction encoded by *Never-ripe. Science,* **270** : 1807-1809.

Worrell AC, Bruneau JM, Summerfelt K, Boersig M and Voelker TA (1991) Expression of a maize sucrose phosphate synthase in tomato alters leaf carbohydrate partitioning. *Plant Cell,* **3** : 1121-1130.

Wunn J, Kloti A, Burkhardt PK, Biswas GCB, Launis K, Iglesias VA and Potrykus I (1996) Transgenic indica rice breeding line IR58 expressing synthetic *cry IA (b)* gene from *Bacillus thurigiensis* provides effective insect pest control. *Bio/ Technology,* **14** : 171-176.

Zagorska NA, Shtereva A, Dimitrov BD and Kruleva MM (1998) Induced androgenesis in tomato (*Lycopersicon esculentum* Mill.). I. Influence of genotype on androgenetic ability. *Plant Cell Rep.,* **17** : 968-973.

Zelcer A, Soferman O and Izhar S (1984) An *in vitro* screening for tomato genotypes exhibiting efficient shoot regeneration. *J. Plant Physiol.,* **115** : 211-215.

Chapter 4

IN VITRO REGENERATION AND TRANSFORMATION OF THE LEAFY VEGETABLES : LETTUCE, CHICORY AND SPINACH

MR Davey★, LC Garratt, FMK Williams and JB Power

Plant Science Division, School of Biosciences, University of Nottingham, Sutton Bonington Campus, Loughborough, Leicestershire LE12 5RD, UK

Summary

Tissue culture based technologies provide an adjunct to conventional breeding approaches for the genetic improvement of lettuce, chicory and spinach. The procedures for shoot regeneration from explants, explant-derived tissues and protoplast-derived tissues, are established in these leafy vegetables. Shoot regeneration in lettuce occurs by organogenesis; both organogenesis and somatic embryogenesis are pathways for plant regeneration in chicory and spinach. The parameters which have been assessed in optimising shoot regeneration include composition of the culture medium, the combination and concentrations of growth regulators and environmental conditions such as the photoperiod and incubation temperature. Somatic hybrid plants have been generated in lettuce following protoplast fusion, while cytoplasmic hybrids (cybrids) have been reported in chicory. Lettuce transformation is reproducible and has enabled several agronomically useful genes to be inserted into this crop. Transgenic plants of chicory have also been produced. In contrast, a transformation procedure that is readily applicable to a range of spinach cultivars is still awaited.

Keywords : *Agrobacterium rhizogenes*, *A. tumefaciens*, chicory, embryo rescue, gene identification, *in vitro*

★Corresponding author : E-mail : mike.davey@nottingham.ac.uk

pollination, lettuce, organogenesis, plant breeding, protoplasts, somaclonal variation, somatic embryogenesis, somatic hybridisation, spinach, transformation, tissue culture, zygotic embryos

Abbreviations : ABA – Abscisic acid; BAC – bacterial artificial chromosome; BAP – benzylamiino purine; 5,6-C12 – 5,6-dichloro-indole-3-acetic acid; CCPU – N-(2-chloro-4-pyridyl)-N-phenyl urea; CaMV – cauliflower mosaic virus; CMV – cucumber mosaic virus; IAA – indole acetic acid; IBA – indole butyric acid; 2,ip – 6-dimethylallylaminopurine; LMV – lettuce mosaic virus; MES – 2-(N-morpholino)ethanesulphonic acid; MS – Murashige and Skoog (1962); PCR – polymerase chain reaction; QTL – quantitative trait loci; RAPD – random amplified polymorphic DNA; RFLP – restriction fragment length polymorphism; SH – Schenk and Hildebrandt (1972); TSWV – tomato spotted wilt virus

1. INTRODUCTION

Lettuce, chicory and spinach are cultivated extensively as leafy vegetables with lettuce production being the most important on a global basis. In the case of lettuce (*Lactuca sativa*), the main production and consumption regions of this annual plant are the USA and Europe. Its annual production as a salad crop is nearly 4 million tonnes (grown on about 1,13,020 hectares) at an estimated value of $1.6 billion (United States Department of Agriculture, 1998) for the USA, with a total European production of 2.3 million tonnes (90,200 hectares; Statistical Office of the European Communities, 1998). Whilst the market for lettuce has expanded during the last 15 years, there has been a change in the varieties under cultivation. Traditional varieties have lost some of their market share to speciality varieties, in line with consumer requirements for exotic produce. For example, in the USA, the consumption of iceberg lettuce declined by 19% between 1989 and 1996, but the demand for romanie and other types increased by 78% during the same period (USDA, 1998). Processed mixed salad, containing different types of lettuce, has gained in popularity and has also stimulated the cultivation of a more extensive range of lettuce varieties.

Chicory (*Chicorium intybus*), like lettuce, is a member of the Asteraceae. It is a biennial monoecious plant which can be divided phenotypically into rosette, Witloof and heading types. Rosette types are open in form with elongated and either divided or entire leaves. Heading varieties produce green, variegated or red leaves, with heads of different morphology. For example, those of Itialian chicory may be round (chioggia), elongate (treviso) or cylindrical (sugar loaf) in shape. Witloof chicory is produced by growing roots of one year-old plants of the rosette type in the dark in either sand or hydroponic culture to produce white buds (chicons). The latter are cooked as a luxury vegetable and command high market prices.

Spinach (*Spinacia oleracea*) is a member of the Chenopodiaceae; the plant is cultivated as an annual for its leaves and as a biennial for seed production. In general, its growth habit is similar to that of lettuce and chicory. The fleshy, upright, wrinkled (savory) or smooth leaves form a compact rosette. An elongate flowering stem is produced in response to temperature and photoperiod (Correll *et al*., 1994). The plant grows well in cool climates and is more tolerant of frost than most other leafy vegetables (Goto *et al*., 1999a). Spinach leaves are used in both fresh and processed (canned and frozen) forms, with about 87,000 tonnes of the vegetable being processed in the USA during 1999. Spinach contains high concentrations of dietary minerals and vitamins (Goto *et al*., 1996). For example, about 93 mg of calcium and 51 mg of vitamin C are present in 100g fresh weight of leaves, together with significant quantities of iron, phosphorus and sodium (Ryder, 1999).

2. BREEDING OF LETTUCE, CHICORY AND SPINACH

As expected, considerably more effort has been invested in the genetic improvement by conventional breeding of lettuce than other leafy vegetables. Current breeding objectives for lettuce include altering leaf shape and pigmentation, primarily to provide the range of novelty that the market requires. Lettuce is an obligate self-fertile species, in which the stamens are fused to form an anther sheath, onto the outside of which pollen collects as the flower opens. As the style elongates and emerges from the anther sheath, pollen is shed onto the stigma. Despite the efficiency of the self-fertilisation system, cross hybridisation is possible involving emasculation prior to cross pollination or, more practically, washing of pollen grains from the stigma with a fine stream of water immediately after the style emerges from the anther sheath (Ryder, 1999). As style emergence occurs at the beginning of the day, sexual crossing can be facilitated by synchronising anthesis by imposing an artificial dawn on the plants. Three principle methods, pedigree, selection and back-crossing, are employed in lettuce breeding (Ryder, 1999). Pedigree breeding involves the generation of a large F_2 population, derived from crossing parental lines exhibiting interesting characteristics. The F_2 generation is screened for plants that exhibit the maximum number of useful traits. Selection breeding relies on the identification of useful variants within an existing cultivar, which arise through mutational changes. Back crossing is used to introduce single allelic characteristics into an existing commercial cultivar. An F_1 plant, derived from a commercial parent and one exhibiting a desired characteristic, is back-crossed with the commercial parent. Repeated back-crossing of subsequent progeny with the original commercial parent results in 99% of the genome of the progeny being derived from the recurrent commercial parent after 6 back-crosses.

Cultivated lettuce has been found to be sexually compatible with only a few of the other 100 or so species within the genus *Lactuca*. These include *L. serriola, L. saligna* and *L. virosa; L. sativa* crosses freely with *L. serriola.* If *L. saligna* is used as the female parent, it can be crossed with *L. sativa* to produce fertile hybrids. *L. virosa* can be used as a male or female parent in crosses with *L. sativa*, but the resulting hybrids are sterile. However, a hybrid resulting from a cross between *L. virosa* and *L. sativa* was made fertile by doubling its chromosome complement via colchicine treatment, creating an amphidiploid (Ryder, 1986). The exploitation of wild *Lactuca* species has led to a number of important traits being introgressed into *L. sativa*, including resistances to LMV, downy mildew, leaf aphid, cabbage looper and tipburn, reduced bitterness and enhanced root systems (Davey *et al.*, 2001).

Chicory is a monoecious plant in which sexual hybridization is hampered by the biennial life cycle, the sporophytic self-incompatability system present in many cultivars (Ryder, 1999) and the difficulty in obtaining synchrony of bolting in different genotypes (Varotto *et al.*, 2000). Flowering occurs when the daylength exceeds 13 hours during the second growing season and requires a period of vernalization between 8 and 18°C. Several traits are targets for both conventional breeding and genetic manipulation technologies. For example, chicory roots contain up to 20% stored carbohydrates (Frulleux *et al.*, 1997). The latter, upon hydrolysis, yield 18% fructose and 2% sucrose, which are of industrial interest. Genetic transformation of cultivars grown for their roots may allow the generation of plants with enhanced fructose production and thus increase their commercial value. One of the stored carbohydrates, inulin, has been found to be a probiotic of bifidiobacteria present in the human gut, where it is thought that bacteria convert inulin to compounds that display anti-carcinogenic activity (Vijn and Smeekes, 1999). Chicory has been introduced as a forage crop in New Zealand (Ryder, 1999). The roots are also cultivated to provide a coffee substitute (Bais *et al.*, 2001). Since the growth of seedlings in the field is slow, weeds are a problem because they out-compete the crop. Consequently, resistance to non-selective herbicides is of interest. The production of high quality chicons is important commercially. During forcing, both the taproot and chicons of Witloof chicory are susceptible to bacterial and fungal infections (Chupeau, 1989), making resistance to these pathogens targets for genetic manipulation. Since quality is also related to the nitrogen source, nitrogen metabolism is an additional target for modification (Abid *et al.*, 1995).

Spinach is dioecious, although monoecious individuals sometimes occur. Consequently, cross fertilisation is the rule, making traditional breeding difficult. Flowers of both male and female plants are formed in clusters of 6 - 12 and develop sequentially. Female flowers have a single ovary. Sex is influenced by X and Y chromosomes (Pandey and Kalloo, 1986), as well a environmental factors including light, temperature and growth regulators. Application of the latter to 10 days-old seedlings alters the 1:1 male to female ratio seen in untreated plants (Chaliakhyan and Khyranin, 1978). Thus, GA_3 increased the percentage of male plants to 79%. BAP, IAA and ABA increased the percentage of female plants, this being to 88% following the application of BAP. Spinach is the target of breeding efforts to improve its nutritional quality, tolerance to elevated temperature and acid soils, resistance to viral and fungal diseases which both cause widespread losses in the field, and a reduction in the high concentraions of oxalate. The latter interferes with Ca^{2+} uptake in humans (Goto *et al.*, 1996).

2.1. Gene identification to assist lettuce breeding

Recently, significant progress has been made in the identification of genes and molecular markers, resulting in the development of a genetic map for lettuce. To date, more than 90 genes have been identified in lettuce, which encode for leaf and floral morphologies and pigmentation, disease and insect resistances, seed and fertility characteristics, together with plant growth and development (Ryder, 1999). Whilst traditional genetic linkage has yielded little information in lettuce, a detailed genetic map has been constructed using isozyme analysis (Kesseli *et al.*, 1994), combined with RFLP and RAPD analyses. This map comprised 319 loci, including 152 RFLPs, 130 RAPDs, 7 isozyme, 19 disease resistance and 11 morphological markers. A further 9 morphological traits were mapped in lettuce relative to RAPD markers (Waycott *et al.*, 1999). The traits for the morphological characteristics of dwarf phenotype (dwf2), white seed (w), brown seed (br), salmon flower colour (sa), pale yellow flower colour (pa), virescent juvenile leaf colour (sa), plump involucre (pl), yellow seed (y) and anthocyanin spotting (Rs) were linked to RAPD loci. The variety of colour and form available in lettuce is extremely diverse, with several of the forms being characterised genetically. Three mutant traits for chlorophyll deficiency in lettuce, namely bleach bud, calico-3 and pale green, were characterised recently as being single recessive alleles (Ryder *et al.*, 1999) and a further six chlorophyll deficient mutants, chlorophyll deficient-3 (cd-3), light green (Ig), chlorophyll deficient-4 (cd-4), virescent (vi), dappled (dap) and sickly (si), have been described (Ryder, 1996) as have additional genes for leaf colour and form (Ryder, 1999). Four recessive alleles, dwf-1, dwf-2, dwf-3 and dwf-4 have been reported (Waycott *et al.*, 1995), controlling plant height from 3 to 30 cm. Recently, Ryder (1999) reviewed the progress made in the identification of genes in lettuce, which govern disease resistance and morphological characteristics. Back-crossing programmes have been enhanced by the identification of two partially dominant genes for flowering, *EF-1 ef-1* and *EF-2 ef-2*, which, in the double-dominant condition, reduce flowering time in lettuce from 140 days to 45 days (Ryder, 1988).

A BAC library has been constructed containing large *Eco*RI and *Hin*dIII genomic fragments of lettuce (Frijters *et al.*, 1997). The library, consisting of over 50,000 clones, was screened with markers linked to disease resistance genes. More recently, QTL were identified for differences between wild and cultivated lettuce with respect to root architecture (Johnson *et al.*, 2000). Undoubtedly, the continued development of a genetic map for lettuce will facilitate the identification of agronomically useful genes for introduction into this crop.

3. TISSUE CULTURE OF LETTUCE, CHICORY AND SPINACH

3.1. Tissue culture of lettuce

Lettuce is highly responsive in culture to a range of growth regulators, with the early studies of the response of explants being discussed in an extensive review by Michelmore and Eash (1988). Subsequently, the effects were investigated of combinations of several growth regulators, media components and environmental parameters, such as light and temperature, on adventitious bud initiation from cultured seedling hypocotyls of the cv. Wayahead (Sasaki, 1975, 1979a,b,c, 1982). The results of these and parallel studies resulted in optimisation of culture conditions for a range of explants with shoot regeneration being achieved for several cultivars. The latter included Greenfields and Summer Gem (Crisphead types), Bronze Mignonette and Green Mignonette (Butterhead types), Salad Bowl (Leaf type) and Cos and Green Cos (Romaine types) (Xinrun and Conner, 1992). An additional 22 lettuce genotypes belonging to different morphological groups were also screened for their shoot regeneration response on SH medium containing 3% (w/v) sucrose, 0.1 mgl^{-1} IAA, 0.5 mgl^{-1} kinetin and 0.05 mgl^{-1} zeatin (Ampomah-Dwamena *et al.*, 1997). Reproducible shoot regeneration was found with Bambino and Iceberg (Crisphead types), Cobham Green and Sweet Butter (Butterhead types), Simpson Elite (Leaf type) and Rosalita and Paris White (Cos types). Whilst the composition of the shoot regeneration media varies between studies, high regeneration rates can be achieved using a regeneration medium with MS salts and vitamins, 3% sucrose, 0.04 mg l^{-1} NAA, 0.5 mgl^{-1} BAP and semi-solidified with 0.8% (w/v) agar at pH 5.8. An incubation temperature of 23 ± 2°C with a 16 hour photoperiod (18 μmol m^{-2} s^{-1}, Daylight fluorescent tubes) is satisfactory. Seedling cotyledons as the source of explants have been favoured in the majority of studies because of their uniformity, their rapid shoot regeneration rates and the ease of generating axenic seedlings. For example, cotyledons from 4 days-old seedlings were reported as being the preferred starting material for root and shoot initiation (Webb *et al.*, 1984). The culture of shoot tips has been exploited for the rapid multiplication of F_1 hybrid lettuce plants and, subsequently, the production of many seedlings, without the need for year-by-year seed production (Takano *et al.*, 1988). Embryo rescue has been exploited to recover hybrids between *L. sativa* and *L. virosa* (Maisonneuve *et al.*, 1995).

3.1.1. Protoplast culture, somatic hybridisation and somaclonal variation in lettuce

Protoplast fusion offers a method of circumventing both pre- and postzygotic barriers to conventional sexual hybridization, thereby permitting the introgression of useful genetic traits into *L. sativa* from other sexually incompatable *Lactuca* species, which would have been impossible using conventional breeding techniques. As a basis for gene transfer by somatic hybridisation, protoplast-to-plant systems have been developed for *L. sativa* and some wild *Lactuca* species, including *L. serriola*, *L. saligna* and *L. virosa* (Berry *et al*., 1982; Engler and Grogan, 1984; Brown *et al*., 1988; Enomoto and Ohyama, 1989; Tanaka *et al*., 1991; Webb *et al*., 1994). Details of the source material, the enzyme mixtures and the culture conditions employed in these investigations can be found in the relevant publications. Likewise, details of the protoplast fusion procedures and subsequent culture conditions used to generate somatic hybrid plants in the genus *Lactuca* can be obtained by consulting the literature. Examples of somatic hybrids in lettuce include *L. sativa* with *L. tartarica* or *L. perennis* (Maisonneuve *et al*., 1995; Chupeau *et al*., 1994), *L. sativa* with *L. virosa* (Matsumoto, 1991), *L. sativa* with *L. deblis* or *L. indica* (Mizutani *et al*., 1989), and *L. sativa* with *L. serriola* (Matsumoto, 1987). The transfer of agronomically useful traits by protoplast fusion include resistances to leaf aphid, powdery mildew, downy mildew and bacterial rot from the wild species into cultivated lettuce Davey *et al*. (2001). However, detailed evaluation of such hybrids, especially under field conditions, is still awaited in most cases. Likewise, the application of this technology to commercial breeding programmes has not been reported.

Somaclonal variation is not uncommon in plants regenerated from cultured cells and tissues. Indeed, the frequency of somaclonal variation in some plants can be high, with 30-40% of regenerated plants showing a least one variant character, and up to 3% of the plants exhibiting variation in a specific trait (Daub, 1986). The phenomenon has been reviewed in detail (Karp, 1995), although the precise mechanisms involved are still not clear. In lettuce, somaclonal variation has been reported as a result of the culture of tissues (Sibi, 1976; Koevary *et al*., 1978; Berry *et al*., 1982; Alconero, 1983; Brown *et al*., 1986) and isolated protoplasts (Engler and Grogan, 1984). Brown *et al*. (1986) reported a somaclone of the cv. Salad Bowl that was more vigorous with an early flowering habit and increased chlorophyll content. In addition, somaclonal variants with reduced susceptibility to both LMV and *Bacillus lactucae* were observed. Engler and Grogan (1984) described the regeneration of protoclones derived from mesophyll protoplasts of the cv. Climax. Mutant phenotypes included variation in pigmentation, ranging from light green to yellow, elongated and crinkled leaves, dwarfism and increased vigour. Unfortunately, to date, most of the somaclonal variants reported in lettuce have exhibited deleterious characteristics, such as reduced vigour, reduced chlorophyll content, reduced fertility and stem fasciation, rather than advantageous traits.

3.2. Tissue culture of chicory

Chicory has been found to be amenable to culture, with cells having a high capacity for shoot regeneration (Frulleux *et al.,* 1997). Plants have been regenerated from cultured tissues by organogenesis and somatic embryogenesis; plants have also been regenerated from protoplast-derived tissues of *C. intybus* and the related species *C. endivia* (endive).

3.2.1. Shoot regeneration in chicory by organogenesis

The regeneration of fertile plants of spinach was reported by Mix (1985), who cultured vein segments from unexpanded leaves on MS-based medium. The author noted that the shoot regeneration response varied between cultivars. Another report in the same year (Bennici and Riepma 1985) demonstrated that storage roots of the cv. Cicoria Dolce de Soncino regenerated shoots on agar-solidified medium. Such shoots, when excised, developed roots on transfer to MS-based medium supplemented with IBA, permitting the recovery of intact, fertile plants.

During the early 1990s, leaf explants of the cv. Witloof were cultured on the medium of Quoirin and Lepoivre (1977), supplemented with BAP and IBA, to induce the formation of nodular tissues (Pieron *et al.*, 1993). The latter were transferred to medium lacking growth regulators to stimulate bud formation. Nodular tissues could also be placed on a Perlite-compost mix to promote bud formation; buds underwent rapid development into shoots and roots. Histochemical studies on the organogenic nodules (Pieron *et al.*, 1998) suggested that they developed in contact with the vascular bundles of the leaves and, within these tissues, vascular centres were initiated which were the sites of bud initiation. The vascular centres were thought to be critical for this form of organogenesis. As others workers have noted (Couillerot *et al.*, 1993), genotypes capable of undergoing organogenesis have shown altered polyamine metabolism compared to non-organogenic cultivars.

Adventitious buds were initiated from leaf explants of the cv. 474, a *C. intybus* x *C. endivia* hybrid (Decout *et al.*, 1994), using medium developed for the induction of somatic embryos (Dubois *et al.*, 1988). Shoot formation was observed at incubation temperatures below 30ºC and was most prolific at 25ºC, with wounding of the leaf tissues being essential for the induction of morphogenesis. An important observation, of possible application to chicory breeding, was that the addition of silver nitrate or putrescine both at a concentration of 40 mM to the culture medium, initiated flowering of regenerated shoots within 28 days of the introduction of explants into culture (Bais *et al.*, 2001).

Three papers reporting genetic transformation of chicory have involved the development of procedures for recovering transgenic plants by organogenesis from cultured leaf tissues (Vermeulen *et al.*, 1992), seedling cotyledons (Abid *et al.*, 1995) and shoot buds (Frulleux *et al.*, 1997). The methods by which plants were regenerated are discussed in more detail in relation to transformation.

3.2.2. Somatic embryogenesis in chicory

Shoots can be regenerated by somatic embryogenesis in chicory. In one of the first reports of somatic embryogenesis in chicory, shoots were regenerated from tissues of the cv. Witloof (Heirwegh *et al.*, 1985). Dubois *et al.* (1988), while attempting to induce androgenesis in chicory, observed the formation of somatic embryos from the sporophytic regions of cultured anthers. In the same study, styles from the cv. 474 were cultured on semi-solid half-strength MS-based medium, in which KNO_3 had been replaced with 10.1 mM KCl and 1.7 mM glutamine, and the inclusion of the microelements of Heller (1953), 5 x 10^{-3} Fe-EDTA, Morel and Wetmore's (1951) vitamins and 0.06 M sucrose. The medium also contained the growth regulators NAA at 0.1 µM and 2,ip at 2.5 µM. Cultures were incubated at temperatures of 15-40°C in the dark. Styles were most embryogenic (98.3% of those cultured) when incubated at 35°C in the dark, with each explant giving rise to 10 - 50 somatic embryos. Interestingly, the age of the explants was found not to be of significance in the response; temperature was the crucial factor in somatic embryo formation from this material. Later, Guedira *et al.* (1989) obtained somatic embryos from anthers excised from micropropagated plants, with anthers at the tetrad and microspore stages being the most responsive. Again, as in the study by Dubois *et al.* (1988), darkness and a temperature of 35°C were essential to induce the development of somatic embryos on medium containing 0.02 mgl^{-1} NAA and 0.5 mgl^{-1} 2,ip. Glutamine and ammonium nitrate provided the most suitable nitrogen sources. A short induction period of 10 days allowed the somatic embryos to develop into plants, the latter exhibiting generally satisfactory phenotypic uniformity, although some variant plants had anthocyanin-pigmented veins and narrow or indented leaves.

The chicory cv. 474 provided suitable source material for several other investigations. For example, Robatche-Claive *et al.* (1992) obtained somatic embryogenesis from leaves excised from 3 months-old plants. The leaves were incubated in the dark (35°C) in agitated liquid medium (Guedira *et al.,* 1989) supplemented with either sucrose alone, or a combination of sucrose with glycerol. Histological studies of material sectioned after 3 days in liquid medium, demonstrated the synchronous development of somatic embryos in explants cultured in medium supplemented with 30 mM sucrose. Increasing the concentration of sucrose to 360 mM increased the number of embryos which developed. The presence of glycerol (330 mM) in the medium inhibited the formation of embryos, which arose, apparently, from single cells.

Couillerot *et al.* (1993) used root segments, also from the cv. 474, to investigate the effect of different carbohydrates on somatic embryogenesis. Cellobiose, fructose, galactose, glucose, lactose, maltose, melibose, raffinose, sucrose and trehalose (each at 120 mM) were added to the embryo induction medium. Most embryos was obtained from explants cultured on medium supplemented with sucrose, although fructose, glucose, maltose and raffinose resulted in satisfactory yields of these structures. The authors correlated the embryogenic potential of explants with polyamines which they could detect in the induction medium after 5 days of culture. In medium which resulted in the induction of somatic embryos, high concentrations were detected of putrescine, spermine and spermidine. Such compounds were not released from explants cultured on a non-inductive medium. The authors noted that, unlike experiments with leaf explants (Robatche-Claive *et al.*, 1988), it was not possible to synchronise the formation of somatic embryos from root explants of chicory.

In 1994, Decout *et al.* published the results of additional investigations on the role of temperature during the incubation of chicory cultures, since earlier work of the same group (Dubois *et al.,* 1988) had demonstrated the importance of temperature in the culture response. Using the cv. 474 and the same medium as Dubois *et al.* (1988), cultures were incubated under continuous light or in darkness at temperatures of 20-35°C. Intact leaves did not respond; wounding of the explants was found to be essential to promote a response. After 5 days, cells near the wound sites began to divide, although the pattern of mitotic division was dependent on the incubation temperature. As had been observed earlier (Dubois *et al.*, 1988), somatic embryogenesis was most pronounced at 35°C, although shoot bud formation by organogenesis was poor at this temperature. At 25°C, the occurrence of somatic embryos was infrequent compared to shoot formation. At an intermediate temperature (30°C) both somatic embryos and shoots developed. No significant differences were found in embryo formation in the light or in darkness, although shoot formation was highest in the light. The authors suggested that temperature may affect the cell cytoskeleton and microtubules in chicory, which may be important in the initiation and subsequent development of somatic embryos.

Cell suspension cultures derived from callus originating from the veins of fully developed leaves of the cv. Witloof, were also used to obtain somatic embryos (Mohamed-Yasseen and Splittstoesser, 1995). Callus was initiated by culturing leaf vein segments on semi-solidified MS-based medium containing 30 gl^{-1} sucrose, 100 mgl^{-1} casein hydrolysate and supplemented with 1.3 μM 2,4-D and 1.3 μM kinetin. Cultures were incubated in the dark for 8 - 12 weeks with transfer to new medium every 4 weeks. Callus tissues were then maintained in liquid medium of the same composition for 4 - 6 weeks. Transfer to liquid MS-based medium with 30 gl^{-1} sucrose supplemented with 1.8 μM BAP, resulted in the formation of somatic embryos on all the calli. Although the somatic embryos were hyperhydric (vitrified), their germination was achieved on half-strength MS-based medium semi-solidified with 8 gl^{-1} agar.

An interesting observation was that high concentrations of glucanases were found in the medium on which explants of the cv. 474 had been incubated in the dark at 35°C (Helleboid *et al.*, 1998) when using the protocol described previously (Dubois *et al.*, 1988) to induce somatic embryogenesis. Glucanase concentrations were up to 3 times higher in the medium following the culture of embryogenic lines compared with non-embryogenic lines. The authors suggested that the proteins may have a role in somatic embryogenesis, possibly by acting upon the walls of cells of the developing somatic embryos. In a follow-up study, Helleboid *et al.* (2000) investigated the presence of proteins in medium conditioned by culturing for 4 days leaf explants of an embryogenic line of cv. 474 or a non-embryogenic line. Three protein bands of 25, 32 and 38 kD were detected in concentrations up to 8 fold higher in the embryogenic line of cv. 474 than in the non-embryogenic line. All 3 proteins were found to be pathogenesis-related proteins. The authors suggested that these proteins, like polyamines, may be involved in somatic embryogenesis.

Other workers have also exploited the same induction procedure as described by Dubois *et al.* (1988), to investigate the effect of doubling the sucrose concentration from 30 mM to 60 mM in the presence of 330 mM glycerol in the culture medium (Bellettre *et al.*, 1999). An increased sucrose concentration reduced the embryo induction period by 1 day, while the number of somatic embryos was increased from about 9 to 14 per mm^{-2} of callus. Embryogenic explants showed modified carbohydrate metabolism and were able to metabolise glycerol.

3.2.3. *In vitro* pollination and culture of immature zygotic embryos of chicory

Doré *et al.* (1996) pollinated the chicory cv. Chioggia and also cvs. expolited for the production of coffee substitute, with pollen from *Lactuca tatarica* and *Cicerbita alpina*. Of the 6 chicory genotypes tested, only the cv. 449-2 produced embryos and plants when crossed with *L. tatarica*, 2 plants being obtained from 55 capitulae. Pollination with *C. alpina* resulted in more embryos. Whether haploid plants or hybrids resulted was dependent on the chicory genotype, with the cv. BO9 producing hybrids and cv. FD1-12 producing haploid plants. However, only 3 plants of the cv. FD1-12 expressed the chicory phenotype and were haploid, despite 522 embryos being produced.

Ovule pollination has been used in several species, including chicory (Castaño and De Proft, 2000), to overcome self-incompatibility. Ovules were isolated from flower heads and flower buds of the cvs. Flash and Carolus and cultured on a wide range of media based on the MS or B5 (Gamborg *et al.*, 1968) formulations with 4-6% (w/v) sucrose. The growth regulators IAA, kinetin and GA_3 were added as supplements. When pollinated ovules were cultured under a 16 h photoperiod at 22°C, low numbers of seedlings developed (7 seedlings from 919 ovules in the cv. Flash). The cv. Carolus gave the best response on MS-based medium supplemented with 7.5 mgl^{-1} glycine, 0.5 mgl^{-1} IAA, 0.5 mgl^{-1} kinetin and 4 mgl^{-1} GA_3. Chromosome analyses confirmed that the plants were diploid.

Traditional methods of haploid plant production, including anther culture, ovule culture, pollen irradiation and ovule irradiation have all been unsuccessful in chicory. However, there has been one example published of microspore culture in this leafy vegetable (Theilerhedtrich and Hunter, 1995). Uninucleate microspores were cultured on MS-based medium supplemented with 0.5 mgl^{-1} 2,4 D, 0.5 mgl^{-1} IAA and 2 mgl^{-1} zeatin for up to 6 months to induce callus. The latter was then sub-cultured on basal medium supplemented with 0.5 mgl^{-1} BAP and 0.5 mgl^{-1} IAA; shoots were induced on medium with a reduced salt concentration in the presence of 0.4 mgl^{-1} kinetin and 0.2 mgl^{-1} IAA. The formation of adventitious roots on the regenerated shoots was achieved on half-strength Quoirin and Lepoivre (1977) medium containing 0.2 mgl^{-1} IBA. About 44 of 450 regenerated plants were haploid.

Recently, Varotto *et al.* (2000) reported plant regeneration from zygotic embryos isolated from the cv. Rosso di Chioggia. Plants were hand-pollinated with pollen from 5 different genotypes in order to overcome self-incompatibility. Embryos were collected 4, 6, 8, 10, 24, 48, 72, 96 and 168 hours after pollination and cultured on semi-solid B5 medium with Morel and Wetmore's (1951) vitamins, 800 mgl^{-1} glutamine and 1% sucrose, under a 14 h photoperiod at 25°C. Embryos collected at the heart or early torpedo stage (3 days after pollination) produced green plantlets within 2 weeks of culture. Less developed embryos either failed to develop or produced abnormal plants. The authors suggested that the culture of zygotic embryos may reduce the duration of chicory breeding programmes and that the torpedo stage is the best time to perform embryo rescue.

3.2.4. Isolation, culture and fusion of chicory protoplasts

In the initial reports of the culture of isolated chicory protoplasts (Crepy *et al.*, 1982), cell colonies often became necrotic with plant regeneration being difficult to stimulate, despite achieving protoplast plating efficiencies of 60 - 80%. These authors isolated protoplasts from young leaves of plants of the cv. Red Witloof grown in the glasshouse under low light conditions. Protoplasts were released using 0.02% (w/v) Macerozyme, 0.1% (w/v) Onozuka Cellulase R10 and 0.05% (w/v) Driselase, with an overnight incubation. Cells were cultured on MS-based medium supplemented with Heller's (1953) microelements, Morel and Wetmore's (1951) vitamins and 5 gl^{-1} sucrose. Ammonium nitrate in the culture medium caused cell necrosis, but increase in the survival of protoplast-derived colonies could be achieved when glutamine (10 mM) was used as the sole nitrogen source. Initially, protoplasts were cultured in the dark before transfer to a 16 hour photoperiod (28°C day and 24°C night temperatures), with 50 - 100 $\mu E\ m^{-2}\ sec^{-1}$of fluorescent illumination. Although callus formation was poor, buds were initiated by culturing protoplast-derived callus on medium with 0.5 mgl^{-1} IAA and 1 mgl^{-1} BAP. However, plant recovery rates were still low.

Saksi *et al.* (1986) isolated protoplasts from leaves of 21 days-old axenically grown seedlings of the Witloof cv. Flash, supplementing the basal medium of Crepy *et al.* (1982) with 1 mgl^{-1} of both BAP and NAA. In agreement with the results of Crepy *et al.* (1982), glutamine was found to be essential for sustained division of protoplast-derived cells. The authors reported that, initially, sucrose at 5 gl^{-1} was advantageous for protoplast growth. Callus formation was stimulated by culturing protoplast-derived cells on semi-solid medium lacking mannitol, the latter being used as an osmotic stabiliser during protoplast culture. Shoots formed on callus cultured on medium in which NH_4NO_3 was replaced by glutamine, and supplemented with 0.05 mgl^{-1} NAA and 0.5 mgl^{-1} BAP. Plants were transferred to the glasshouse after a culture period lasting 4 or 5 months.

In general, young leaves have been the preferred starting material for the isolation of protoplasts, as other tissues such as those from roots, stems, bracts and callus, resulted in poor yields (Chupeau, 1989). Optimum plant growth conditions were found to be critical in maximising yields of viable protoplasts of 5 - 7.5 x 10^6 g^{-1} fresh weight of tissue (Chupeau, 1989; Rambaud *et al.*, 1990). The latter authors, using procedures described in the literature, isolated and cultured protoplasts from the chicory var. Magdebourg cv. Pévèle.

The cv. 474 which was known to be highly embryogenic from previous culture studies, has also featured in the development of protoplast-to-plant systems. Mesophyll protoplasts were cultured on half-strength MS-based medium lacking NH_4NO_3, but containing Heller's (1953) microelements, Morel and Wetmore's (1951) vitamins, 375 mgl^{-1} glutamine, 150 mgl^{-1} myoinositol, 5 gl^{-1} sucrose and 90 gl^{-1} mannitol, together with with 10.7 μM NAA and 5 μM 2,ip. Somatic embryos were induced by 3 different procedures (Sidikou-Seyni *et al.*, 1992). Firstly, 7 days-old protoplast-derived callus were transferred to semi-solid medium with 500 mgl^{-1} glutamine, 50 gl^{-1} mannitol and 20 gl^{-1} sucrose and supplemented with 0.5 μM NAA and 2.5 μM 2,ip. Secondly, 1-2 mm protoplast-derived microcolonies were transferred to medium already optimised for somatic embryo induction with 2.5 μM 2,ip and 0.1 μM NAA (Dubois *et al.*, 1988). Thirdly, well-developed callus was transferred to the same somatic embryo induction medium. Embryos developed when transferred to agitated liquid Heller's (1953) medium containing 0.15 μM GA_3, followed by germination on semi-solid Heller medium lacking growth regulators. Embryos did not develop from single protoplasts as a callus stage was essential. It was noted that the procedure was not as rapid as somatic embryogenesis from leaf explants. However, regenerated plants were found to be morphogenically normal and fertile.

Several Italian red chicory types have been regenerated from protoplasts (Varotto *et al.*, 1997), with the genotype CH 363 resulting in 10 plantlets per protoplast-derived callus. Later, Nenz *et al.* (2000) investigated the behaviour of leaf mesophyll-derived protoplasts of the cv. Rosso di Chioggia. A comparison was made of culture in liquid medium, stratification in semi-solid medium and embedding in alginate droplets. Protoplasts embedded in calcium alginate droplets exhibited the highest division and plating frequencies. Plants could be regenerated in 2 months, compared with 4 - 5 months reported by Saksi *et al.* (1986) for protoplasts of the cv. Flash. The plating efficiency was highest (56.2%) on medium containing mannitol (8 gl^{-1}) with 10 mgl^{-1} NAA and 4.5 mgl^{-1} BAP. The same medium also resulted in the highest division frequency (35.9%) of protoplast-derived cells.

To date, there is one report of attempts to transfer genetic material into chicory by inter-generic protoplast fusion. Thus, Varotto *et al.* (2001) generated male sterile asymmetric hybrids (cybrids) by fusing iodoacetic acid inactivated mesophyll protoplasts from a red chicory genotype CH363 with gamma irradiated hypocotyl protoplasts of a male sterile line of sunflower (CMSHA89). After mixing protoplasts in a 3:1 or 1:1 ratio (chicory to sunflower), the protoplasts were fused using polyethylene glycol (PEG). Following chemical fusion treatment, protoplasts were immobilised in alginate droplets (Nenz *et al.*, 2000) and incubated at 23°C under a 14 hour photoperiod. Thirty three plants were regenerated of which 18 were found to have the same chromosome complement as chicory ($2n = 2x = 18$). Three plants were found to be asymmetric hybrids following DNA analysis by Southern blotting. These 3 cytoplasmic hybrids were morphologically similar to plants of the chicory parent, but their anthers contained less pollen at anthesis than plants of the genotype CH363 and failed to undergo self pollination.

3.3. Tissue culture of spinach

Spinach has been found to be relatively totipotent *in vitro*, although there is no universal procedure for plant regeneration from cultured tissues. Several general parameters relate to the culture of this leafy vegetable. The germination of spinach seeds is normally performed in the dark on half-strength MS-based medium lacking growth regulators. Germination is often erratic; the incubation temperature is important, with most workers using temperatures of 20 - 25ºC. Since seeds are often contaminated by micro-organsms, it may be necessary to remove the pericarp manually (Sasaki, 1989; Molvig and Rose, 1994), or to use long periods of sterilisation such as 2 - 3% sodium hypochlorite for 2 hours (Komai *et al.*, 1996a) prior to germination. Explants are normally derived from 5 - 10 days-old seedlings, although Al-Khayri *et al.* (1991a) used leaves from 3 months-old plants as source material. Knoll *et al.* (1997) employed explants taken from the roots of shoots that had been in culture for up to 24 months.

The culture conditions have varied widely between the studies reported. Incubation temperatures have ranged from 20°C (Al-Khayri *et al.* 1991a) to 25°C (Komai *et al.*, 1996b). The majority of workers have used a photoperiod of 16 hours for studies of organogenesis and somatic embryogenesis, with a light intensity of 35 – 90 µmol m^{-2} sec^{-1}. However, callus has also been initiated under continuous illumination (Mii *et al.*, 1992) and in continuous darkness (Al-Khayri *et al.*, 1991a; Molvig and Rose, 1994). In general, MS medium has been employed most extensively as the basal medium, although Knoll *et al.* (1997) utilised medium based on the formulation of Nitsch and Nitsch (1969). Some researchers (Zdravkovic-Korac and Neskovic, 1999) have supplemented basal media with additional vitamins. Sucrose (10 – 30 gl^{-1}) has been used as the main carbon source, although Komai *et al.* (1996b) reported fructose (5 gl^{-1}) to stimulate somatic embryogenesis from cultured root explants. Although spinach grown *in vivo* is intolerant of acid soils, a pH of 5.8 is generally used to culture tissues and plants of spinach. However, other workers have used media of higher pH. For example, Satoh *et al.* (1992) maintained the pH of the culture medium at 6.5, while Goto *et al.* (1996) found a pH of 6.3 increased root formation on shoots derived from mesophyll protoplasts. The duration of the period from the begining of culture to acclimatization of regenerated plants in the glasshouse varies considerably. Al-Khayri *et al.* (1991a) obtained plants by organogenesis in 29 weeks, whereas Zhang and Zeevaart (1999) succeeded in reducing this time for transgenic plants to 6 weeks. Somatic embryogenesis has been reported to require a culture period of about 13 weeks (Xiao and Branchard, 1993).

3.3.1. Shoot regeneration in spinach by organogenesis

Adventitious shoot formation was described initially in spinach by Neskovic and Radojevic (1973). Shoots were induced from callus derived from seedling apices on MS-based medium supplemented with 4.65 µM kinetin and 2.89 μM GA_3 under diffuse light conditions. Sixteen years later, Sasaki (1989) reported callus formation and organogenesis from the 5 spinach cvs. Wakakusa, Nippon, Toko, Hoyo and Tokai by culturing hypocotyl segments from 6 days-old seedlings on MS-based medium containing 20 gl^{-1} sucrose with various combinations of GA_3 (0, 10^{-7}, 10^{-6}, 10^{-5} and 10^{-4} M) and 15 mgl^{-1} IAA under continuous light at 25°C. Callus development was dependent on the cultivar, with each cultivar requiring a very specific concentration of GA_3 in the medium. Adventitious shoot buds were observed in all cultivars except Tokai, callus of the cv. Nippon giving the highest frequency of shoot regeneration. Interestingly, there was an inverse correlation between callus initiation and regeneration, cultivars capable of shoot regeneration, such as Nippon, exhibited poor callus development. Shoots were excised after 5 - 6 weeks and rooted on MS-based medium supplemented with 10 mgl^{-1} IBA. Regenerated plants flowered and set seed.

In a series of publications, Al-Khayri *et al.* (1991a) attempted to optimise organogenesis from cultured tissues of spinach. They used leaf explants from 3 months-old plants of the cvs. High Pack, Grandstand, Baker and Kent. Explants were cultured on MS-based medium supplemented with 2 mgl^{-1} kinetin in combination with 0.1, 0.5 or 1 mgl^{-1} 2,4-D in the dark, or exposed to a 10 hour photoperiod to induce callus formation. After 4 weeks, callus was transferred to medium with the same concentration of kinetin, but with 0.1 mgl^{-1} 2,4-D, 1 mgl^{-1} GA_3 and 100 mgl^{-1} ascorbic acid to optimise shoot regeneration. Fourteen weeks later, shoots (each 1.5 - 2 cm in height) were rooted on medium with 1 mgl^{-1} IBA, the latter being more effective than the other auxins (NAA, IAA, IBA and 2,4-D) assessed. Although explants of the cv. Grandstand cultured on medium with 0.5 mgl^{-1} 2,4-D in the dark readily formed callus, shoot regeneration was not as good as in the cv. High Pack. These results were in agreement with those of Sasaki (1989), who also found that cultivars that readily produced callus exhibited poor shoot regeneration. Callus induced in the dark regenerated more shoots than callus initiated in the light. Al-Khayri *et al.* (1991a) stressed that the system was highly cultivar dependent and, therefore, could not be considered as a general protocol. An interesting observation was that by utilising shoots derived from these studies, Al-Khayri *et al.* (1991c) reported flowering *in vitro* in the cv. High Pack when the shoots were incubated under long day conditions on medium lacking growth regulators. In contrast, shoots remained vegetative when maintained under short days. Shoots rooted readily and it was envisaged that such an experimental system might provide a simple procedure for maintaining selected, elite plants.

Subsequently, the same research group (Al-Khayri *et al.*, 1992a) targeted the cvs. High Pack and Baker, which had shown the highest shoot regeneration frequencies in their previous investigations. They employed the same culture protocol and basal medium, but added 0, 5, 10, 15 and 20% (v/v) coconut milk. Fifteen percent coconut milk increased callus production by explants of both cvs. and reduced the time for callus initiation by 7 days. Shoot formation was also stimulated and was more rapid on this medium, leading to a reduction in the overall culture duration of 8 - 7 weeks. The cv. High Pack responded better, in terms of shoot regeneration, than the cv. Baker.

Having determined the optimum cultivar and the most suitable basal medium, Al-Khayri *et al.* (1992b) investigated the role of GA_3 in the culture response. Noteworthy was the fact that shoot regeneration was not observed on explants cultured on medium lacking GA_3. The optimum concentration of GA_3 was 7.2 μM, which resulted in 70% of the explant-derived calli regenerating shoots. Interestingly, high concentrations of GA_3 reduced shoot height, which was unexpected, since GA_3 normally promotes stem elongation. The authors concluded that whilst 7.2 μM GA_3 was optimum for callus formation, transfer of tissues to medium with GA_3 reduced to 1.4 μM was recommended for shoot multiplication.

Other early reports of organogenesis in spinach were those of Kondo *et al.* (1991), who induced shoot primordia of the cvs. Minsterland and Jiromaru to produce multiple shoots when cultured in liquid MS-based medium with IAA and BAP both at 0.2 - 4 mgl^{-1}under a 16 hour photoperiod at 22°C. The optimum concentrations of growth regulators were found to be 0.02 - 1 mgl^{-1} IAA with 2 mgl^{-1} BAP. Roots were formed on shoots excised from cultured explants on medium either with IBA at 0.2 mgl^{-1} or lacking growth regulators. In agreement with the results of other workers (Komai *et al.*, 1996c; Zdravkovic-Korac and Neskovic, 1999), plant regeneration was observed only in cultures with red pigmentation. Explants cultured on a medium with a pH 5.8 exhibited the highest frequency of shoot regeneration.

Other workers used hypocotyl explants from 7 days-old seedlings of the cv. Sunlight (Satoh *et al.*, 1992) to induce callus on MS-based medium supplemented with 7 mgl^{-1} NAA, with the pH of the medium increased from 5.8 to 6.5. In a further report of shoot regeneration from hypocotyl segments, Mii *et al.* (1992) cultured 7 days-old explants of the same age from 9 spinach cultivars on Nitsch (1969) medium containing 10 mgl^{-1} IAA, the cvs. Solomon, Hoyo and Jiromaru responding best in terms of callus formation. This work again reflected cultivar differences, a re-occurring theme throughout reports on the culture of spinach tissues. The effects of pre-culture of explants on medium containing a range of auxins was investigated using the cv. Hoyo, since explants of this cv. showed the highest rates of callus initiation. Pre-culture on medium suplemented with 5,6-Cl2-IAA for 20 days before transfer to medium lacking growth regulators, was beneficial in stimulating 81.7% of the hypocotyl-derived calli to regenerate shoots after 10 weeks of culture.

Using cotyledons, roots and hypocotyls from the cv. Hybrid 102, Molvig and Rose (1994) modified the procedure of Neskovic and Radojevic (1973) in order to regenerate plants. Callus was initiated from all explants in the dark on MS-based medium supplemented with 4.6 μM kinetin and 4.5 μM 2,4-D. Callus growth was enhanced by subsequent culture under a 16 hour photoperiod on medium with the same concentration of kinetin, but with 2.9 μM GA_3 replacing 2,4-D. Shoot and root formation was observed on the explants after several culture cycles, each of 4 - 6 weeks duration. Subsequently, shoots were excised and transferred to hormone-free medium for development into plants.

Both hypocotyl and root explants have also been used as source material in other studies. For example, Xiao and Branchard (1995) cultured such explants on MS-based medium containing 85.62 μM IAA and 100 μM GA^3. These high concentrations of growth regulators induced more than 90% of the root explants to form adventitious shoots. In contrast, hypocotyl segments were less responsive, with only 75% of the explants regenerating shoots. Ninety two % of the shoots excised and cultured on medium containing 2.85 - 5.71 μM IAA developed roots.

More recently, Zhang and Zeevart (1999) reported a rapid and efficient procedure for regenerating transgenic plants from 5 days-old cotyledons of the cv. Longstanding Bloomsdale Dark Green. This method, together with a shoot regeneration protocol exploiting the ability of root segments to undergo shoot regeneration (Knoll *et al.*, 1997), focused on the generation of transgenic plants of spinach. These procedures are discussed in a later section on spinach transformation.

3.3.2. Somatic embryogenesis in spinach

One of the first papers to report somatic embryogenesis in spinach was by Xiao and Branchard (1993). Seven days-old hypocotyl explants of the cv. Carpo were cultured on semi-solid MS-based medium supplemented with 3% sucrose, 0.01 mgl^{-1} biotin, 250 mgl^{-1} glutamine, 15 mgl^{-1} IAA and 34.6 mgl^{-1} GA_3. Callus was induced by incubation in the dark for 7 days, followed by 14 days with a 10 hour photoperiod preceeding continuous light. Somatic embryos were induced from the callus by transfer of the latter to liquid medium with 0.5 mgl^{-1} IAA and 3.46 mgl^{-1} GA_3 in continuous light. The decrease in IAA concentration was found to be essential for the induction of somatic embryogenesis and subsequent embryo development. The somatic embryos developed into plants when cultured in liquid medium containing 0.5 mgl^{-1} IBA under continuous light.

Two years later, Komai *et al.* (1995) reported somatic embryogenesis from root explants taken from 10 days-old seedlings of the cv. Jiromaru. Again, callus was induced on MS-based medium, but with 20 gl^{-1} sucrose in combination with 10 or 30 μM NAA and 0 - 100 μM GA_3. A nitrate : ammonium ratio of 2 : 1 in the medium was beneficial for callus induction. After 28 days exposure to a 16 hour photoperiod, cell masses were excised from the calli and cultured in liquid or on semi-solid MS-based medium lacking growth regulators to induce the formation and development of somatic embryos. Most plants were regenerated from cell masses in liquid medium. Histological studies confirmed the presence of somatic embryos of all stages of development after 28 days on callus induction medium. As in the case of organogenesis, the presence of GA_3 in the callus induction medium was found to be essential for somatic embryogenesis.

In subsequent experiments from the same laboratory, Komai *et al.* (1996c) performed a more detailed study of somatic embryogenesis in spinach. Explants of cotyledons, roots, hypocotyls and leaves from the 8 cvs. Jiromaru, Nihon, Hoyo, King of Denmark, Ujo, Minsterland, Viroflay and Nobel were cultured on 6 different media [Nitsch (1969), half-strength MS, full strength MS, White's (1943), B5 and SH] supplemented with 10 gl^{-1} sucrose, 0.1 μM GA_3 and 10 μM NAA. Explants were incubated as in their earlier report (Komai *et al.*, 1995). Root explants of the cvs. Jiromaru, Hoyo and Nihon produced the most embryogenic tissues, while half-strength MS-based medium was the most suitable for callus induction and somatic embryo formation.

A further study by Komai *et al.* (1996d) targeted root explants of the cv. Jiromaru which the authors exposed to different concentrations of growth regulators. IAA, IBA, NAA, 2,4-D (all at 0, 0.1, 1, 10, 30 or 100 μM), purine, 2,ip, kinetin, BAP, CPPU and GA_3 (all at 0.01, 0.1, 1 or 10 μM) were used either individually, or in combination. Callus was induced on medium supplemented with all of the auxins evaluated, the optimum response being on medium containing 30 or 100 μM IBA with 10 or 30 μM NAA. However, callus induced on medium containing only auxins failed to produce somatic embryos; GA_3 was found to be essential for the formation of embryogenic callus.

Komai *et al.* (1996b) also investigated the effect of the carbohydrate source on somatic embryogenesis in spinach. Again using the same cv. Jiromaru and the culture conditions described previously, they modified the medium with 29, 87 or 145 mM glucose, galactose, mannose, fructose, sorbose, maltose, cellobiose, lactose or raffinose to replace sucrose. Although embryos developed on medium with most of the carbohydrates assessed, somatic embryogenesis was optimal on medium with 29 mM fructose. Subsequent analysis of sugar concentrations in the medium suggested that this response may have been related to fructose being metabolised in preference to other carbohydrates.

An interesting observation, in contrast to the results of Komai *et al.* (1996c), was that Zdravkovic-Korac and Neskovic (1999) reported the presence of GA_3 and IAA in the medium suppressed embryo initiation, whereas abscisic acid (4 μM) increased the number of embryos produced per gram fresh weight of callus. Discs from fully expanded young leaves of the cv. Matador were cultured on MS-based medium containing 20 gl^{-1} sucrose, 100 mgl^{-1} myo-inoistol, 2 mgl^{-1} thiamine hydrochloride, 2 mgl^{-1} pyridoxine hydrochloride, 2 mgl^{-1} adenine, 5 mgl^{-1} nicotinic acid, 4.4 μM 2,4-D and 4.6 μM kinetin. Embryo induction was achieved by transferring the explants to medium lacking 2,4-D, but supplemented with ABA, GA_3 or IAA. All cultures were incubated under a 8 hour photoperiod. Embryos formed only from red pigmented callus, in agreement with the results of Komai *et al.* (1996c). Somatic embryos appeared as white organised structures and were easily distinguished from the parental callus and any developing bud initials. Shoot formation also occurred on many of the calli. The authors also reported that the growth regulators determine the ratio of globular to polar embryos in the cultures, leading them to suggest that specific growth regulators influence cell division and expansion during the early stages of somatic embryogenesis.

More recently, Ishizaki *et al.* (2000) investigated the role of ethylene in somatic embryo formation from root explants of the cv. Nippon. The addition of 10 μM ethephron to callus induction medium resulted in a 50% increase in the frequency of embryogenic callus initiation compared to controls lacking this compound. However, addition of the same concentration of ethephron to medium used for later stages of somatic embryo development reduced the final number of somatic embryos produced. Silver nitrate (an ethylene inhibitor) at the same concentration had the opposite effect. Thus, this compound decreased embryogenic callus formation, but when used at a later stage it increased the number of somatic embryos that developed from embryogenic callus. Therefore, the authors argued that ethylene may be critical in embryogenic callus initiation in spinach, but was inhibitory to the later development of somatic embryos.

3.3.3. Isolation, culture and fusion of spinach protoplasts

Spinach protoplasts were first isolated by Otsuki and Takebe in 1969. Initially, such protoplasts were used only in chloroplast studies and were first cultured for longer periods by Rose (1980) on MS-based medium supplemented with 0.5% sucrose, 0.4M sorbitol and 0.4M mannitol. The latter author employed leaves from 3 weeks-old seedlings and found that protoplasts isolated in the presence of Ca^{2+} and cultured under dim light conditions had a 50% survival rate after 8 days of culture. Darkness was found to inhibit cell wall regeneration by cultured protoplasts.

A period of 23 years elapsed before Goto and Miyazaki (1992) became the first investigators to succeed in regenerating plants from protoplasts derived from leaf material of the cv. Jiromaru, with most mitotic division of protoplast-derived cells leading to cell colony formation being optimal on half-strength MS-based medium supplemented with 5 mgl^{-1} BAP, 1 mgl^{-1} 2,4-D and CPW salts (Frearson *et al.*, 1973). In common with the work of Rose (1980), protoplasts were cultured at 1 x 10^5 ml^{-1} at 25°C under dim light for 7 days, before transfer to more intense light conditions. Adventitious shoot formation was induced on MS-based medium with 1 mgl^{-1} kinetin, 1 mgl^{-1} zeatin or various combinations of BAP and NAA; shoots were rooted in liquid MS-based medium supplemented with 1 mgl^{-1} IBA or 1 mgl^{-1} IAA. Initially, shoot regeneration levels were low, but subsequent work by the same authors (Goto *et al.*, 1996) demonstrated an increase in organogenesis using the same basal medium, cultivar and isolation procedure. The addition of KM8P vitamins (Kao and Michayluk, 1975) and 0.5M glucose as osmoticum increased the protoplast plating efficiency, although mannitol at 0.5M decreased this efficiency. Again, protoplasts were maintained in the dark for the first 7 days of culture. Goto *et al.* (1996) also obtained evidence that rooting of protoplast-derived shoots was stimulated when the pH of the shoot regeneration medium was adjusted to 6.3. In subsequent studies, Goto *et al.* (1998a) analysed the regeneration potential of protoplast-derived tissues of 7 spinach cultivars using their established culture conditions. Protoplasts isolated from the cvs. Ujo, Jiromaru, Hojo and Minsterland exhibited higher plating efficiencies and rates of cell colony formation than the cvs. Sapporo-aba, Viroflay and Viking. Shoot regeneration from callus was not observed in the later 3 cultivars, again emphasising cultivar-specific responses. These workers also suggested that smooth seeded spinach cvs. may be less amenable to culture than prickly seeded varieties.

In the same year as Goto *et al* (1996) published their studies on spinach protoplasts, Komai *et al.* (1996a) also reported that they had obtained somatic embryos from protoplasts isolated from cotyledons, roots and hypocotyls from 10 days-old seedlings and leaves from 30 days-old plants of the cv. Jiromaru. These studies were based on their demonstration of somatic embryogenesis from cultured root explants (Komai *et al.*, 1996b). Isolated protoplasts were cultured on a medium similar to that employed by Goto and Miyazali (1992), but with 1 μM 2,4-D or 10 μM NAA combined with 10 μM zeatin. An extended dark period of 21 days at 25°C was used, followed by transfer to dim light conditions for a further 21 days, during which time the cultures were diluted with new medium every 3 days. Callus was induced from protoplast-derived cells by supplementing the basal medium with 30 mM sucrose, 30 mM glucose, 500 mgl^{-1} casein hydroxylate, 0.3 μM 2,4-D and 3 μM zeatin, whilst maintaining the cultures under a 16h photoperiod. Somatic embryos formed on protoplast-derived callus when the growth regulators were changed to 0.1 μM GA_3 with 10 μM NAA. Interestingly, adventitious roots were also induced from protoplast-derived callus. These roots could be stimulated to produce multiple somatic embryos on MS-based medium supplemented with 30 mM fructose.

More recently, in attempts to maximise the efficiency of the protoplast-to-plant system for spinach, Goto *et al.* (1999a) evaluated the effect of organic acids on cell division and shoot multiplication. The addition of 0.1 M citric acid and other organic acids to growth medium proved beneficial in the presence of 0.5 M glucose, with the division of protoplast-derived cells, callus formation and shoot regeneration all being enhanced.

In medium supplemented with 0.5 M mannitol, other organic acids such as sodium pyruvate, fumaric acid, malic acid, malonic acid and succinic acid, either alone or in combination, had no effect or a negative influence. An interesting observation was that in the presence of 0.1 M citric acid, cell division was recorded in protoplast cultures at plating densities as high as 20 x 10^4 ml^{-1}. The authors found that leaves from 25 days-old seedlings were the preferred source material for the isolation of viable, totipotent protoplasts.

4. TRANSFORMATION OF LETTUCE, CHICORY AND SPINACH

4.1. Transformation of lettuce

Although there are reports of lettuce transformation by electroporation-mediated gene transfer into isolated protoplasts (Chupeau *et al.*, 1989), few genotypes have been transformed using this procedure which has been superseded by *Agrobacterium* based methods. Indeed, the use of *Agrobacterium* mediated gene delivery as a means of generating transgenic lettuce plants is now routine in many laboratories. A reliable genotype-independent protocol for lettuce transformation was published by Curtis *et al.* (1994, 1995) involving seedling cotyledons as target material. The technique has also been described in detail (Davey *et al.*, 2001), together with the merits of using supervirulent strains of *Agrobacterium*, such as strain 1065, for gene transfer into lettuce. In general, shoot regeneration occurs from 50% of excised, inoculated cotyledons.

4.1.1. Introduction of agronomically important genes into lettuce

Efforts have been made to introduce specific characteristics into lettuce, some of which are considered in this discussion, using a recombinant DNA-transformation technology approach. For example, the iron storage protein, ferritin, is responsible for sequestering the intracellular iron present within plants, and has therefore been a target for manipulation in tobacco (Deak *et al.*, 1998; Goto *et al.*, 1998b), rice (Goto *et al.*, 1999b) and, more recently, lettuce (Goto *et al.*, 2000). The overproduction of ferritin reduces oxidative stress and improves tolerance to pathogens in tobacco (Deak *et al.*, 1998) by sequestering free-iron which would normally produce iron-mediated Fenton oxidants, such as hydroxyl radicals ($OH^{\bullet}$). Transgenic lettuce plants have been generated which exhibit enhanced iron accumulation and increased growth (Goto *et al.*, 2000). In the latter studies, the binary plasmid BG1, carrying the CaMV 35S promoter driving the soybean ferritin cDNA with the *npt*II gene as a selctable marker, was transferred into *A. tumefaciens* strain LBA4404. Lettuce plants transformed with this construct contained iron concentrations which were 1.2 - 1.7 fold higher than in wild-type plants. Enhanced growth rates were also observed in the transgenic plants. While the enhanced growth rate was attributed to a reduction in iron-mediated Fenton oxidant production, free-radical production and antioxidant scavenging potential has not been assessed in these plants.

Male sterility has been described by Curtis *et al.* (1996) in *Agrobacterium* transformed lettuce plants of the cv. Lake Nyah. Thus, transgenic plants expressing a pathogenesis-related β-1,3-glucanase gene linked to a tapetum-specific promoter, A9, exhibited dissolution of the microspore callose wall, resulting in male sterility. The possible exploitation of male sterile plants in lettuce breeding programmes was discussed in the work reported.

Lettuce is a vegetable which undergoes rapid leaf deterioration following harvesting. Consequently, attempts have been made to delay this process of senescence. For example, an *ipt* gene encoding isopentenyl phosphotransferase, under the control of the senescence-specific promoter SAG12 from *Arabidopsis thaliana* (P_{SAG12}-*IPT*), significantly delayed developmental and post-harvest leaf senescence in transgenic lettuce (cv. Evola) homozygous for the transgene (McCabe *et al.*, 2001). Isopentenyl phosphotransferase catalyses the rate limiting step for *de novo* cytokinin biosynthesis. The P_{SAG12}-*IPT* gene is activated only at the onset of senescence, resulting in cytokinin biosynthesis which inhibits leaf senescence and, consequently, attenuates activity of the P_{SAG12}-*IPT* gene, preventing cytokinin overproduction. Apart from retardation of leaf senescence, mature 60 days-old plants exhibited normal morphology with no significant differences in head diameter or fresh weight of leaves or roots. Importantly, harvested lettuce heads from transgenic plants expressing the P_{SAG12}-*IPT* transgene exhibited a significant increase in shelf life, with their outer leaves showing delayed senescence and retention of chlorophyll for up to 7 days longer than those of lettuce heads derived from non-transformed plants.

Since viruses and weeds cause severe crop losses, it is not surprising that they have been targets for genetic manipulation in lettuce and, indeed, other plants. Virus resistance has been introduced into lettuce using a coat protein-mediated approach. Thus, the nucleocapsid (N) protein gene of the lettuce isolate of the tospovirus TSWV was overexpressed in lettuce (Pang *et al.*, 1996), resulting in suppression of N protein accumulation through transgene silencing and subsequent TSWV resistance. Similarly, the coat protein gene from LMV strain O (LMV-O) was introduced into three virus-susceptible lettuce cvs. Girelle, Jessy and Cocarde (Dinant *et al.*, 1997). The effectiveness of this strategy for inducing resistance to LMV was shown to be related to the developmental stage of the transgenic plants at the time of inoculation. Whilst some plants exhibited complete resistance, others did show, however, viral symptoms during later development. Resistance to the broad-spectrum, glufosinate ammonium based herbicide (*Challenge*®), was introduced into lettuce (cv. Evola) by *Agrobacterium* mediated transformation (Mohapatra *et al.*, 1999). The *bar* gene encoding phosphinothricin acetyltransferase, was shown to be integrated into the lettuce genome and stably expressed over several seed generations, conferring resistance to 300 mg l^{-1} glufosinate ammonium. While conventional plant breeding has also been used to transfer sulfonylurea resistance into cultivated lettuce from the wild species *L. serriola* (Mallory-Smith *et al.*, 1990), sexual incompatibility and lack of availability of resistance genes in the lettuce gene pool restricts further introgression of herbicide resistance genes to *Agrobacterium* or allied gene transfer systems.

Although gene insertion by *Agrobacterium*-mediated transformation is reproducible in lettuce, the data currently available indicate that transgene expression can be inconsistent in several cultivars. Additionally, the choice of promoters for long-term gene expression is crucial. Gene copy number and the extent of DNA methylation also influence transgene expression in this leafy vegetable. These parameters, together with their possible limitations, are discussed in detail in relation to the genetic modification of lettuce by Davey *et al* (2001). These authors also provide reference to detailed background information from the relevant publications.

4.2. Transformation of chicory

In an early report of genetic manipulation in chicory, Sun *et al.* (1991) transformed root and pedicel explants of the cv. Hybrid Flash using 2 strains of *A. rhizogenes*, namely A4RSII and 8169. Of interest was the fact that regenerated plants exhibited a change from biennial to annual flowering, which was passed to seed progeny in a Mendelian fashion. It was suggested that this change in flowering habit may have been associated with the integration into the plant genome of *rol* genes from the T-DNA of the Ri plasmid of *A. rhizogenes*, since *rol* genes had been demonstrated to have a similar effect on flowering in other biennial plants, such as carrot.

Later, Vermeulen *et al.* (1992) introduced into chicory a mutant acetolactate synthase gene (*csr1-1*) from *Arabidopsis*, which conferred resistance to the herbicide chlorsulfuron. Leaf discs from 2 months-old plants of an unspecified cultivar were inoculated by immersion for 1 minute in an overnight culture of *A. tumefaciens* strain LBA4404 harbouring the recombinant vector pGH6, the latter carrying the gene of interest and the *neo* gene (from pBIN19) conferring kanamycin resistance on transformed plant cells. Following co-cultivation, leaf discs were transferred to semi-solid MS-based shoot induction medium with Morel and Wetmore's (1958) vitamins, 58 mM sucrose, 3.5 mM MES, 160 μM iron-citrate-ammonium, 0.015 μM bromocresol purple and supplemented with a range of concentrations of IAA, IBA, BAP and NAA. Microelements of Heller (1953) or the B5 formulation were also included in the medium. Optimum shoot regeneration occurred after 12 – 15 days in the presence of 0.1 mgl^{-1} IAA and 1 mgl^{-1} BAP with B5 microelements. Since the addition of kanamycin sulphate at 100 mgl^{-1} to the shoot induction medium inhibited growth, the selection of transgenic plants was delayed until rooting of the regenerated shoots. The medium for rooting was similar to that for shoot induction, but with MS components reduced to half strength and sucrose increased to 87 mM. Growth regulators were omitted and the medium was supplemented with 100 mgl^{-1} of kanamycin sulphate. Five independently transformed shoots were selected and micropropagated in the presence of kanamycin. Kanamycin resistant plants were acclimatised to glasshouse conditions and flowering was induced by a vernalisation period of 2 months at 6°C. Two plants were self-pollinated and the seeds were germinated on medium containing kanamycin and chlorsulfuron; a 2:1 rather than the expected 3:1 segregation was found for kanamycin and chlorsulfuron tolerances compared to sensitivity to these compounds. The authors suggested that this ratio may have been due to the chimaeric nature of the transgenic plants as selection had not been carried out until after shoots had developed. Nevertheless, despite this difficulty, the transformation protocol of Vermeulen *et al.* (1992) was also exploited to introduce a fructan:fructan 6G-fructosyltransferase (6G-FFT) from onion into chicory (Vijn *et al.*, 1997). In these experiments, the authors demonstrated that it was possible to alter the type of fructan synthesised in a fructan-producing plant. Transgenic plants were found to accumulate the fructan encoded by the transgene, together with the endogenous fructans.

In two publications, tap-roots and leaf discs, as well as cotyledons from etiolated seedlings of the Witloof cv. Flash were transformed using *A. tumefaciens* strains carrying either pGSGLUC1 or pTDE4, both encoding *npt*II and *uidA* (*gus*) genes under different promoters (Abid *et al.*, 1995, 2001). The genes on pGSGLUC1 were under the control of TR1' and TR2' promoters, while on pTDE4 the *uid*A and *npt*II genes were driven by the CaMV35S and *nos* promoters, respectively. Explants were inoculated with a bacterial suspension (OD_{600} = 1) in which the Agrobacteria were cultured in medium supplemented with 100 µM acetosyringone. Tap-roots were incubated on Heller's (1953) medium supplemented with 75 mgl^{-1} kanamycin; leaves and cotyledons were wounded manually prior to inoculation before culture on half-strength MS-based medium in the presence of 0.1 mgl^{-1} BAP for 2 days followed by selection in the presence of 75 µg ml^{-1} kanamycin. After 2 - 3 weeks, shoots were transferred to agar-solidified Heller's (1953) medium containing 10 gl^{-1} sucrose and 75 μgl^{-1} kanamycin to induce root formation. Cotyledons inoculated with *Agrobacterium* carrying pTDE4 resulted in the highest transformation efficiency (15%). The choice of appropriate culture medium was important in these experiments. As shoot formation is normally rapid from tap-roots, it was necessary to maintain explants of these organs on a medium which was relatively poor in nutrients in order to prevent the production of chimaeric shoots. Leaf and cotyledon tissues required a more nutrient rich medium to induce adventitious shoots. Seven independently transformed plants were subjected to Southern analysis; only 1 plant contained a single T-DNA insert. Other plants had 1 - 5 copies of T-DNA per 2C genome, although the majority of the T-DNA sequences were incomplete or re-arranged. In contrast to the studies of Vermeulen *et al.* (1992), the authors did not encounter problems with kanamycin selection inhibiting growth or shoot regeneration from cultured explants.

Excised buds have also been used in chicory transformation studies (Frulleux *et al.*, 1997). For example, explants of the 6 cvs. Hicor, Inula, Tilda, VBF, VBG and VAX were cultured on B5 medium with 0.9 mM BAP and semi-solidified with 7 gl^{-1} agar. In transient gene expression experiments, strain GV2260:35SGUSINT was used, containing a *gus* gene with a plant intron to preclude gene expression in Agrobacteria. Strain EHA101 carrying pGV1531 with *nos-npt*II and CaMV 35S-*uid*A genes was used for stable transformation. Shoot buds (each 2 - 3mm in size) were inoculated with an overnight bacterial suspension (OD_{600} 1.1), before co-cultivation with the bacteria for 2 days. Explants were washed with sterile water containing 500 mgl^{-1} carbenicillin and cultured in the presence of the same concentration of carbenicillin to remove bacteria, with 100 mgl^{-1} kanamycin for selection. Four weeks later, shoots were transferred to medium in which the concentration of kanamycin was increased to 200 mgl^{-1}. Shoots were exposed to this selection pressure for 4 months, before rooting by exposure to 0.98 mM IBA in the presence of carbenicillin and kanamycin. Only in the cvs. Hicor and Inula did about 10% of the shoots survive kanamycin selection, despite 60% of the inoculated buds exhibiting transient *gus* expression. However, more than 95% of leaf discs excised from the putatively transformed plants regenerated shoots when cultured on medium supplemented with 100 mgl^{-1} of kanamycin. Southern analysis confirmed the presence of the *npt*II gene in kanamycin resistant plants, although no mention was made of transgene copy number. Transformed plants were self-pollinated after vernalisation treatment (6°C); segregation analysis showed a 3:1 ratio of kanamycin resistant to sensitive seedlings.

There are reports of the effects on chicory of the expression of genes on the T-DNA of the Ri plasmid of *Agrobacterium rhizogenes*, the first of which was by Sun *et al.* (1991). Plants regenerated from roots transformed by *A. rhizogenes* strain A4SRII exhibited modifications in morphology and physiology, including a change from biennial to annual flowering, stunted growth, wrinkled leaves, reduced development of tuberised roots, but proliferic development of lateral roots. Although the plants still had a requirement for an inductive period to promote flowering, a vernalisation period was unnecessary. Phenotypic alteration of growth habit in these transgenic plants was at least partially associated with expression of the *rol A-C* genes on the Ri T-DNA. Based on this initial report by Sun *et al.* (1991), Limami *et al.* (1998) further investigated the switch from biennial to annual flowering. Back-crossing of transformed with non-transformed plants produced a line that retained annual flowering in the absence of other phenotypic abnormalities. Molecular analysis indicated that annualism was associated with a truncated T-DNA insertion into the plant genome. When transgenic plants were harvested at the end of a 5 months period, the inulin concentration in their roots was found to be the same as that in non-transformed plants, ranging from 71 - 75% of the root dry weight. However, at the end of the vegetative period, the free fructose content increased 8 - 10 fold in the roots of transformed plants because of increased inulin hydrolysis in the transgenic plants at this time. Unlike the situation in non-transformed plants, chicon induction in transgenic plants did not require a vernalisation period of 0 – 2°C for 14 days.

In other studies using *A. rhizogenes*, Bais *et al.* (2001) employed strains of mannopine-type Agrobacteria (LMG-150, A20/83 and AZ/83) to transform wounded hypocotyls from 2 – 10 days-old seedlings of the cv. Lucknow Local. After co-cultivation on MS-based medium for 3 days, explants were exposed to 500 mgl^{-1} of carbenicillin, with subculture to new medium every 3 days. Roots which developed were transferred to liquid MS-based medium in the dark on a rotary shaker to obtain axenic hairy root cultures. For all strains of *Agrobacterium* tested, a bacterial concentration of 10^8 cells ml^{-1} was optimal for infection; strain LMG-150 resulted in the highest percentage of inoculated explants producing transformed roots. This response declined with explants taken from seedlings more than 2 days old. Shoots were induced from transformed roots on MS-based medium with 4 mgl^{-1} BAP and 1 mgl^{-1} NAA; such shoots were micropropagated on medium with 0.5 mgl^{-1} of GA_3, 2 mgl^{-1} 2,ip and 0 – 50 mM putrescine, with the latter compound at 40 mM being optimal for shoot induction. Interestingly, this concentration of putrescine also stimulated flowering *in vitro* of regenerated shoots. Silver nitrate at 40 μM could also be used to replace putrescine to stimulate shoot multiplication. The presence of the *rol* A gene in hairy roots and plants arising from transformation with *Agrobacterium* strain LMG-150 was confirmed using PCR.

At present, chicory is the only example of a genetically modified leafy vegetable to be approved for field release in Europe and the USA. The Agro-Chemical Company Bejo Zaden generated 3 lines of the cv. Radicchio Rosso using *Agrobacterium*-mediated transformation which were both male sterile and glufosinate tolerant (BiotechKnowledgeCentre, European Commission, Ref. No. 1004). Male sterility was achieved using the anther-specific promoter TA29 to drive expression of the barnase gene from *Bacillus amyloliquefaciens*, which encodes an extracellular ribonuclease, in anthers of transgenic plants. A *bar* gene, encoding herbicide tolerance, was placed under the control of the *Arabidopsis* promoter ssuAra and the *nos* termination sequence from *A. tumefaciens*. All transgenic plants were selected by their kanamycin resistance. Expression of the *bar* gene in transgenic plants was used solely for breeding purposes, as only 50% of the crop was herbicide tolerant. Taste studies indicated that there was no alteration in the palatability of the transgenic chicory compared to non-transformed plants.

4.3. Transformation of spinach

Transgene expression in spinach was first reported by Al-Khayri (1995). Using a culture system developed previously in the same laboratory (Al-Khayri *et al.,* 1991a), leaf-discs from 3 months-old plants, hypocotyl segments from 14 days-old seedlings and cell suspensions from leaf-derived callus of the cv. High Pack, were targeted using *A. tumefaciens* strain A208 carrying the disarmed binary vector pMON9749. The latter carried CaMV 35S-*gus* and *nos-npt*II genes, together with a gene for spectinomycin resistance. Leaf explants and hypocotyl segments were co-cultivated with *Agrobacterium* on MS-based medium supplemented with 2 mgl^{-1} kinetin and 0.5 mgl^{-1} 2,4-D to induce callus. Subsequently, the explants were transferred to medium with 100 mgl^{-1} cefotaxime and 100 mgl^{-1} carbenicillin (to remove the bacteria) and with 50 mgl^{-1} kanamycin to select transformed callus and transgenic shoots. Callus formed in the dark on 3-8% of the inoculated leaf explants and on 4-20% of the hypocotyl segments. In the case of suspensions, the cells were inoculated with 50-200 μl aliquots of an overnight bacterial suspension. Two days later, the cells were washed with liquid MS-based medium to remove the bacteria and the cultures maintained in liquid medium with antibiotics, including kanamycin, for 21 days in the light. Cells were plated onto kanamycin-containing selection medium for 21 days in the dark, prior to culture on semi-solid shoot regeneration medium. Al-Khayri (1995) reported that up to 100% of the cells from the suspensions were GUS positive and, consequently, they argued that the approach using cell suspensions was preferable for transformation of spinach. It was thought that differences in cell wall development in the cell systems may have been crucial in eliciting the transformation response. Kanamycin resistant plants transferred to the glasshouse appeared morphologically normal. The main deficiency of this study was that molecular analysis was not performed on the regenerated plants.

Three publications during 1997 reported spinach transformation. Thus, Daniels *et al.* (1997) targeted leaf and cotyledon explants of spinach using *A. tumefaciens* strain AGLO with a binary vector carrying the *npt*II gene and a gene for green fluorescent protein (*gfp*). The latter, under the control of the CaMV 35S promoter, had been engineered to target the endoplasmic reticulum. The expression of the *gfp* gene permitted the non-destructive monitoring of transformed plant material. Explants were inoculated with overnight bacterial cultures of optical densities of 0.3 or 1.0, before culture in the dark at 24°C for 84 days on callus induction medium (Al-Khayri *et al.*, 1991a). Stable expression was recorded after 21 weeks in culture. An *Agrobacterium* suspension of optical density 1.0 resulted in a higher percentage of calli expressing *gfp* than a more dilute bacterial suspension with an optical density of 0.3. Although transformed callus was obtained from leaf explants, cotyledon explants could not be transformed using the same approach.

In the same year, Knoll *et al.* (1997) reported a shoot regeneration system from spinach root explants, the latter being induced on hypocotyl segments taken from axenic seedlings of the cvs. Longstanding Round and RS No. E. This regeneration system was used as a basis for transformation. *Agrobacterium tumefaciens* strains 0065 and 1065, both carrying the binary vector pMOG23 with *npt*II and *gus* genes, were used for inoculation of explants, strain 1065 being supervirulent because of the presence of pTOK47 carrying copies of the *vir B*, *C* and *G* genes. Explants were either dipped in a 1:10 (v:v) dilution of an overnight culture for 2 seconds or immersed in a 1:1 dilution of the bacterium for 10 minutes. Following 2 days of co-cultivation with Agrobacteria, explants were transferred to Nitsch and Nitsch (1969) medium supplemented with 100 mgl^{-1} cefotaxime, 100 mgl^{-1} carbenicillin and 50 mgl^{-1} kanamycin sulphate for selection. Explants cultured in the light for 56 days after inoculation were assayed histochemically for GUS activity. GUS positive shoot buds were observed on explants taken from roots that had been cultured for 20 months prior to inoculation with strain 1065.

During 1997, Yang *et al.* attempted to introduce a gene of potential agronomic interest into the cv. High Pack and also demonstrated transient gene expression in the spinach cv. Fall Green. *Agrobacterium tumefaciens* strain LBA4404 was used with the binary vector pB1121 modified so that the *gus* gene of the original construct was replaced by the coat protein gene of the CMV isolates, SP103 and SP104. Leaf strips and hypocotyl segments from 6 weeks-old plants were inoculated with overnight cultures of *Agrobacterium* to which 0.05 mM acetosyringone had been added 4 hours prior to explant inoculation. After 2 days of co-cultivation on semi-solid MS-based medium, the explants were transferred to callus induction medium (Al-Khayri *et al.*, 1991a) supplemented with 237 μM carbenicillin, 209 μM cefotaxime and 86 μM kanamycin for 12 weeks in the dark, with transfer to new medium every 4 weeks. Calli were then transferred to shoot regeneration and root induction media (Al-Khayri *et al.*, 1991a), both supplemented with cefotaxime and carbenicillin, in the light. Although the cv. Fall Green gave the highest rates of transformed calli, shoot regeneration was not achieved in this case. In contrast, the cv. High Pack responded well to regeneration medium and plants were recovered from shoots initiated on both leaf and hypocotyl-derived calli. Twelve percent of the regenerated plants were found to carry the CMV coat protein gene, which was detected using PCR. Southern and Northern blots confirmed the presence of the gene and its protein product in transgenic plants. However, the latter were not assessed for their response to inoculation with CMV. Consequently, it was not known whether the presence and expression of the coat protein gene resulted in increased tolerance/resistance to the virus.

In the most recent report of spinach transformation, Zhang and Zeevaart (1999) claimed to have developed a rapid and efficient procedure for generating transgenic plants using cotyledons excised from 5 days-old seedlings of the cv. Longstanding Bloomsdale Dark Green as starting material. *Agrobacterium tumefaciens* strain LBA4404 harbouring the binary vector pB1121 was used, in which the CaMV 35S-*gus* region of the original construct was replaced by 35S-*smgfp*, a gene encoding a more soluble version of a codon-modified green fluorescent protein. The basis for the use of the *gfp* gene as a marker in this study was provided by the earlier investgations of Yang *et al.* (1997) and Daniels *et al.* (1997), who both used constructs carrying the *gfp* gene for spinach transformation. *Agrobacterium* was cultured overnight with 0.2 mM acetosyringone, before being diluted 1:10 by volume with new medium and culture for a further 48 hours prior to use for inoculating explants. Explants were inoculated for 25 minutes and co-cultured with Agrobacteria on MS-based medium supplemented with 1 mgl^{-1} BAP and 0.4 mgl^{-1} NAA, prior to transfer to selection medium supplemented with 100 mgl^{-1} cefotaxime, 200 mgl^{-1} carbenicillin and 50 mgl^{-1} kanamycin. Shoots formed on 86% of the inoculated explants cultured in the presence of kanamycin. An amber filter was used during culture, although it was not confirmed if this was critical for shoot regeneration. Regenerated shoots were excised from the parental explants and immersed in liquid MS-based medium supplemented with 20 mgl^{-1} IBA for 2 hours, before transfer to semi-solid MS-based medium containing 1 mgl^{-1} IBA. The regenerated, kanamycin resistant plants (T_0 generation) were hybridised with wild-type plants as the dioecious nature of spinach precluded self-fertilisation. When seeds from the T_0 plants were germinated on selection medium containing kanamycin sulphate at 50 mgl^{-1}, a 1:1 segregation was found of resistant to susceptible T_1 generation seedlings. However, seedlings of the subsequent T_2 generation were found to segregate 3 :

1 (resistant to susceptible), suggesting that the *npt*II gene had been inherited in a Mendelian fashion. Northern and Southern analyses established the transgenic nature of the T_2 generation. To date, this investigation by Zhang and Zeevaart (1999) has advanced considerably spinach transformation, although the applicability of this procedure to a range of cvs. remains to be confirmed by workers from other laboratories.

In general, the transformation efficiency of spinach is low and cultivar dependent, with various approaches being attempted to increase the transformation rate. For example, Daniels *et al.* (1997) found that the addition of the surfactant Pluronic F-68 to the medium used for bacterial inoculation, increased transformation rates from 5.1% to 19.7%, while dilution of the *Agrobacterium* culture prior to use and the time of inoculation have been found to be important. In this respect, Knoll *et al*. (1997) noted that a 1:10 (v:v) dilution of an overnight *Agrobacterium* culture resulted in a higher transformation frequency than either an undiluted culture, or a bacterial suspension diluted 1:1 by volume to provide the inoculum. Similarly, the *Agrobacterium* strain used for inoculation of plant tissues is also important, with supervirulent strains, such as 1065, giving the highest rates of transformation. The source and age of explants are also critical in maximising transformation.

5. CONCLUSIONS AND FUTURE PROSPECTS

To date, considerably less effort has been invested in developing somatic cell technologies for the genetic improvement of lettuce, chicory and spinach compared to research directed to some of the major crops, such as cereals and oilseeds. Nevertheless, significant progress has been made during the last decade in these leafy vegetables. For example, protoplast fusion to generate somatic hybrid and cybrid plants offers a means of introgressing genetic material from wild into cultivated species without the need for complex recombinant DNA technology. Indeed, this approach has already been applied to lettuce and chicory. In terms of transformation, the ease of generating transgenic plants of lettuce using an *Agrobacterium*-based procedures may enable this vegetable to become a "model" system for gene introduction and expression. Similarly, chicory responds to *Agrobacterium* inoculation, although subsequent exploitation of transgenic plants in breeding programmes may be more difficult in this case, because of the biennial life-cycle of this leafy vegetable. Despite the transformation systems reported for spinach, such procedures still necessitate considerable refinement before they can be applied to a range of commercially important germplasms.

REFERENCES

Abid M, Huss B and Rambour S (2001) Transgenic Chicory (*Cichorium intybus* L.) In: *Biotechnology in Agriculture and Forestry, Vol. 47: Transgenic Crops II* (Ed Bajaj YPS), Springer-Verlag, Berlin, pp. 102-123.

Abid M, Plams B, Derycke R, Tissier JP and Rambour S (1995) Transformation of chicory and expression of the bacterial *uid*A and *npt*II genes in the transgenic regenerants. *J. Exp. Bot.*, **46** : 337-346.

Alcohero R (1983) Regeneration of plants from cell suspensions of *Lactuca saligna, Lactuca sativa* and *Lactuca serriola*. *HortSci.* **18** : 305-307.

Al-Khayri JM (1995) Genetic transformation in *Spinacia oleracea* L. (spinach). In: *Biotechnology in Agriculture and Forestry, Vol. 34: Plant Protoplasts and Genetic Engineering* (Ed Bajaj YPS), Springer-Verlag, Berlin, pp. 279-288.

Al-Khayri JM, Huang FH and Morelock TE (1991a) Regeneration of spinach from leaf callus. *HortSci.,* **26** : 913-914.

Al-Khayri JM, Huang, FH and Morelock TE (1991b) *In vitro* flowering in regenerated shoots of spinach. *HortSci.,* **26** : 1422.

Al-Khayri JM, Huang FH, Morelock TE and Busharar TA (1992a) Stimulation of shoot regeneration in spinach callus by gibberellic acid. *HortSci.,* **27** : 1046.

Al-Kharyi JM, Huang FH, Morelock TE and Busharar TA (1992b) Spinach tissue culture improved with coconut water. *HortSci.,* **27** : 357-358.

Ampomah-Dwamena C, Conner AJ and Fautrier AG (1997) Genotype response of lettuce cotyledons to regeneration *in vitro*. *Sci. Hort.,* **71** : 137-145.

Bais HP, Venkafesh RT, Chandrashkar A and Ravishankar GA (2001) *Agrobacterium rhizogenes* mediated transformation of Witloof chicory - *In vitro* shoot regeneration and induction of flowering. *Curr. Sci.,* **80** : 83-87.

Bellettre A, Couillerot JP and Vasseur J (1999) Effect of glycerol on somatic embryogenesis in *Cichorium* leaves. *Plant Cell Rep.,* **19** : 26-31.

Bennici AR and Riepma P (1985) *In vitro* clonal multiplication of chicory (*Cichorium intybus* L.). Rivista della Ortoflorofrutti *Coltura Italiana,* **69** : 235.

Berry SF, Lu DY, Pental D and Cocking EC (1982) Regeneration of plants from protoplasts of *Lactuca sativa* L. *Z. Pflanzenphysiol.,* **108** : 31-38.

Brown C Lucas JA Crute IR Walker DGA and Power JB (1986) An assessment of genetic variability in somaclonal lettuce plants (*Lactuca sativa*) and their offspring. *Ann. Appl. Biol.,* **109** : 391-407.

Castãno CI and De Proft MP (2000) *In vitro* pollination of isolated ovules of *Cichorium intybus* L. *Plant Cell Rep.,* **19** : 616-621.

Chaliakhyan MKH and Khyranin VN (1978) Effect of growth regulators and role of roots in sex expression in spinach. *Planta,* **142** : 207-210.

Chupeau Y (1989) Regeneration of plants from chicory (*Cichorium intybus* L.) protoplasts. In: *Biotechnology in Agriculture and Forestry, Vol. 8. Plant Protoplasts and Genetic Engineering I* (Ed Bajaj YPS), Springer-Verlag, Berlin, pp. 206-216.

Chupeau M, Maisonneauve B, Bellec Y and Chupeau Y (1994) A *Lactuca* universal hybridizer, and its use in creation of fertile interspecific somatic hybrids. *Mol. Gen. Genet.,* **245** : 139-145.

Correll JC, Morelock TE, Black MC, Koike ST, Brandenburger LP and Dainello FJ (1994) Economically important diseases of spinach. *Plant Dis.,* **78** : 653-660.

Couillerot JP, Decout E, Warnot F, Dubois J and Vasseur J (1993) Free polyamines evolution in relation to carbohydrate sources and somatic embryogenesis in a *Cichorium* hybrid. *C.R. Acad. Sci. Paris Série III,* **316** : 299-305.

Crepy L, Chupeau MC and Chupeau Y (1982) The isolation and culture of leaf protoplasts of *Cichorium intybus* and their regeneration into plants. *Z. Pflanzenphysiol.,* **107** : 123-131.

Curtis IS, Davey MR and Power JB (1995) Leaf disk transformation. In: *Methods in Molecular Biology, Vol. 44: Agrobacterium Protocols.* (Eds Gartland KMA and Davey MR), Humana Press Inc., Totowa, NJ, pp. 59-70.

Curtis IS, He CP, Scott R, Power JB and Davey MR (1996) Genomic male sterility in lettuce, a baseline for the production of F_1 hybrids. *Plant Sci.,* **113** : 113-119.

Curtis IS, Power JB, Blackhall NW, de Laat AMM, Davey MR (1994) Genotype-independent transformation of lettuce using *Agrobacterium tumefaciens. J. Exp. Bot.,* **45** : 1441-1449.

Daniels SJ, Zelcer A, Ball LF and Goldsbrough AP (1997) GFP as a reporter gene for spinach transformation. In: *Spinach Report,* Plant Breeding International, Cambridge, UK, pp. 20.

Daub M (1986) Tissue culture and the selection of resistance to plant pathogens. *Annu. Rev. Phytopathol.,* **24** : 159-186.

Davey MR, McCabe MS, Mohapatra U and Power JB (2002) Genetic manipulation of lettuce. In: *Transgenic Plants* (Eds Khachatourians GG, McHughen A, Scorza R, Nip W-K, Hui YH). Marcel Dekker Inc., New York. (in press).

Deak M, Horvath GV, Davletova S, Torok K, Sass L, Vass I, Barna B, Kiraly Z and Dudits D (1998) Plants ectopically expressing the iron-binding protein, ferritin, are tolerant to oxidative damage and pathogens. *Nature Biotech.,* **17** : 192-196.

Decout E, Dubois T, Guedira M, Dubois J, Audran J-C and Vasseur J (1994) Role of temperature as a triggering signal for organogenesis or somatic embryogenesis in wounded leaves of chicory cultured *in vitro. J. Exp. Bot.,* **45** : 1859-1865.

Dinant S, Maisonneauve B, Albouy J, Chupeau Y, Chupeau MC, Bellec Y, Gaudefroy F, Kusiak C, Souche S, Robaglia C and Lot H (1997) Coat protein-mediated protection in *Lactuca sativa* against lettuce potyvirus strains. *Mol. Breed.,* **3** : 75-86.

Doré C, Prigent J and Desprez B (1996) *In situ* gynogenetic haploid plants of chicory (*Cichorium intybus* L.) after intergeneric hybridisation with *Cicerbita alpina* Walbr. *Plant Cell Rep.,* **15** : 758-761.\

Dubois T, Dubois J, Guedira M and Vasseur J (1988) Embryogenése somatique directe sur les styles de *Cichorium*: effets de la temperature et origine des embryïdes. *C.R. Acad. Sci. Paris Série III,* **307** : 669-675.

Dubois T, Dubois J, Guedira M and Vasseur J (1990) Direct embryogenesis in roots of *Cichorium*. Is callose an early marker? *Ann. Bot.,* **65** : 539-545.

Engler DE and Grogan RG (1984) Variation in lettuce plants regenerated from protoplasts. *J. Hered.,* **75** : 426-430.

Frearson EM, Power JB and Cocking EC (1973) The isolation, culture and regeneration of petunia leaf protoplasts. *Dev. Biol.,* **33** : 130-137.

Frijters ACJ, Zhang Z, Van Damme M, Wang GL, Ronald PC and Michelmore RW (1997) Construction of a bacterial artificial chromosome library containing large *Eco*RI and *Hin*dIII genomic fragments of lettuce. *Theor. Appl. Genet.,* **94** : 390-399.

Frulleux F, Weyens G and Jacobs M (1997) *Agrobacterium tumefaciens*-mediated transformation of shoot-buds of chicory. *Plant Cell Tiss. Org. Cult.*, **50** : 107-112.

Gamborg OL, Miller RA and Ojima K (1968) Nutrient requirements of suspension cultures of soybean root cells. *J. Exp. Res.*, **50** : 151-158.

Goto T and Miyazaki M (1992) Plant regeneration from mesophyll protoplasts of *Spinacia oleracea* L. *Plant Tiss. Cult. Letts.*, **9** : 15-21.

Goto T, Miyazaki M and Oku M (1996) Improved procedure for protoplast culture and plant regeneration of spinach (*Spinacia oleracea* L.). *J. Japan. Soc. Hort. Sci.*, **65** : 349-354.

Goto T, Miyazaki M and Oku M (1998a) Varietal variations in plant regenerative potential from protoplasts in spinach (*Spinacia oleracea* L.). *J. Japan. Soc. Hort. Sci.*, **67** : 503-506.

Goto T, Miyazaki M and Oku M (1999a) Stimulation of protoplast division of spinach (*Spinacia oleracea* L.) by addition of citric acid. *J. Japan. Soc. Hort. Sci.*, **68** : 762-767.

Goto F, Yoshihara T and Saiki H (1998b) Iron accumulation in tobacco plants expressing soybean ferritin gene. *Transgenic Res.*, **7** : 173-180.

Goto F, Yoshihara T, Shigemoto N, Toki S and Takaiwa F. (1999b) Iron fortification of rice seed by the soybean ferritin gene. *Nature Biotech.*, **17** : 282-286.

Goto F, Yoshihara T and Saiki H (2000) Iron accumulation and enhanced growth in transgenic lettuce plants expressing the iron-binding protein ferritin. *Theor. Appl. Genet.*, **100** : 658-664.

Guedira M, Dubois-Tylski T, Vasseur J and Dubois J (1989) Embryogénèse somatique directe de cultures d'anthères de *Cichorium* (Asteraceae). *Can. J. Bot.*, **67** : 970-976.

Helleboid S, Bauw G, Belingheri L, Vasseur J and Hilbert J-L (1998) Extracellular ß-1,3-glucanases are induced during early somatic embryogenesis in *Cichorium. Planta,* **205** : 56-63.

Helleboid S, Hendriks T, Bauw G, Inze D, Vasseur J and Hilbert J (2000) Three major somatic embryogenesis related proteins in *Cichorium* identified as PR proteins. *J. Exp. Bot.*, **51** : 1189-1200.

Heller R (1953) Reserches sur la nutrition minérale des tissus végétaux cultivés *in vitro. Ann. Sci. Nat. Bot. Biol. Vég.*, **14** : 1-223.

Heirwegh KMG, Banerjee N, Van Nerum K and De Langhe E (1985) Somatic embryogenesis and plant regeneration in *Cichorium intybus* L. (Witloof, Compositae). *Plant Cell Rep.*, **4** : 108

Ishizaki T, Komai T, Msuda K and Megumi C (2000) Exogenous ethylene enhances formation of embryogenic callus and inhibits embryogenesis in cultures of explants in spinach roots. *J. Amer. Soc. Hort. Sci.*, **125** : 21-24.

Johnson WC, Jackson LE, Ochoa O, van Wijk R, Peleman J, St Clair DA and Michelmore RW (2000) Lettuce, a shallow-rooted crop, and *Lactuca serriola*, its wild progenitor, differ at QTL determining root architecture and deep soil water exploitation. *Theor. Appl. Genet.* **101** : 1066-1073.

Kao K and Michayluk MR (1975) Nutritional requirements for *Vicia hajastana* cells and protoplasts at a very low plating density in liquid medium. *Planta,* **126** : 105-110.

Karp A (1995) Somaclonal variation as a tool for crop improvement. *Euphytica,* **85** : 295-302.

Kesseli RV, Paran I and Michelmore RW (1994) Analysis of a detailed genetic linkage map of *Lactuca sativa* (lettuce) constructed from RFLP and RAPD markers. *Genetics,* **136** : 1435-1446.

Knoll KA, Short KC, Curtis IS, Power JB and Davey MR (1997) Shoot regeneration from cultured root explants of spinach (*Spinacia oleracea* L.): a system for *Agrobacterium* transformation. *Plant Cell Rep.,* **17** : 96-101.

Koevary K, Rappaport L and Morris LL (1978) Tissue culture propagation of head lettuce. *HortSci.,* **13** : 39-41.

Komai F, Kiyoshi K, Harada T and Okuse I (1996a) Plant regeneration from adventitious roots of spinach (*Spinacia oleracea* L.) grown from protoplasts. *Plant Sci.,* **120** : 89-94.

Komai F, Okuse I and Harada T (1995) Histological identification of somatic embryogenesis from excised root tissues of spinach (*Spinacia oleracea* L.). *Plant Tiss. Cult. Letts.,* **12** : 313-315.

Komai F, Okuse I and Harada T (1996b) Effective combinations of plant growth regulators for somatic embryogenesis from spinach root segments. *J. Jpn. Soc. Hort. Sci.,* **65** : 559-564.

Komai F, Okuse I and Harada T (1996c) Somatic embryogenesis and plant regeneration in culture of root segments of spinach (*Spinacia oleracea* L.). *Plant Sci.,* **133** : 203-208.

Komai F, Okuse I, Saga K and Harada T (1996d) Improvement on the efficiency of somatic embryogenesis from spinach root tissues by applying various sugars. *J. Japan. Soc. Hort. Sci.,* **65** : 67-72.

Kondo K, Nadamitsu S, Tanaka R and Taniguchi K (1991) Micropropagation of *Spinacia oleracea* L. through culture of shoot primordia. *Plant Tiss. Cult. Letts.,* **8** : 1-4.

Limami MA, Sun L-Y, Douat C, Helgeson J and Tepfer D (1998) Natural genetic transformation by *Agrobacterium rhizogenes* – annual flowering in two biennials, Belgian endive and carrot. *Plant Physiol.,* **118** : 543-550.

Maisonneuve B, Chupeau MC, Bellec Y and Chupeau Y (1995) Sexual and somatic hybridisation in the genus *Lactuca. Euphytica,* **85** : 281-285.

Mallory-Smith CA, Thill DC, Dial M.J and Zemetra RS (1990) Inheritance of sulfonylurea resistance in *Lactuca* spp. *Weed Technol.,* **4** : 787-790.

Matsumoto E (1987) Production of somatic hybrids between *Lactuca sativa* and *L. serriola* by cell fusion. *Jpn. J. Breed.,* **35** : 134-135.

Matsumoto E (1991) Interspecific somatic hybridisation between lettuce (*Lactuca sativa*) and wild species (*L. virosa*). *Plant Cell Rep.,* **9** : 531-534.

McCabe MS, Garratt LC, Schepers F, Jordi WJRM, Stoopen GM, Davelaar E van Rhijn JHA, Power JB and Davey MR (2002) Effect of P_{SAG12}-*IPT* gene expression on development and senescence in transgenic lettuce. *Plant. Physiol.* (in press).

Michelmore RW and Eash JA. (1988) Tissue culture of lettuce. In: *Handbook of Plant Cell Culture, Vol. 4* (Eds Evans DA, Sharp WR and Amirato PV), Collier MacMillan, London, pp. 512-551.

Mii M, Nakano K, Okuda K and Iizuka M (1992) Shoot regeneration from spinach hypocotyl segments by short term treatment with 5,6-dichloro-indole-3-acetic acid. *Plant Cell Rep.,* **11** : 58-61.

Mix G (1985) *In vitro* regeneration from leaf vein segments of *Cichorium intybus* L. Landbau. *Volkenrode,* **35** : 59-62.

Mizuttani T, Liu XJ, Tashiro Y, Miyazaki S and Shimanasaki K (1989) Plant regeneration and cell fusion of protoplasts from lettuce cultivars and related wild species in Japan. *Bull. Faculty Agric. Saga University,* **67** : 109-118.

Mohamed-Yasseen Y and Splittstoesser WE (1995) Somatic embryogenesis from leaf of witloof chicory through suspension culture. *Plant Cell Rep.,* **14** : 804-806.

Mohapatra U, McCabe MS, Power JB, Schepers F, Van der Arend A and Davey MR (1999) Expression of the bar gene confers herbicide resistance in transgenic lettuce. *Trans. Res.* **8** : 33-44.

Molvig L and Rose RJ (1994) A regeneration protocol for *Spinacia oleracea* using gibberellic acid. *Aust. J. Bot.,* **42** : 763-769.

Morel G and Wetmore RH (1951) Fern callus tissue culture. *Am. J. Bot.,* **38** : 141-143.

Murashige T and Skoog F (1962) A revised medium for rapid growth and bioassays with tobacco tissue cultures. *Physiol. Plant.* **15** : 473-497.

Nenz E, Varotto S, Lucchin M and Parrini P (2000) An efficient and rapid procedure for plantlet regeneration from chicory mesophyll protoplasts. *Plant Cell Tiss. Org. Cult.,* **62** : 85-88.

Neskovic M and Radojevic L (1973) The growth of and morphogenesis in tissue cultures of *Spinacia oleracea. Bull. Inst. Jardin Bot. Univ. Béograd VIII, Series* **1-4** : 35-37.

Nishio T, Sakata Y, Yamagishi H, Tabei and Takayanagi K (1989) Protoplast culture in vegetable crops development and improvement of culture procedure. *Bull. Natl. Res. Inst. Veg. Orn. Plants,* **3** : 67-96.

Nishio T, Sato T, Mori K and Takayanagi K (1988) Simple and efficient protoplast culture procedure of lettuce, *Lactuca sativa* L. *Jpn. J. Breed.,* **38** : 165-171.

Nitsch JP (1969) Experimental androgenesis in *Nicotiana. Phytomorphology,* **19** : 389-404

Nitsch JP and Nitsch C (1969) Haploid plants from pollen grains. *Science,* **163** : 85-87.

Otsuki Y and Takebe I (1969) Isolation of intact mesophyll cells and their protoplasts from higher plants. *Plant Cell Physiol.,* **10** : 917-921.

Panday SC and Kalloo G (1986) Spinach, *Spinacia oleracea* L. In: *Genetic Manipulation in Plant Breeding.* Proc. Internatl Symp. (Eds Horn W, Jensen CJ, Odenbach W and Schieder O), Walter de Gruyter, Berlin, pp. 325-336.

Pang SJ, Jan FJ, Carney K, Stout J, Tricoli DM, Quemada HD and Gonsalves D (1996) Post-transcriptional transgene silencing and consequent topovirus resistance in transgenic lettuce are affected by transgene dosage and plant development. *Plant J.,* **9** : 899-890.

Pieron S, Belaizi M and Boxus P (1993) Nodule culture, a possible morphogenetic pathway in *Cichorium intybus* L. propagation. *Sci. Hort.,* **53** : 1-11.

Pieron S, Boxus P and Dekegel D (1998) Histological study of nodule morphogenesis from *Cichorium intybus* L. leaves cultivated *in vitro. In Vitro Cell. Dev. Biol.-Plant,* **34** : 87-93.

Quoirin M and Lepoivre P (1977) Improved media for *in vitro* culture of *Prunus* sp. *Acta Hort.,* **78** : 437-442.

Rambaud C, Dubois J and Vasseur J (1990) Some factors related to protoplast culture and plant regeneration from leaf mesophyll protoplasts of Magdeburg chicory (*Cichorium intybus* L. var. Magdeburg). *Agronomie,* **10** : 767-772.

Robatche-Claive AS, Couillerot JP, Dubois J, Dubois T and Vasseur J (1992) Embryogenése somatique directe dans les feuilles du *Cichorium* hydride 474: synchronisation de l'induction. *C.R. Acad. Sci., Série III,* 314.

Rose RJ (1980) Factors that influence the yield, stability in culture and cell-wall regeneration of spinach mesophyll protoplasts. *Aust. J. Plant Physiol.,* **7** : 713-725.

Ryder EJ (1983) Inheritance, linkage and gene interaction studies in lettuce. *J. Amer. Soc. Hort. Sci.,* **108** : 985-991.

Ryder EJ (1986) Lettuce Breeding. In: *Breeding Vegetable Crops* (Ed Bassett MJ). AVI Publishing company, Westport, USA, pp. 433-474.

Ryder EJ (1988) Early flowering in lettuce as influenced by a second flowering time gene and seasonal variation. *J. Amer. Soc. Hort. Sci.,* **113** : 456-460.

Ryder EJ (1996) Inheritance of chlorophyll deficiency traits in lettuce. *J. Hered.,* **87** : 314-318.

Ryder EJ (1999) *Lettuce, Endive and Chicory - Crop Production Science in Horticulture Series.* CABI Publishing, Wallingford, UK.

Ryder EJ, Kim ZH and Waycott W (1999) Inheritance and epistasis studies of chlorophyll deficiency in lettuce. *J. Amer. Soc. Hort. Sci.,* **124** : 636-640.

Saksi N, Dubois J, Millecamps J-L and Vasseur J (1986) Régénération de plantes de chicoré Witloof cv. Zoom à partir de protoplasts: influence de la nutrition glucidique et azotée. C. R. Acad. Sci. Paris Série III, **302** : 165-170.

Sasaki H (1975) Physiological and morphological studies on development of vegetable crops. III. Adventitious bud formation of callus tissue derived from lettuce hypocotyls. *J. Japan. Soc. Hort. Sci.,* **44** : 138-143.

Sasaki H (1979a) Physiological and morphological studies on development of vegetable crops. IV. Effect of various media on the adventitious bud formation of lettuce hypocotyls tissue cultured *in vitro. J. Japan. Soc. Hort. Sci.,* **47** : 479-484.

Sasaki H (1979b) Physiological and morphological studies on development of vegetable crops. VI. Effect of several auxins, cytokinins and cytokinin-ribosides on the adventitious bud formation of lettuce hypocotyl tissue cultured *in vitro. J. Japan Soc. Hort. Sci.,* **48** : 67-72.

Sasaki H (1979c) Physiological and morphological studies on adventitious bud formation of lettuce hypocotyl tissue cultured *in vitro. J. Japan. Soc. Hort. Sci.,* **48** : 67-72.

Sasaki H (1982) Effect of temperature and light on adventitious bud formation of lettuce hypocotyl tissue culture *in vitro. J. Japan. Soc. Hort. Sci.,* **51** : 187-194.

Sasaki H (1989) Callus and organ formation in tissue cultures of spinach (*Spinacia oleracea* L.). *J. Japan. Soc. Hort. Sci.,* **58** : 149-153.

Satoh T, Abe T and Sasahara T (1992) Plant regeneration from hypocotyl-derived calli of spinach (*Spinacia oleracea* L.) and anatomical characteristics of regenerating calli. *Plant Tiss. Cult. Letts.,* **9** : 176-183.

Schenk RU and Hildebrandt AC (1972) Medium and techniques for induction and growth of monocotyledonous and dicotyledonous plant cell cultures. *Can. J. Bot.,* **50** : 199-204.

Sibi M (1976) La notation de programme genetique chez les vegetaux superieurs. II- Expt I. Aspect - Production of variants by *in vivo* tissue culture of *Lactuca sativa* L. Increase in vigour in outcrosses. *Ann. Amelior. Plantes.,* **26** : 523-547.

Sidikou-Seyni R, Rambaud C, Dubois J and Vasseur J (1992) Somatic embryogenesis and plant regeneration from protoplasts of *Cichorium intybus* L. x *Cichorium endivia* L. *Plant Cell Tiss. Org. Cult.*, **29** : 83-91.

Sun LY, Tourand G, Charbonnier C and Tepfer D (1991) Modification of phenotype in Belgian endive (*Cichorium intybus*) through genetic transformation by *Agrobacterium rhizogenes*: conversion from biennial to annual flowering. *Transgenic Res.*, **1** : 14-22.

Takano T, Kabeya H and Isono K (1988) Multiplication of lettuce F_1 hybrids by tissue culture. *Sci. Rep. Fac. Agric. Meijo Univ.*, **24** : 17-28.

Tanaka T, Matsumura T and Morinaga Y (1991) Studies on the protoplast culture 1: Procedure of protoplast culture of lettuce *Lactuca sativa* L. *Proc. Fac. Agric. Kyushu Tokai Univ.*, **10** : 29-36.

Theilerhedtrich R and Hunter CS (1995) Regeneration of dihaploid chicory (*Cichorium intybus* L. var. Hegi) via microspore culture. *Plant Breed.*, **114** : 18-23.

Varotto S, Lucchin M and Parrini P (1997) Plant regeneration from protoplasts of Italian red chicory (*C. intybus*). *J. Genet. Breed.*, **51** : 17-22.

Varotto S, Lucchin M and Parrini P (2000) Immature embryo culture in Italian red chicory. *Plant Cell Tiss. Org. Cult.*, **62** : 75-77.

Varotto S, Nenz E, Lucchin M and Parrini P (2001) Production of asymmetric somatic hybrid plants between *Cichorium intybus* L. and *Helianthus annuus* L. *Theor. Appl. Genet.*, **102** : 950-956.

Vermeulen A, Vaucheret H, Pautot V and Chupeau Y (1992) *Agrobacterium* mediated transfer of mutant *Arabidopsis* acetolactate synthase gene confers resistance to chlorsulfuron in chicory (*Cichorium intybus* L.). *Plant Cell Rep.*, **11** : 243-247.

Vijn I and Smeekens S (1999) Fructan: more than a reserve carbohydrate? *Plant Physiol.*, **120** : 351-359.

Vijn I, van Dijken A, Sprenger N, van Dun K, Weisbeek P, Wiemken A and Smeekens S (1997) Fructan of the inulin neoseries is synthesized in transgenic chicory plants (*Cichorium intybus* L.) harbouring onion (*Allium cepa* L.) fructan:fructan 6G-frctosyltransferase. *The Plant J.*, **11** : 387-398.

Waycott W, Fort SB and Ryder EJ (1995) Inheritance of dwarfing genes in *Lactuca sativa* L. *J. Hered.*, **86** : 39-44.

Waycott W, Fort SB, Ryder EJ and Michelmore RW (1999) Mapping morphological genes relative to molecular markers in letrtuce (*Lactuca sativa* L.). *Heredity*, **82** : 245-251.

Webb CL, Davey MR, Lucas JA and Power JB (1994) Plant regeneration from mesophyll protoplasts of *Lactuca perennis*. *Plant Cell Tiss. Org. Cult.* **38** : 77-79.

Webb DT, Torres LD and Fobert P (1984) Interactions of growth regulators, explant age and culture environment controlling organogenesis from lettuce cotyledons *in vitro*. *Can. J. Bot.*, **62** : 586-590.

White PR (1943) *A Handbook of Plant Tissue Culture*. Jacques Cottell Press, Lancaster, PA, USA.

Xiao XG and Branchard M (1993) Embryogenesis and plant regeneration of spinach (*Spinacia oleracea* L.) from hypocotyl segments. *Plant Cell Rep.*, **13** : 69-71.

Xiao X-G and Branchard M (1995) *In vitro* high frequency plant regeneration from hypocotyls and root segments of spinach by organogenesis. *Plant Cell Tiss. Org. Cult.*, **42** : 239-244.

Xinrun Z and Conner AJ (1992) Genotype effects on tissue culture response of lettuce cotyledons. *J. Genet. Breed.,* **46** : 287-290.

Yang Y, Al-Khayri JM and Anderson EJ (1997) Transgenic spinach plants expressing the coat protein of cucumber mosaic virus. *In Vitro Cell Dev. Biol.-Plant,* **33** : 200-204.

Zdravkovic-Korac S and Neskovic M (1999) Induction and development of somatic embryos from spinach (*Spinacia oleracea*) leaf segments. *Plant Cell Tiss. Org. Cult.* **55** : 109-114.

Zhang HX and Zeevaart JAD (1999) An efficient *Agrobacterium tumefaciens*-mediated transformation and regeneration system for cotyledons of spinach (*Spinacia oleracea* L.). *Plant Cell Rep.,* **18** : 640-645.

Chapter 5

TRANSGENIC CAULIFLOWER PLANT

Ching-Yeh Hu★ and Claire M Leonard

Biology Department, Wm. Paterson University of New Jersey, Wayne, NJ 07470, USA

Summary

In vitro culture techniques, transformation procedures, transgenes used and transgenic status confirmation measures for cauliflower were reviewed and evaluated. Explant pre-incubation, ethylene-inhibitor supplementation in culture medium and culture ventilation are special culture practices found to improve the success rate of cauliflower in vitro regeneration and should also be built in the procedure of the cauliflower transformation. Leaf protoplasts, seedling explants, leaf disks of cauliflower were subjected to direct and/or Agrobacterium-mediated transformation procedures. Transgenic plantlets were either directly regenerated from these explants or indirectly from the resultant tumor/hairy root tissues that were developed out from the Agrobacterium-inoculated wounds on intact stems or hypocotyls. Marker genes had been used to develop cauliflower transformation procedures by a number of laboratories. Some of the agronomic potentially useful target genes, such as trypsin inhbitor, CaMV capsid, antisense CaMV gene VI, and anther-specific Bcp1 genes, were also explored in transforming the cauliflower crop. After reviewed and analyzed the merits and drawbacks of the reported cauliflower tissue culture and transformation procedures, we are endorsing the seedling explant-based Agrobacterium-mediated transformation for cauliflower genetic engineering.

Keywords : *Agrobacterium*, *Brassica oleracea* var. *botrytis*, cauliflower, genetic engineering, *in vitro* culture, regeneration, transformation

---

★Corresponding author : E-mail : huc@wpunj.edu

Abbreviations : 2,4-D: 2,4-dichlorophenoxyacetic acid; 35S promoter: a promoter from cauliflower mosaic virus; AVG: aminoethoxyvinylglycine; B5: Gamborg et al. (1968) medium; BAP: N^6-benzylaminopurine (= BA: N^6-benzyladenine); *bar*: gene coding for the enzyme PAT; CaMV: cauliflower mosaic virus; GA_3: gibberellic acid; *gus*: β-glucuronidase; *hpt*: hygromycin phosphotransferase; IAA: indole-3-acetic acid; IBA: Indolebutyric acid; KIN – kinetin: MS: Murashige and Skoog (1962) medium; NAA: α-naphthaleneacetic acid; *nos*: nopaline syntase; *neo* or *npt*II: neomycin phosphotransferase II; PAT: phosphinotocin acetyltransferase; PCR: polymerase chain reaction; TI: trypsin inhibitor.

1. INTRODUCTION

Cauliflower (*Brassica oleracea* var. *botrytis* L.) is a common vegetable been cultivated worldwide. Based on FAOSTAT Database (FAO, 2000), it's world cultivation area increased steadily since 1961 from less than 0.24 million ha to more than 0.77 million ha in 1999 and accounted for over 1.9% of world vegetable and melon cultivation area. The world cauliflower production in 1999 was over 13.8 million Mt. It is a favorable vegetable in Asia with the combined production in China and India in 1999 exceeded 70% of world production. The importance and tissue culture of cauliflower were reviewed by Grout (1988). Recent development in crop genetic engineering, which has the potential of re-designing crop plants, has stimulated active investigations in cauliflower transformation and its associated *in vitro* culture methods.

2. *IN VITRO* CULTURE

In vitro regeneration is essential for crop genetic transformation and for quick clonal multiplication of the resultant transgenic plants. Organogenesis of seedling explants is the main *in vitro* procedure used in cauliflower genetic engineering. Curd culture is the preferred method in cauliflower clonal multiplication (Bhalla and Weerd, 1999).

Standard or modified MS (Murashige and Skoog, 1962) or B5 (Gamborg *et al.*, 1968) medium with 2-3% sucrose and, for solid medium, agar or phytagel was used as the basal media in all the reported cauliflower *in vitro* works. Although De Block *et al.*, (1989) reported that MS medium is superior compared to B5 medium for cauliflower regeneration, both basal media served well based on a survey of the literatures. When different solidifying agents were tested, phytagel was preferred by Bhalla and Smith (1998a) as callus might turn black on agar-based medium. However, agar was used in most of the works.

2.1. Type of cultures

In vitro cauliflower curd, seedling explant, leaf disk, protoplast, and anther cultures have been performed. Here culture procedures applicable to genetic engineering will be reviewed and their merits be discussed.

2.1.1. Curd culture

In view of the fact that curd culture is the best-developed procedure among cauliflower *in vitro* techniques (Bhalla and Weerd, 1999), we shall review this procedure first. Curd is a unique type of explant available to cauliflower only. The market-ready curd consists of a main shoot from which many first-order branches develop in acropetal succession. The second-order branches are initiated from the apices of the first-order branches. Each successive branching order develops similarly up to the tenth-order (Sadik, 1962; Torres *et al.*, 1980). Higher order branches contain simple meristematic apices that, depending on environmental conditions, can either develop into flowers or into vegetative shoots *in vitro*. Phytohormone(s), such as cytokinin alone (Torres *et al.*, 1980) or auxin alone (Kumar *et al.*, 1993) or the combination of both (Kumar *et al.*, 1993), can lead toward *in vitro* vegetative shoot development probably from the not-yet-determined lateral meristems. While the more advance-developed terminal (apical) meristem do not revert into a vegetative phase but aborted *in vitro* (Margara and David, 1978). Pre-incubate curd explants in osmoticum can replace the hormonal effects and leads toward shoot development in hormone-free medium (Vandemoortele, 1999), while only floral development would take place in hormone-free medium without such pretreatment. The best medium for shoot development from curd reported by Vandemoortele *et al.* (1999) was MS supplimented with 2.5 μM BA, 5 μM NAA and 168 mM (5.75%) of sucrose.

Based on Torres *et al.* (1980) anatomical observations, development of true callus from the curd meristems cultured *in vitro* was not likely to happen. Thus, tissue culture induced mutation (somaclonal variation) is not likely to take place among the curd meristem regenerated shoots.

Grout (1988) stated in his review article: "The vast majority of micropropagation of cauliflower has been from floral meristem (curd) explants, and there have been no report of genetic instability in the literature to date. In the author's own laboratory over 10,000 plants have been regenerated in 10 years, with no evidence of heritable abnormality". Lack of tissue culture induced mutation is one of the two key advantages of employing curd culture in cauliflower clonal propagation. The other advantage is the gigantic quantity of shoots that can be produced from the compact meristematic masses.

Despite of the above-mentioned advantages, curd culture has not been used in cauliflower transformation manipulations. This is likely due to the recalcitrancy of organized-meristem to genetic transformation. Such recalcitrancy is testified by the extreme rarity of meristem-based transformation reports appearing in the enormous volume of plant transformation literature. Organogenesis-based *in vitro* procedures seem suit better in supporting genetic transformation. However, there is no question that curd culture can be successfully adopted for rapid clonal multiplication of the transgenic cauliflower plants obtained from genetic engineering.

2.1.2. Leaf-disk culture

Although the leaf-disk is the chief explant type for organogenesis-based transformation manipulation for numerous plant species, its value in cauliflower transformation is not at all clear. In fact, one of the first two successful cauliflower genetic engineering works reported in 1988 used leaf-disk explants (Srivastava *et al.*, 1988). However, there was no quantitative figure given throughout this report. Without information on its regeneration and transformation efficiencies, there is no way to evaluate the procedure's usefulness. Eimert and Siegemund (1992) was the only other group tested leaf disc transformation method on cauliflower and obtained transgenic calli which were incapable of regenerating shoot even on the non-selective medium.

Srivastava *et al.* (1988) obtained their leaf disk explants from mother plants grown on peat-sand (1:3) mixture. The experiment will be difficult to be repeated, since no specifics on raising the mother plants or explant selection were given. Shoot regeneration of explants was induced with a medium containing 0.5 mg/l BAP and 1 mg/l NAA. There was no phytohormone supplemented for shoot elongation and rooting medium. Eimert and Siegemund (1992) maintained their axenic shoot cultures on basal medium supplemented with 0.1 mg/l 2,4-D. They attempted to regenerate buds on a medium containing 1 mg/l BAP and 0.4 mg/l NAA.

2.1.3. Seedling explants culture

Cauliflower seedling parts are powerful tissues for *in vitro* bud regeneration. Usually explants were obtained from sterile *in vitro* germinated seedlings. Seeds can be surface disinfected with sodium hypochlorite (0.5%) for around 30 minutes. A brief (up to two minutes) pre-soaking seeds with 70% ethanol before sodium hypochlorite treatment may be helpful. Disinfected seeds, after washing, can be germinated on a basal medium. Four-day old seedlings were used as the explant sources by Ding *et al.* (1998). Nine- to 14-day old seedlings were used by other workers (Bhalla and Smith, 1998a,b; De Block *et al.*, 1989; David and Tempe, 1988).

Seedling cotyledon, hypocotyl and, less frequentely, root explants were commonly used. Seedlings of three Taiwan cauliflower cultivars were cut into hypocotyl, cotyledon petiole and cotyledon sections by Ding *et al.* (1998). After three-week incubation, the hypocotyl-explants of all three tested cultivars expressed the highest adventitious bud regeneration capacity. Little, if any, adventitious buds appeared on the cotyledon explants. Eleven Australian cauliflower genotypes were tested by Bhalla and Smith (1998a). In addition to hypocotyl and cotyledon, root explants were also tested. Their results showed that root and hypocotyl explants were more responsive to shoot regeneration than that of cotyledon explants. De Block *et al.* (1989) used hypocotyl explants only from two cultivars on their cauliflower transformation work.

Cytokinin is the most critical phytohormone group for stimulating bud organogenesis from cauliflower explants. A low concentration of auxin is also required to complement the cytokinin. Sometimes a small amount of gibberellin was used to stimulate internodal elongation of the regenerated shoots. The optimum cytokinin concentration varies with genotype and explant types. For the three cultivars and three explant types used by Ding *et al.* (1998), they found that the optimum BAP concentration ranged from 5.0 mg/l to 10.0 mg/l when combined with 0.2 mg/l IAA. However many workers frequently use only one cytokinin concentration. For example, Bhalla and Smith (1998a,b) used 3 mg/l BAP complemented with 0.2 mg/l NAA and 0.001 mg/l GA_3 for all the tested genotypes and explant types. De Block *et al.* (1989) used 1 mg/l BAP complemented with 0.1 mg/l NAA and 0.01 mg/l GA_3.

On cytokinin containing shoot induction (regeneration) medium, callus will develop from the explants. Adventitious buds will subsequently develop from the callus. Shoot elongation and rooting of cauliflower adventitious buds can occur on basal medium without hormonal supplementation (Srivastava *et al.*, 1988; Ding *et al.*, 1998). However, some authors used specific media for shoot elongation/rooting stages. Bhalla and Smith (1998a,b) and Bhalla and Weerd (1999) added 40 mg/l adenine hemisulfate and 1.25 mg/l BAP to the shoot outgrowth medium and 0.2 mg/l IBA to a sucrose-free rooting medium. De Block *et al.* (1989) reduced BAP concentration from 1 mg/l of the shoot induction medium to 0.0025 mg/l along with 40 mg/l adenine sulfate in the shoot outgrowth medium. Their rooting medium used ½ MS salts supplemented with 0.1 mg/l IBA and vermiculite, instead of agar, was used as the supporting material.

2.1.4. Protoplast culture

Cauliflower protoplast culture and relatively high efficiency of plantlet regeneration were reported by Jourdan *et al.* (1990). However, problems of high degree of genotypic dependence, low regeneration frequency and inconsistent response of protoplasts in culture still limit the application of this technology for cauliflower improvement. Since no transgenic plant has been resulted from cauliflower protoplast transformation to date, we will not review cauliflower protoplast isolation, culturing and plantlet regeneration procedures here.

2.2. Beneficial culture practices

The following special culture practices were found to improve the success rate of cauliflower *in vitro* regeneration. Therefore, such culture practices should also be included in the procedure of the cauliflower transformation.

2.2.1. Explant pre-incubation

Pre-incubation is frequently used to re-condition the physiological conditions of explants in order to improve regeneration or transformation rates. Ding *et al.* (1998) pre-incubated cauliflower hypocotyl-explants in a Callus-Inducing (CI) Medium containing 1 mg/l 2,4-D and 0.5 mg/l KIN for 3 days before transferring into the shoot regeneration medium. The resultant regeneration rates for all three tested cultivars improved significantly, exceeding 95%.

2.2.2. Ethylene inhibitors

Ethylene gas frequently accumulates in the atmosphere of the culture containers. To the sensitive species, like cauliflower, *in vitro* ethylene accumulation may lead to excess callus formation, tissue/organ hyperhydricity (vitrification) and inhibition of shoot regeneration. The hyperhydric status of cauliflower, as described by Vandemoortele (1999), consists of thick, easy-to-break stems, reduced leaf area and curled leaf blades. It results in poor survival when the hyperhydric shoots are transferred *ex vitro*. The hypothesis that ethylene produced by explants often inhibits *in vitro* shoot regeneration was experimentally proved by Pua *et al.* (1999) using cauliflower, cabbage and broccoli. Added proof of ethylene regulating *in vitro* shoot regeneration was provided via constructing transgenic *Brassica* plants with antisense ACC oxidase gene (Pua and Lee, 1995).

In vitro ethylene accumulation can be reduced either by supplementing ethylene inhibitors in culture medium or increase ventilation (see I.B.3.) of the culture containers. Silver ion and AVG (aminoethoxyvinylglycine) are commonly used *in vitro* ethylene inhibitors. Bhalla and Smith (1998b) used 0.5 mg/l $AgNO_3$ during shoot regeneration of their successful cauliflower transformation experiment. De Block *et al.* (1998) used 2-10 mg/l $AgNO_3$ in their cauliflower transformation experiment and considered the use of $AgNO_3$ was a prerequisite for efficient shoot regeneration under selective conditions. Ding *et al.* (1998) used 29.4 μM silver thiosulfate (STS) and significantly increased shoot regeneration percentage over the control. Higher STS concentrations (upto 117.6 μM) tested showed a similar degree of stimulation. We recommend the use of STS over $AgNO_3$. According to the Merck Index $AgNO_3$ is "caustic and irritating to the skin and mucous membranes. Swallowing can cause severe gastroenteritis and may be fatal." In solution it is light sensitive and will darken the medium when exposed to light over a period of time. STS, on the other hand, can be used more safely. To prepare STS, 1:3 molar concentration solutions of silver nitrate and sodium thiosulfate are mixed and added to culture medium via filter sterilization. However, De Block *et al.* (1998) cautioned about the use of STS because of its toxic effects on potato upon prolonged use. Such toxic effects was not noticed by Ding *et al.* (1998) on the few-week-long cauliflower regeneration period. Pua *et al.*, (1999) reported that lower ethylene was emanated by cauliflower, cabbage and broccoli explants during the three-week culture period when 10 μM AVG was supplemented to the culture medium. Beneficial effects of AVG supplementation on shoot regeneration were observed.

2.2.3. Culture ventilation

Culture ventilation is a more efficient way of preventing *in vitro* ethylene accumulation than ethylene-inhibitor supplementation. Zobayed *et al.* (1999) investigated the effects of various methods of ventilations (sealed, diffusive and humidity-induced convective throughflow {HICT} ventilation) on cauliflower shoot regeneration. Throughout the 30-day experiment, high ethylene accumulation at 1.75 μl/l was observed in sealed system and the diffusive system accumulated 0.55 μl/l, while no accumulation was detected under the HICT-ventilation system. The HICT-ventilation system resulted in the best growth of the regenerated shoots. On the contrary the sealed vessels showed very poor growth with little leaf and shoot number, fresh weight and callus volume. The diffusive treatment is being intermediate. Bhalla and Smith (1998a) found that sealing of Petri dishes with porous surgical tape (3M) rather than non-porous plastic (Nescofilm or Parafilm) was essential for shoot initiation and growth. This was likely due the defusing out of ethylene through the porous tape.

The most valuable application of *in vitro* ventilation is to improve the survival rates of *ex vitro* transplantation. This is accomplished through a complex physiological/morphological benefits resulted from reduced RH in the culture atmosphere, including reduced tissue/organ hyperhydricity, higher photosynthetic rate, elevated accumulation of epidermal wax, more complete cuticle deposition and normalized stomatal function (Short and Warburton, 1987). Thus, *in vitro* ventilation can also be considered as *in vitro* acclimatization (hardening).

The effects of in vitro ventilation during the rooting stage on photosynthetic, growth and transpiration rates of *in vitro* cauliflower plantlets were investigated by Kanechi *et al.* (1998). Plantlets were established in airtight or ventilated vessels for 15 days before transplanting *ex vitro*. Leaves in ventilated vessels had a higher chlorophyll content and photosynthetic rate than those produced in airtight vessels. Greater leaf expansion and shoot/root dry matter accumulation during *in vitro* culture and acclimatization stages of the ventilated plantlets were observed. Fifteen days after transplanting *ex vitro*, the high photosynthetic ability and stomatal resistance to transpiratory water loss of ventilated plantlets contributed greatly to successful rooting and acclimatization.

Abe *et al.* (1997) had also reported that the *in vitro* generated cauliflower leaves in a ventilated vessel had higher rates of photosynthesis than those in an airtight vessel. The enhanced rates of photosynthesis contributed to promote the growth *in vitro* and the acclimation of plantlets to *ex vitro*.

Reduction in culture atmospheric RH can also be achieved by elevating medium gel concentration or supplementation of osmoticums such as PEG or sugars. Since too low an *in vitro* atmospheric RH would induce osmotic stress and lead to reduced plantlet growth, particularly root, high mortality and slow *ex vitro* growth rate following transferring to soil (Wardle *et al*, 1983; Short and Warburton, 1987). Short and Warburton (1987) had determined that culture atmospheric RH of 80%, which can be accomplished by partly closed screw top lids, was optimum for cauliflower. They reported that an 80% culture atmospheric RH would bring about a significant increase in wax bloom production to a level characteristic of brassicas. The resultant plantlets could be transferred to soil without humidity protection and wilting did not occur.

3. TRANSFORMATION

Cauliflower is known to be recalcitrant to genetic transformation and very few successful transformation experiments have been reported. We will base this portion of review on two themes: (A) tissue/cell sources and the associated transformation procedures used to produce transgenic plants and (B) transgenes been used, confirmation of transformation of transgenic status and expression of transgenes in the host plants.

3.1. Tissue sources and transformation procedures

The methodology on how the cauliflower transformation experiments were performed and its effectiveness will be reviewed at this first portion of the "Transformation" section. Leaf protoplasts, seedling explants, leaf disks of cauliflower were subjected to direct and/or *Agrobacterium*-mediated transformation procedures. Transgenic plantlets were either directly regenerated from these explants or indirectly from the resultant tumor/hairy root tissues that were developed out from the *Agrobacterium*-inoculated wounds on intact stems or hypocotyls.

One common factor for the majority of reported cauliflower transformation works was the use of selective medium. When *npt*II gene was used as the selectable marker, antibiotic kanamycin (25-50 mg/l) or G418 (15-30 mg/l) was added to the selective medium. When bar gene was used, herbicide D+L-phosphinotricin (NH_4^+-salt; 20 mg/l) was added.

3.1.1. Protoplasts

Eimert and Siegemund (1992) demonstrated transcient expression of *npt*II gene using cauliflower protoplasts via either *Agrobacterium*-, or PEG-/electroporation-mediated transformation. Although the procedures were not efficient enough to produce transgenic plants, interested readers can use them as starting points for further optimizing.

Enzymatic isolated cauliflower leaf protoplasts were cultivated in a modified MS0 medium. Transformation procedures were applied at this stage. After two weeks, regenerated cell clusters were transferred to an agarose-bead-type culture medium of the same composition under full illumination. Phytohormone supplemented solid MS0 media were used for shoot regeneration and rooting.

For *Agrobacterium* transformation, the regenerating protoplasts were inoculated with *Agrobacterium* (100 cells per protoplast) and cocultivated for 24-48 h at 20°C in the dark. After washing they were incubated in medium containing 500 mg/l cefotaxime or carbinicellin. Selection started in agarose-bead-type culture by adding 50 mg/l kanamycin.

For direct gene transfer, 10 μg plasmid DNA and 50 μg calf thymus DNA were used to transform aliquots of 10^6 protoplasts. DNA-protoplast mixture subjected to PEG-assisted transformation following Pröls *et al.* (1988) or Negrutiu *et al.* (1987) procedure. For electroporation-assisted transformation, DNA-protoplast mixture was re-suspended in 1 ml MEM solution. Square pulses (20 μs) of different field strength (800-1500 V/cm) were applied. After a reconstitution period of 10 min, the protoplasts were washed and cultivated.

3.1.2. Seedling explants

Most of the successful cauliflower transformation reports were based on *Agrobacterium*-mediated transformation of seedling explants using binary plasmid harboring *A. tumefacience* (De Block *et al.*, 1989; Ding *et al.*, 1998; Bhalla and Smith, 1998b).

For transformation works, Bhalla and Smith (1998b) found that explants from 4-day old seedlings were better than that from 7- to 14-day old seedlings as very few green shoots recovered when older seedlings were used. The same authors tested hypocotyl and cotyledon explants from three cauliflower cultivars in their transformation work. Although the hypocotyl produced as many as twice the number of shoots per explant compared to cotyledon explants, higher numbers of transformed plants were obtained from cotyledon explants. They have concluded that cotyledon from young seedling was the best target for transformation of cauliflower, presumably such explants are "more amenable" to genetic transformation. Of course, such claim has to be confirmed by other laboratories.

In general, *Agrobacterium* was grown in kanamycin-containing culture medium at 27-28°C under rapid shaking for 18 - 36 hours. The suspension were washed and diluted with kanamycin-free medium. Explants were inoculated by soaking in diluted bacterium suspension, with or without shaking, for 30 minutes to 1 hour. Co-cultivation usually lasted for 2-3 days in darkness before transferred onto carbenicillin containing regeneration (shoot induction) medium. De Block *et al.* (1989) and Bhalla and Smith, (1998b) used 500 mg/l carbenicillin, while Ding *et al.* (1998) used 150 mg/l after washed the explants with 1500 mg/l carbenicillin. The same concentration of carbenicillin was supplemented to all subsequent culture media to inhibit bacterial growth. Selection started immediate (De Block *et al.*, 1989) or one week after co-cultivation (Ding *et al.*, 1998; Bhalla and Smith, 1998b). Explants were transferred to fresh selective medium every 1-2 weeks and the green shoots that formed were removed and placed into shoot outgrowth and, then, rooting media.

Here we are presenting Ding *et al.* (1998) transformation procedure, which resulted in over 100 G418/kanamycin-resistant plantlets, in more details:

A single colony bacterium was transferred into liquid YEP medium containing 100 mg/l kanamycin and shaken at 240 rpm under 28°C for 2 days. The suspension were then mixed with sterile glycerol (1:1 vol/vol) and stored at -80°C until use. Eighteen hours before inoculation, the bacterium suspension was mixed with the YEP medium (1:25 vol/vol) and shake-incubated at the above conditions. Acetosyringone was added to the suspension after 10 h incubation at a concentration of 50 μM. By the end of 18 h incubation the suspension reached an OD_{600} reading of 0.8 with approximately 3×10^8 cells/ml. The bacterial medium in the suspension was, then, replaced with an equal volume of Inoculation Medium (liquid Callus Induction Medium [CI medium; see Sect I.B.1] + 10 mM D-glucose) before use. The CI medium pre-incubated hypocotyl explants were inoculated with diluted bacterial suspension. Based on the frequencies of explants regenerating green shoots under selective pressure, the optimum inoculation conditions were 100 x dilution of bacterium suspension incubated at 20-25°C for 1 h. After blotting dry with sterile filter paper the inoculated explants were incubated on a solid CI Medium at 25°C in darkness for a 3-day co-cultivation. The co-cultivated explants were washed with 1.5% D-mannitol with 1500 mg/l carbenicillin, blotted dry and placed on a 150 mg/l carbenicillin containing regeneration medium.

Six antibiotic selection regimes (Fig. 1) were tested. The highest percentage of antibiotic-resistant green buds resulted in regime "f", which had no selective pressure at the first two weeks following co-cultivation. But many of the resultant plantlets were confirmed as "escapes" or "chimeras". On the other hand, there was no green bud regeneration when antibiotic selective pressure began right after co-cultivation in regimes a, b, and c. The above results suggested that the buds of cauliflower, even the *npt*II gene transformed ones, were quite sensitive to the selective agents at the early stages of development. They became progressively more tolerant to the selective antibiotics, even without the resistant gene. Thus, the lack of selective agent during the first week after co-cultivation, followed by a progressively increasing the selective pressure, *i.e.* the regime "e", provided the optimum conditions for selecting the *npt*II gene containing transgenic cauliflower buds. Results also indicated that G418 selection leaded to a higher number of regenerated buds than that of kanamycin. In regime "e", for example, the percentages of explants regenerating green buds from the three tested cultivars in the presence of G418 were 8.7, 10.3, and 33.3 (see Fig. 1B), while from kanamycin selection the percentages were 0, 2.0, and 0, respectively (data not shown on Fig. 1B). Thus, G418 selection provided a higher probability of obtaining putative transgenic plants than that of kanamycin.

3.1.3. Leaf disks

Srivastava *et al.* (1988) inoculated cauliflower leaf disks, which were pre-conditioned on BAP and NAA containing medium for 24 h, with a diluted (1:20) culture of an 48 h grown oncogenic strain of *A. tumefaciens* for 5-10 minutes. Without the standard co-cultivation, explants were transferred onto a carbenicillin or cefortaxime (500 mg/l) supplemented hormone-free medium for a 2-week incubation. Explants were transferred onto hormone-free kanamycin containing selection medium afterwards. Calli and shoots regenerated and developed into transgenic plantlets (no

A. Regimes Flowchart

Week[1]	a	b	c	d	e	f
1	H[2]	L	L	N	N	N
⇓	⇓	⇓	⇓	⇓	⇓	⇓
2	H	H	L	H	L	N
⇓	⇓	⇓	⇓	⇓	⇓	⇓
3	H	H	H	H	H	H
⇓	⇓	⇓	⇓	⇓	⇓	⇓
4	H	H	H	H	H	H

B. Effects of Selection Regimes

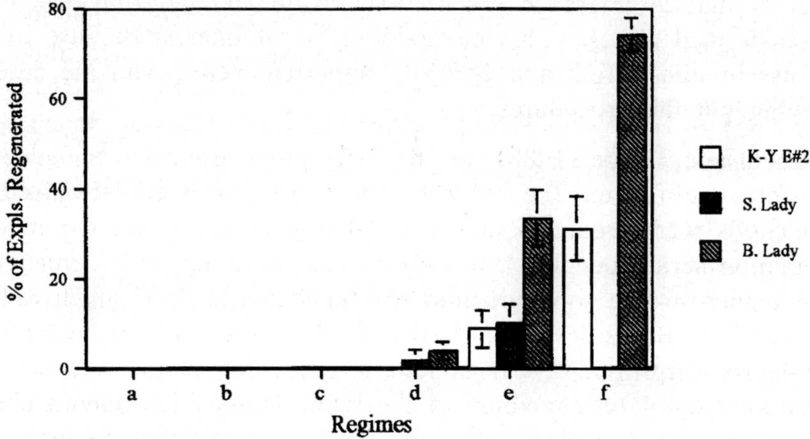

Figure 1 : The effects of timing and concentration of antibiotic selective regimes on resistant cauliflower plant regeneration following co-cultivation with pBI121/TI vector containing *Agrobacterium.* A. The flow chart of antibiotic selection regimes. [1] Week after co-cultivation. [2] Selective media: N = No selection, using SI-1 medium (SI medium + 150 mg/l carbenicillin); L = Low level selection, using SI-2 medium (SI-1 + 25 mg/l kan or 15 mg/l G418); H = High level selection, using SI-3 medium (SI-1 + 50 mg/l kan or 30 mg/l G418). B. Percentages of explants of 'Known You Early #2', 'Snow Lady', and 'Beauty Lady' producing putative transgenic plants under different G418 selection regimes. Verticle bar = standard error. ND = Not determined.

quantitative data were given). Eimert and Siegemund's (1992) transformation work using leaf disks only resulted in transgenic calli.

3.1.4. Tumors and hairy roots

Passelegue and Kerlan (1996) wounded the stem of one-month-old *in vitro* seedlings with forceps dipped in a mixture of two *A. tumefaciens* strains (A_{600} adjested to 1). The bacterial strains were disarmed C58pMP90 and a wild type strain 82.139 at a 10:1 concentration ratio, respectively. The first tumors appeared in 20 days. On an average, 67% treated plants displayed tumors. Tumors were excised one month later, shaken for 6-8 h at 25°C in a liquid medium supplemented with 72 mg/l adenine sulfate, 4 mg/l tetracycline and cultivated on solid medium of the same composition with an addition 500 mg/l carbenicillin supplemented. Three weeks later, regenerated shoots were individually subcultured on the same medium until rooted. Only 60% of the resultant 750 shoots were analyzed because of abnormalities in the other 40%. No selection pressure was exerted in this work because a significant level of natural tolerance toward kanamycin and hygromycin of the untransformed plants. This methodology is of interest because of the success in transferring agonomically important genes, and the ease of carrying out the procedures.

David and Tempe (1988) was the only group used *A. rhizogenes* to transform cauliflower. Ten-day-old *in vitro* dark grown, etiolated seedling hypocotyls were wounded and inoculated with bacterium strain 8196 containing virulence pRi8196. Inoculated seedlings were transferred onto a sucrose-free solid medium and incubated at 25°C under a 16 h photoperiod. Roots developed in 8 to 24 days at the site of inoculation. Single root tips (ca. 2 cm in length) were repeatedly excised and transferred onto fresh medium of the same composition and incubated under the same conditions. In some cases, carbenicillin (50 µg/l) was added to the medium for the two or three first transfers. Regeneration occurred spontaneously from both normal and transformed roots. Buds developed earlier from normal roots (40 days after root isolation) than from hairy roots (2 to 8 months according to the root lines). Root cultures re-isoloated from regenerated plantlets had more pronounced tendency in bud neoformation than primary cultures.

3.2. Transgene types, confirmation of transgenic status and transgene expression

The molecular aspects of the cauliflower transformation experiments will be reviewed in this second portion of the "Transformation" section.

3.2.1. Transformation using marker genes

Marker genes had been used to develop cauliflower transformation procedures by a number of laboratories. Selectable markers were used to reduce the frequency of nontransgenic plantlets production and reporter genes were used to discriminate between the transgenic plantlets or tissue sectors and the nontransgenic ones among the regenerated plantlets. In order to positively confirm the transgenic status, gene integration has to be established using molecular procedure. The commonly applied molecular steps are:

- Extraction of the total DNA from the putative transgenic plantlet/ tissue.
- Complete restriction of plant DNA material
- Transfer of the restricted DNA onto a filter support
- Hybridization analysis with labeled DNA derived from the marker gene pieces for verification of size and intact integration of the marker gene in the plant genome

3.2.1.1. In 1988, David and Tempe reported the successful transformation of cauliflower with several different strains of *A. rhizogenes* carrying virulence Ri plasmid DNA that, in addition to the oncogenes, encoding gene for opine synthesis. The putative transgenic plants were analyzed for the presence of mannopine or mannopine and agropine to report the transgenic status of the tissues. In addition, transformation was verified by Southern blot analysis of restriction digested DNA from different plant parts (roots, leaves and curds) using a radiolabelled probe to T-DNA. The authors reported regenerates with normal and modified phenotypes that carried similar or the same T-DNA content. The phenotypic variations were supposed to be due to the differential expression of the oncogenes in the regenerated plants.

3.2.1.2. Selectable marker gene, conferring resistance to aminoglycoside drugs was employed by Srivastava *et al.* (1988). They used an *A. tumefaciens* strain with an oncogenic Ti plasmid vector containing a chimaeric *npt*II gene in its T-DNA. The presence of the marker gene were analyzed in restricted DNA from regenerated plants, first in a quick screen for transformants in a protein dot blot, then by Southern blot analysis on a putative transgenic plantlet. Although the transformation efficiency can not be evaluated in this work due to only one putative transgenic plantlet was analyzed using Southern blot and only a single lane for putative transgenic plantlets was shown on both dot blot analyses. However, the analyses clearly demonstrated the presence of true transgenic individual(s) among the putative regenerates.

The interesting fact of this work was that, when leaf and petiole explants were taken from the resultant transgenic plantlets and incubated on a "hormone-free", kanamycin-containing medium, clonal plantlets regenerated. The authors used such hormonal autonomous regeneration as their evidence that auxin/cytokining-like Ti-oncogenes were inserted onto the genomes of the resultant transgenic plantlets. Unfortunately, no quantitative value was given on the number of such clonal progenies obtained and no photograph was presented that clearly showing the formation of such clonal plantlets. In addition, wide morphological variations in the resultant transgenic plants were observed and the authors also contributed this as hormonal imbalance resulted from the expression of the Ti-oncogenes. Since seed progenies of these transgenic plantlets were not studied, one cannot rule out the possibility that some of such morphological variations were tissue-culture-induced epigenetic variants.

3.2.1.3. De Block *et al.* (1989) used *neo* (*npt*II) and *bar* as selectable marker genes. In the chimaeric plasmid, the *bar* gene, coding for the enzyme PAT, an inactivator of the herbicide phosphinotricin, under the 35S promoter and the *neo* gene was under the *nos* promoter.

The molecular analysis of transformants followed the generalized procedure listed above: fully restricted transformed plant DNA was analyzed using Southern blot by probing with a radiolabelled DNA fragment either from the *neo* or from the *bar* gene. The copy number of the chimeric gene in the transformants were determined. On the average, each transgenic plant contained one to three copies of the chimeric *neo* or *bar* genes. There was no correlation between the copy number and the resistance level in the leaf disc assay. In addition, up to 25% of the transformants showed no detectable PAT or *npt*II enzyme activities although Southern blotting demonstrated that these plants were indeed transformed.

3.2.1.4. Bhalla and Smith (1998b) used a binary vector containing the selectable marker *npt*II gene under the control of the *nos* promoter and terminator sequences. Integration and expression of the introduced transgene were analyzed by DNA gel blot and PCR analysis and *npt*II expression assays. A PCR fragment identical to the *npt*II gene was identified in seven of the transgenic plants. No amplification was observed in untransformed cauliflower. This was also confirmed in Southern blot of several putative transgenic plants, indicating that the transgene has been integrated in the genome of these cauliflower plants.

3.2.2. Transformation using target genes

In addition to the marker genes used to assist the transformation procedure *per se,* some agronomic potentially useful target genes were also tested in transforming the cauliflower crop.

3.2.2.1. Until 1996, there was no report of cauliflower transformation by genes of agronomic interest. Passelegue and Kerlan (1996) described the successful transformation of cauliflower with a disarmed plamsids included the capsid gene and the antisense gene VI of CaMV along with the marker genes *nptII*, *hpt* and *gus.* CaMV can cause extensive damage and losses in *Brassica* crops and no known natural resistance had been clearly identified. The regenerated shoots were screened for GUS-like activity, which was not detected in the control tissue. While the authors offered no data on the behavior of the transgenic plant lines towards CaMV infection, they did offer extensive molecular analysis of the transformants. The procedures used to verify the presence of transgenes included:

PCR analysis : Either the capsid gene or the whole/truncated forms of gene VI was detected in all transformants .

Southern blot analysis : The copy number of introduced genes was estimated by this analysis as 2-4 copies. The analysis was performed on restricted genomic DNA using a labelled probe specific for the *ntpII* gene.

Quantitative analysis of the transgene expressions was achieved by employing the following methodologies :

Northern blot analysis : The presence of transcripts of *hpt* and capsid genes as well as that of the antisense gene VI was detected in the total mRNA population from the putative transformed plants.

PCR analysis : Reverse transcription PCR (RT-PCR) was used to provide more sensitive detection of DNAs that were corresponding to the predicted size fragments of the capsid gene and gene VI transcripts.

The expression data revealed variability in transgene expression among independently transformed plants and lack of correlation between copy number of transgene and mRNA accumulation. Possible reasons leaded to such expression observations, as suggested by the authors, were:

Multiple copies of T-DNA influenced the expression of introduced genes and high copy number often associated with reduction of transgene activity.

The "position effect" due to random T-DNA insertion sites within the host genome.

Both maker genes and the CaMV genes on the expression plasmid were under the 35S promoters and the presence of homologous sequences is believed as an important factor for transgene inactivation.

Numerous reports were cited by the authors in supporting the above proposed mechanisms. The low accumulation of CaMV gene transcripts could result either from reduced transcriptional activity, or RNA instability. No detectable capsid protein in transgenic plants was evident by Western blot analysis. The authors contended that the capsid protein was not essential to obtain viral protection, and analysis the progenies resistance to virus induction is pending.

3.2.2.2. Genetic transformation was first shown by Ding *et al.* (1998) to be of agronomic beneficial to cauliflower crop. Insect-resistant transgenic cauliflower plants, that expressed the TI gene, were developed. TI is a subgroup of protease inhibitors that are thought to be part of the natural plant defense system against insect infestation. It makes up over 80% of the total soluble proteins in the storage roots of Taiwan sweet potato cultivars from where the gene was isolated. Authors were able to demonstrate *in planta* insect resistance in transgenic plants but not in control plants.

On the expression plasmid used, the TI gene was placed under the control of 35S promoter next to the *gus* and *npt*II marker genes. All the tested putative transgenic plants showed significantly higher TI activity versus control plants demonstrated by a cell-free, *in vitro* TI assay. Conformation of the TI transgenic status of the resultant plants was made by PCR analysis. The PCR products produced corresponded to the flanking sequences of the *TI* gene, and were present in all the putative transgenic plants but not in control plants. The same sequences were used as a probe in a Southern blot analysis of the transgenic plant DNA. Finally TI protein was detected in the putative transgenic plants by Western blot analysis using a rabbit anti-TI serum with suitable conjugated secondary antibody. The TI polypeptide detected was 24 kD, and present only in the transgenic plants.

Thus, the expression of the inserted TI gene was demonstrated on the transcriptional and translational levels in the transformed cauliflower plants and not in the control counterparts. The functional integrity of the TI gene products in the transformed plants was demonstrated by an *in vitro* TI assay and also by *in planta* bioassays. The *in vitro* TI assay was a cell free assay for suppression of trypsin by the TI gene product measured by changes in optical density. The *in planta* bioassays included force feeding (Table 1) and open infestation (Fig. 2). Taken together, this represents the first report of the insertion of an agronomic important gene into the cauliflower plant, showing the full functioning of the inserted gene product.

3.2.2.3. In many instances, it will be highly desirable to confine the expression of transgenes to a specific time and location by using specialized promoters. Bhalla and Smith (1998b) transformed cauliflower with an antisense *Bcp1,* an anther-specific gene isolated from *Brassica campestris*, linked to promoter *Lat52*, a pollen-specific promoter active post-meiotically in haploid pollen. The *Bcp1* gene is essential for the production of functional pollen. The introduction of antisense *Bcp1* was expected to produce sterility in pollen carrying this transgenic construct. Northern blot analysis proved the presence of the appropriate sized antisense transcripts of *Bcp1.* A specific reduction in the endogenous *Bcp1* mRNA in pollen of transgenic plants was observed. One half of the pollen grains from anthers of the transgenic plants were non-viable indicated by fluorochromatic reaction assay. This methodology provides

Table 1. *In planta* feeding bioassys using 2 clonal plantlets of a TI transgenic cauliflower 'Snow Lady' plant #1296 1-1 and non-transgenic control (CK) plants. Experimant was carried out once without replication.

Insect species	Host Plant	% Larvae survived	% Reached pupa stage	% Reached adult stage
Spodoptera litura	1-1	56.7	22.7	13.2
	CK	86.7	63.6	49.1
Plutella xylostella	1-1	60.0	36.0	21.6
	CK	100.0	100.0	100.0

Figure 2 : In *Planta* open infestation test carried out in a greenhouse with opened windows. Left : TI transgenic cauliflower plants *in vitro* cloned from the R_0 plants. Right : Control plants. The insects were identified as *Pieris conidia* (A) and *Plutella xylostella* (B).

an opportunity to create nuclear male sterility for hybrid seed production, of obvious commercial value.

4. CONCLUSIONS AND FUTURE PROSPECTS

Most of the tissue culture and transformation procedures commonly applied to crop genetic engineering have been tested on cauliflower transformation. Here we are attempting to evaluate the future prospects of these procedures for cauliflower crop improvement based on the available data on this crop and current state of knowledge in crop genetic engineering.

Tissue culture procedures applied to cauliflower can be roughly divided into protoplast *vs.* tissue regeneration. Protoplast regeneration is highly laborious in addition to low regeneration frequency, inconsistent response and high degree of genotypic dependence. Since the availability of the simpler, more efficient and more dependable tissue regeneration procedures, few laboratories concentrate on developing cauliflower protoplast technology. Thus, it is not likely that the cauliflower protoplast culture technology will mature to the extent like that in tobacco or cereals can be efficiently applied to genetic transformation. Cauliflower tissues used in regeneration can be divided to seedling explants, leaf disks, pathogenic (tumor/hairy root) and the unique curd tissues. Regeneration from pathogenic tumor and hairy root tissues containing wild-type *Agrobacterium* oncogenic phytohormone gene(s) are likely to lead to high percentages of resultant transgenic plants expressing abnormal morphology/physiology as described by David and Tempe (1988). Such fact will lead to workload increase on selection. That means that high rate of transformation is required in order to produce sufficient number of transgenic plantlets to be selected from. Therefore, this source to regeneration for obtaining transgenic plantlets should be avoided when regeneration procedures capable of producing higher percentages of normal plantlets are available. To compare seedling explant and leaf-disk transformations, since seedling explants containing younger tissues that, as a rule, physiologically more totipotant (regenerable) than that of the physiologically older leaf disk tissues, and since the transformation efficiency is positive correlated with the tissue regeneration efficiency, it is expected that the seedling explant transformation will be more successful. The actual experimental outcomes supported such prediction: the leaf-disk transformation reports were either devoid of any quantitative data (Srivastava *et al.*, 1988) or stated that no transgenic plantlet resulted (Eimert and Siegemund, 1992). The unique crud explant tissue, that only available to cauliflower, is a highly sensible explant source to be adapted to cauliflower transformation. This is meanly due to it's gigantic regenerating quantity and high genetical stability, in addition to the fact as the best developed cauliflower *in vitro* cloning procedure. Unfortunately, transformation from organized meristems, in general, is difficult and the reason behind is poorly understood. But in view of the current rapid advancement on crop transformation mechanisms and procedures, it is foreseeable that this problem will be solved in the near future.

Commonly applied transformation processes can be divided into the indirect *Agrobacterium*-mediated transformation and the direct plasmid DNA uptake via microprojectile bombardment or PEG-/electroporation-mediated transformation. The PEG-/electroporation-mediated direct DNA transformation procedures are specifically for the protoplast works and are not suitable for cauliflower as stated in the previous paragraph. As far as *Agrobacterium*-mediated transformation *vs.* microprojectile bombardment, each has its merits and limitations. Without go into the detail comparisons, the major limitations of microprojectile bombardment for low budget laboratories are the expensive instrumentation and high operation costs. Its main drawbacks for all the laboratories are the non-specific starting/ending points and tandemly multiple-copied gene insertion. For *Agrobacterium*-mediated transformation, the experiment can be carried out in a standard bio-laboratory less costly and whole gene at single or low-copy number insertions are expected. Such advantages even made numerous laboratories in the highly active area of cereal transformations, a field exclusively using protoplast-based direct DNA transformation in previous years, shifting emphases toward tissue-based *Agrobacterium*-mediated transformation in recent years.

In view of all the forgoing considerations, we are endorsing the seedling explant-based *Agrobacterium*-mediated transformation for cauliflower genetic engineering. We are also closely following the development on meristem transformation and hoping that the crud-based transformation procedure will one day become a reliable technology for cauliflower crop improvement.

In conclusion, cauliflower genetic transformation was commonly being considered as recalcitrant by many workers. Breakthrough had been made in recent years by the pioneer works cited in this review. Now, using genetic engineering to re-design the cauliflower crop is no longer just an idea or a dream, it becomes an achievable goal.

REFERENCES

Abe M, Kanechi M, Inagaki N and Maekawa S (1997) Effects of the natural ventilation and light intensity on the growth, photosynthesis and acclimation of *in vitro* cultured cauliflower plantlets under photomixotrophic conditions. *Environ. Control Biol.*, **35** : 47-53.

Bhalla PL and Smith N (1998a) Comparison of shoot regeneration potential from seedling explants of Austatralian cauliflower (*Brassica oleracea* var. *botrytis*) varieties. *Australian J. Agr. Res.*, **49** : 1261-1266.

Bhalla PL and Smith N (1998b) *Agrobacterium tumefaciens*-mediated transformation of cauliflower, *Brassica oleracea* var. *botrytis*. *Mol. Breed.*, **4** : 531-541.

Bhalla PL and Weerd N de (1999) *In vitro* propagation of cauliflower, *Brassica oleracea* var. *botrytis* for hybrid seed production. *Plant Cell Tiss. Org. Cult.*, **56** : 89-95.

David C and Tempe J (1988) Genetic transformation of cauliflower (*Brassica oleracea* L. var. B*otrytis*) by *Agrobacterium rhizogenes*. *Plant Cell Rep.*, **7** : 88-91.

De Block M, Brouwer D and De Tenning P (1989) Transformation of *Brassica napus* and *Brassica oleracea* using *Agrobacterium tumefaciens* and the expression of the *bar* and *neo* genes in the transgenic plants. *Plant Physiol.*, **91** : 694-701.

Ding LC, Hu CY, Yeh KW and Wang PJ (1998) Development of insect-resistant transgenic cauliflower plants expressing the trypsin inhibitor gene isolated from local sweet potato. *Plant Cell Rep.*, **7** : 854-860.

Eimert K and Siegemund F (1992) Transformation of cauliflower (*Brassica oleracea* L. var. *botrytis*) – an experimental survey. *Plant Mol. Biol.*, **19** : 485-490.

FAO (2000) http://apps.fao.org/lim500/nph-wrap.pl? Production. Crops. Primary & Domain = SUA & servlet = 1

Gamborg OL, Miller RA and Ojima K (1968) Nutrient requirements of suspension cultures of soybean root cells. *Exp. Cell Res.*, **50** : 151-158.

Grout BWW (1988) Cauliflower (*Brassica oleracea* var. *botrytis* L.). In *: Biotechnology in Agriculture and Forestry* (Ed Bajaj YPS), Springer-Verlag, Berlin pp 211-225.

Jourdan PS, Earle ED and Mutschler MA (1990) Improved protoplast culture and stability of cytoplamic traits in plants regenerated from leaf protoplasts of cauliflower *Brassica oleracea* spp. *botrytis*. *Plant Cell Tiss. Org. Cult.*, **21** : 227-236.

Kanechi M, Ochi M, Abe M, Inagaki N and Maekawa S (1998) The effects of carbon dioxide enrichment, natural ventilation, and light intensity on growth, photosynthesis, and transpiration of cauliflower plantlets cultured *in vitro* photoautotrophically and photomixotrophically. *J. Amer. Soc. Hort. Sci.*, **123** : 176-181.

Kumar A, Kumar VA and Kumar J (1993) Rapid *in vitro* propagation of cauliflower. *Plant Sci.*, **90** : 175-178.

Margara J and David C (1978) Les étapes morphologiques du développement du méristème de chou-fleur, *Brassica oleracea* L. var. *botrytis*. *C. R. Acad. Sci.*, Paris, **287** : 1369-1371.

Murashige T and Skoog F (1962) A revised medium for rapid growth and bioassays with tobacco tissue cultures. *Physiol. Plant.*, **15** : 473—498

Negrutium I, Shillito R, Potrykus I, Biasini G and Sala F (1987) Hybrid genes in the analysis of transformation conditions I. Setting up a simple method for direct gene transfer in plant protoplasts. *Plant Mol. Biol.* **8** : 363-373.

Passelegue E and Kerlan C (1996) Transformation of cauliflower (*Brassica oleracea* var. *botrytis*) by transfer of cauliflower mosaic virus genes through combined cocultivation with virulent and avirulent strains of *Agrobacterium. Plant Sci.*, **113** : 79-89.

Pröls M, Töpfer R, Sehell J and Steinbiss HH (1988) Transient gene expression in tobacco protoplasts. I. Time course of CAT appearance. *Plant Cell Rep.*, **7** : 221-224.

Pua EC, Deng XY and Koh ATC (1999) Genotypic variability of de novo shoot morphogenesis of *Brassica oleracea in vitro* in response to ethylene inhibitors and putrescine. *J. Plant Physiol.*, **155** : 598-605.

Pua EC and Lee J (1995) Enhanced de novo shoot morphogenesis *in vitro* by expression of antisense 1-aminocyclopropane-1-carboxylate oxidase gene in transgenic mustard plants. *Planta,* **196** : 69-76.

Sadik S (1962) Morphology of the curd of cauliflower. *Amer. J. Bot.*, **49** : 290-297.

Short KC and Warburton J (1987) *In vitro* hardening of cultured cauliflower and chrysanthemum plantlets to humidity. *Acta Hort.,* **212** : 329-334.

Srivastava V, Reddy AS and Guh-Mukherjee S (1988) Transformation and regeneration of *Brassica oleracea* mediated by an oncogenic *Agrobacterium tumefaciens. Plant Cell Rep.,* **7** : 504-507.

Torres AC, Paviani TI, Caldas LS and Vecchia PTD (1980) Anatomy of shoot production *in vitro* from explants of cauliflower curd (*Brassica oleracea* L. var. *botrytis* subvar. *cauliflora* DC.) *Pesquisa agropecuaria brasileira,* **15** : 435-440.

Vandemoortele JL (1999) A procedure to prevent hyperhydricity in cauliflower axillary shoots. *Plant Cell Tiss. Org. Cult.*, **56** : 85-88.

Vandemoortele JL, Billard PB, Boucaud J and Gaspar T (1999) Evidence for an interaction between basal medium and plant growth regulators during adventitious or axillary shoot formation of cauliflower. *In Vitro Cell. Dev. Biol. – Plant*, **35** : 13-17.

Wardle K, Dobbs EB and Short KC (1983) *In vitro* acclimatization of aseptically cultured plantlets to humidity. *J. Amer. Soc. Hort. Sci.*, **108** : 386-389.

Zobayed SMA, Armstrong J and Armstrong W (1999) Cauliflower shoot-culture: Effects of different types of ventilation on growth and physiology. *Plant Sci.*, **141** : 209-217.

Chapter 6

GENETIC TRANSFORMATION IN TWO IMPORTANT VEGETABLE CROPS : HOT CHILLI (*CAPSICUM ANNUUM*.L) AND EGGPLANT (*SOLANUM MELONGENA* L.)

G Lakshmi Sita★ and G Franklin

Department of Microbiology and Cell Biology, Indian Institute of Sciences, Bangalore - 560 012, India

Summary

Plant genetic engineering took birth in mid eighties, when for the first time, plants were successfully engineered for improved virus resistance. Plant genetic engineering techniques could be effectively used to exploit some of the untapped potentials to increase the harvestable crop yield. In nineties virus resistant squash seeds followed soon by bio-engineered cantaloupes, potatoes, and papaya reached to market. Subsequently Bt cotton (Insect resistant) reached the market and it was one of the first bio-engineered crop seeds to become available commercially. Since then around a dozen of other crop plants have been genetically modified to resist pests and diseases many with genes derived from basic research findings reported decades earlier. In spite of several successes there are many vegetable crops, which need genetic manipulation for crop improvement in general and yield in particular. This review attempts to look at the progress made in the crop improvement of two important crops namely chilly and eggplant towards developing virus resistant and insect resistant plants, with special emphasis on the work done in the author's laboratory.

Keywords : Chilli, Egg plant, Transformation NPTII

★Corresponding author : E-mail : sitagl@mcbl.iisc.ernet.in

1. INTRODUCTION

For thousands of years, farmers have battled against, insects, micro-organisms and weeds that destroy or compete with the main crop. Many major events in history have resulted from devastating plant diseases epidemics or insect infestations. The Irish potato famine of the mid1800s, which was caused by *Pytophthora* infestation, is one such example killing more than a million people. Farmers have used plant management techniques coupled with plant breeding to eliminate these problems. Although pesticides and herbicides have been helpful to some extent, continuous usage results in environmental problems and residual toxicity. Advent of genetic engineering in plants has resulted in improved seeds that are genetically endowed, not only to resist damage from insects and micro-organisms but also to be resistant to herbicides. Such bio-engineered seeds have the potential to revolutionise agriculture and improve environmental quality by making it possible to reduce the use of pesticides and others. Bio-engineered seeds although seen as a recent outcome in plant breeding efforts, they are actually results of 50 years of research by many scientists. Incrementally many scientists paved the way for isolating genes that protect a particular organism from pests and for transforming these genes into a wide variety of plants. Careful selection and repetitive crossing of progeny can eventually generate varieties that are both high yielding and resistant to some pests. But the process is extremely time consuming and can take more than fifteen years to bring a new variety. Generally only closely related species of plants can be crossbred. If no varieties are naturally resistant to a particular fungus or insect, traditional breeding has no way to create resistance to that fungus or insect. Further more, breeders frequently face a situation in which a resistance gene is closely linked to a gene that adversely affects the quality of a crop, that is where the two traits are always inherited together. Despite the best efforts of plant breeders to improve the pest and disease resistance in 1990, farmers world over lost nearly one fourth of their crops.

The phytopathogenic bacteria *Agrobacteriun tumefaciens* genetically transforms plants by transferring a portion of resident Ti plasmid, the T-DNA, to the plant. Recent developments in cell biology to regenerate plants from single cells and organized tissues and the discovery of methodology to isolate and clone genes in *Agrobacterium* provided the basic pre-requisite for the practical use of genetic engineering to introduce agronomically important traits into plants (Lakshmi Sita and Manoharan, 2001). With the exception of some species (tobacco, tomato, and potato) work has been limited in Solanaceae. Many important vegetable crops belong to the family Solanaceae. Among these tobacco, tomato. potato, brinjal and capsicum are the most important crops. Tobacco and tomato are widely used as model systems to study plant genetic transformation with marker genes as well as important desirable genes. Extensive literature is available in both these species and hence not discussed here. Potato also has been extensively studied. Potato has already been engineered for disease and pest resistance, for increased adaptability to biotic and abiotic stresses, for oral vaccines (Haq *et al.*, 1995) and also for increased nutritional value (Chackraborty *et al.*, 2000). This review attempts to look at the progress made in the crop improvement of two important crops namely chilli and eggplant towards developing virus resistant and insect resistant plants, with special emphasis on the work done in the author's laboratory.

2. GENETIC TRANSFORMATION IN CHILLI

Chilli is a spice cum vegetable of commercial importance. Besides imparting pungency and colour to the dishes it is also a rich source of vitamins (A, C and E) which are having a high medicinal value. Considering the culinary importance, medicinal value and vitamins content the demand for chillies in the world market is increasing progressively. Hence it is essential to increase the per hectare production for economic returns to the farmer to meet the local demands of the growing population and to have enough quantity for foreign export. In India, the production of chilli has declined from 779 thousand tonnes during 1992-93 to 730 thousand tonnes during 1993-94, the overall percentage in decrease is -6.25. Consequently, the export of chilli witnessed a fall in quantity during the year 1994-95 (Peter, 1995). There are many factors, which contributed to this decline and most important among them are the diseases caused by viruses, bacteria, fungi and insects. Spraying of fungicides and pesticides can control the diseases to some extent however, effective resistance against several destructive pathogens is still lacking. Hence, efforts are being made to produce disease resistant plants through genetic engineering. Recent advances of genetic engineering of plants have evoked great interest in developing modern technology for crop improvement. In many countries, efforts have been made on cloning genes potentially useful in agriculture and introducing them stably into plants of important crop species. In consequence, many transgenic plants have been successfully produced which expressed remarkable effects such as resistance to chemicals, pests and diseases. Genetic transformation through *Agrobacterium tumefaciens* is now a routine procedure for introducing foreign genes into many plant species including several vegetable crop plants such as tomato, *Brassica* etc. (McCormick *et al.*, 1986; Metz *et al.*, 1995). The two most important pre-requisites for the success of the method are the availability of a plant regeneration system from the explants and suitable method for transformation. Although chilli belongs to the family Solanaceae, whose members are easily amenable to tissue culture and transformation practices, it proves to be highly recalcitrant. Several reports of plant regeneration either through organogenesis (Gunay and Rao, 1978; Fari and Czako, 1981; Philips and Hubstenberger, 1985; Sripichitt *et al.*, 1987; Agarwal *et al.*, 1989; Arroyo and Revilla, 1991; Valera-Montero and Ochoa-Alejo, 1992; Ebida and Hu, 1993; Christopher and Rajam, 1994; Szasz *et al.*, 1995; Hyde and Philips, 1996) or embryogenesis (Harini and Lakshmi Sita, 1993) are available. However, these reports are genotype-specific and consequently, the regeneration protocol as well as viable transformation has to be established for each commercial cultivar for exploiting the potential of genetic engineering. Although there are one or two reports on transformation in sweet pepper with viral coat protein gene (Zhu, 1996) and herbicide resistant gene (Tsafataris, 1996), there are no reports of transformation in any of the hot chillies with desirable genes.

2.1. Protocol development for the regeneration of plantlets from explant

Several combinations of cytokinins and auxins were tried to induce direct regeneration of shoot buds from cotyledons and hypocotyls and first leaves, from two to three week old seedling. Among the three explants tested cotyledons were found to be the best. Almost 90-97 % regeneration was obtained on MS media (Murashige and Skoog, 1962) augmented with 10 mg/L BAP and 1mg/L IAA (Med 1) from cotyledons followed by first leaves and hypocotyls. However this medium did not support the elongation of the shoot buds. Hence further modification of the regeneration medium was required to obtain shoots which could be rooted for proper plantlet formation. A series of media combinations were tried with cytokinins and GA. Media combinations with BAP and GA (1 mg/L each) was found to be more appropriate for successful proliferation and elongation. (Med 2). Transfer from medium 1 to medium 2 was found to be absolutely essential for successful elongation of 2-3 mm shoot buds to 20-25 mm shoots. In about four to six weeks, elongation was accomplished. These shoots could be excised from the clumps and rooted on half strength MS medium supplemented with 0.1 NAA within 10-15 days (Med 3). In brief the regeneration of plantlets is a three-stage process. Growth regulator requirement was almost same in both varieties used. Regeneration protocol established as above was used routinely for transformation.

2.2. Transformation of explants and regeneration of transgenics

For transformation, thousands of cotyledonary leaves were infected and co-cultivated with *Agrobacterium* for 48 h. The shoot bud induction was observed after 15-20 days of inoculation on the selection medium (Fig 1A). The explants with shoot buds were subcultured in the fresh selection medium. After 30-45 days culture, the putatively transformed shoot buds were transferred to the elongation medium containing 50 mg/L kanamycin and 400 mg/L cefotaxime. Transformation efficiency is calculated as the number of regenerating explants over total number of explants in each plate on selection medium. About 30-40 % transformation efficiency in sweet pepper and 10 % in hot chilli variety Pusa jwala was observed.

Efficient and high percentage shoot bud induction was observed in our studies from cotyledons, compared to hypocotyls and first leaves. Age of the explant is also critical. If the explants are more than three weeks regeneration percentage was quite low, or even nil. Elongation of shoot buds into long shoots has been a consistent problem with the previous workers. The shoot buds, which formed at the cut end of the distal part of the cotyledonary leaves after 30-40 days of culture were transferred to shoot elongation medium (Med 2). The transfer of shoot buds to the medium containing GA (1 mg/L) and Kinetin (1 mg/L) leads to excessive callus growth and did not support shoot bud elongation and rooting. On the other hand transfer of shoot buds to MS medium supplemented with GA and BAP resulted in excellent elongation without any callus formation. At least two passages of four weeks on Med 2 were found to be desirable. By transferring the shoot buds to Med 2 large number of plantlets could be obtained. Normally one and occasionally two elongated shoots were produced from the total of ten to thirty shoot buds per explant (observed under stereo microscope) if the shoot buds were not transferred to the elongation medium. Elongation step is necessary for proper plantlet development. Earlier workers failed to induce rooting whereas Ebida and Hu (1993) rooted the rosette buds and transferred to soil where the shoots were elongated. . Rooting of the shoots was obtained on Med 3 in the presence of kanamycin. In the present study rooting was obtained with the single shoots separated from the clumps on Med 3 before transferring to soil. Rooting of

Figure 1 : Different developmental stages of transgenic hot chilli plants. A: Regeneration of shoot buds from cotyledonary explants. B: rooted shoots C: A well developed plant in flowering. D: fully ripened fruits from R0 plants.

transgenic shoots in Sweet pepper and Jwala varieties was carried resulting in rooted plants. Rooting of a large number of the shoots was carried on the selection medium. Successful rooting was obtained clearly indicating the transgenic nature of the plants. Almost 100% rooting of the shoots was obtained on half strength MS medium supplemented with 0.1 mg/L NAA (Fig. 1B). The plants were transferred to soil after hardening. However induction of rooting was taking much longer time compared to the controls. It is well known that roots are generally much more sensitive to antibiotics, and thus the ability of shoots to root on selection medium containing higher levels of selective agent (kanamycin 25-50 mg/L) is therefore a very strong indication of its transformed nature. In the majority of the cases only transformed shoots will root successfully on the medium containing kanamycin. It should again be noted that it is not at all uncommon for some of the regenerated shoots taken randomly failed to root in the presence of the selective agent. They could be escapes thus reducing the final number of transgenic plants. Putative transgenic shoots were transferred to antibiotic free medium before the analysis of GUS activity and PCR analysis to make sure that there is no contamination with *Agrobacterium*, which might interfere with the analysis.

Transgenic nature of these plants was confirmed by histochemical localization of GUS gene and PCR analysis. Histological tests carried with leaves from transgenic plants showed the characteristic blue colour. Several plants were tested and confirmed by molecular analyses such as PCR and southern hybridization. PCR analysis showed the expected 0.7kb amplified fragment of DNA with the NPTII primers used. The amplified DNA samples were electrophoresed on 0.8% agarose gel along with controls from plasmid DNA (pBI121) 0.7 kb fragment co-running with the amplified product from pBI121 could be detected from the transgenic plants but not from non-transformed plant. In some cases GUS primers were also used for PCR analyses. Southern hybridization was also carried in order to confirm the T-DNA integration. A 0.7kb PCR fragment of the *nptII* gene was used as a probe. DNA was digested with Hind 111, which cuts once in the plasmid. Hybridization showed positive signal in the undigested positive control at 13kb (Two bands corresponding two forms of plasmid). However, undigested DNA from transgenic plants gave signal at about 23 kb size indicating the integration of *nptII* gene into the genome of transgenic plants. Hybridization results clearly show variation in pattern among transgenic plants, indicating independent transformation events, and differed from the 13 kb band corresponding to the plasmid DNA digested with Hind111. No hybridization signals were observed in genomic DNA isolated from non-transformed plants. Thus, genomic DNA blot hybridization data confirmed the integration of NPT11 gene in the genome of transgenic chilli plants (Manoharan *et al.*, 1998)

In both the varieties (Californian Wonder and Pusa Jwala) plants were established in pots. Established transgenic plants have flowered and developed into fully mature fruits (Figs. 1C and D). Fruits were collected and the seeds were planted to study R_1 generation. R_1 plants were also found to positive in PCR analysis. The results presented demonstrate the potential of genetic transformation, for the introduction of desirable traits such as disease resistance, in *Capsicum annuum* L var. Pusa Jwala and insect resistance. In sweet pepper gene transfer with viral coat protein gene and also with herbicide resistant gene, however more work needs to be done before commercialization could be achieved.

3. GENETIC TRANSFORMATION IN EGGPLANT (*SOLANUM MELONGENA* L.)

Solanum melongena (eggplant) is an important vegetable crop widely cultivated in countries like Spain, France, Italy, Greece, North Africa, India and China. Eggplant fruit is an integral part of food for the major vegetarian population in India. The annual production of this important vegetable is severely limited by several soil-born pathogens and viruses. Although, sources of resistance exist in several resistant solanaceous species, the incorporation of these traits into eggplant gene pool by breeding has been slow, laborious and inadequate.

Recently effort has been made to transform eggplant with *Agrobacterium tumefaciens*. Rotino and Gleddie (1990) achieved transformation of eggplant using a disarmed binary vector system containing the *nptII* gene as the selectable marker and chloromphenicol acetyl transferase (CAT) as reporter gene. GUS gene was successfully introduced into eggplant by Fari *et al.* (1995) through *Agrobacterium* strain. Recently stable transformation of luciferase (*luc*) gene in eggplant was reported by Hanyu *et al.* (1999). An agronomically important (insect resistant) gene was introduced in eggplant by Kumar *et al.* (1998). They introduced a lepidopteran-specific, synthetic *cry1AB* gene extensively modified for higher expression in plant cells. The transgenic brinjal fruits were demonstrated to be significantly resistant to the larvae of

Leucinodes orbonalis. In view of the limited information available on genetic transformation in the commonly used varieties for commercialization several Indian varieties were screened for establishing suitable protocols. A successful protocol for high efficiency transformation has been established. Results obtained are briefly discussed here (unpublished).

3.1. Protocol development for the regeneration of plantlets from explant

An important pre-requisite for transferring genes into plants is the availability of efficient regeneration system. It is therefore important to identify conditions conducive to high rate of regeneration for each species. The critical conditions for regeneration of plants obviously varies between species and even between varieties. Concentrations, combinations and the type of plant growth regulator in the media critically affect regeneration potential. Among the different combinations of TDZ and NAA tested, 0.1 mg/L of TDZ and 0.2 mg/L NAA in the medium influenced efficient regeneration of shoots via indirect organogenesis. Callus induction and shoot regeneration occurred subsequently in this media. Eventhough, several types of cytokinins were used for regeneration in different species of Solanaceae, at present TDZ is the most widely used for successful regeneration. TDZ at low concentrations was generally found to be a better growth regulator for regeneration compared to other cytokinins. Magioli *et al.* (1998) also obtained similar results in *S. melongena* cotyledonary and leaf explants with TDZ at low concentrations. In *S. integrifolium*, the highest frequency of shoot regeneration was obtained with TDZ regardless of the concentration used and a mean value of 33 shoot buds per leaf disc was obtained with 2 mg/L TDZ (Rotino *et al.*, 1992). Rotino and Gleddie (1990) were successful in transforming eggplant when they used young leaves obtained from aseptic plants. While, they failed to achieve the same with explants obtained from green house grown plants due to excessive colonization of *Agrobacterium*, which resulted in the death of co-cultivated explants. All the explants tested namely cotyledonary leaf, first leaf, hypocotyl, shoot tip and root are morphogenetically competent in producing shoots, however the frequency and percentage of regeneration varied depending upon the explant type. Cotyledonary leaves followed by first leaves are more efficient in terms of percentage and frequency of regeneration. In eggplant most of the previous studies as well as the present study show that *in vitro* grown material was superior to material grown *in vivo* as the source of initial explants. Protocol thus developed was used for transformation studies.

3.2. Transformation and regeneration of transgenics

Young leaves excised from 15 days old aseptic seedlings were made into two longitudinal pieces and pre-cultured on regeneration media (0.1mg/L TDZ and 0.2mg/LNAA) for 1-2 days. Pre-cultured leaf pieces were infected with *Agrobacterium* strain LBA4404 containing a binary vector pBAL2. The binary plasmid vector pBAL2 harbors neomycin phosphotransferase (NPT11) gene driven by nopaline synthase promoter and terminator conferring resistance to antibiotic kanamycin as plant selection marker. *uidA* gene interrupted with intron (GUS INT) driven by the CAMV 35S promoter (CAMV35S) and nopaline synthase terminator as reporter gene. One tenth dilution (v/v) of overnight *Agrobacterial* culture was taken and the explants were infected for 5 minutes, blot dried and transferred to the plates in which they were precultured, for co-cultivation for two days. After repeated washes, co-cultivated explants were transferred to selection medium containing growth regulators for regeneration (MS medium containing 0.1 mg/L TDZ, 0.2 mg/L NAA, 100mg/L kanamycin and 500 mg/L cefotaxime). After three weeks of incubation, green callus grew along the cut surface of the infected leaves whereas the negative control leaves turned brown and died. The frequency of transgenic calli formation was better with cotyledonary explants compared to first leaves. After 30 days on selection medium, the calli were separated from the parent explant and transferred to fresh selection media (same as above) where differentiation of shoot buds was observed (Fig. 2A). Shoot buds elongated in the same media in 3-4 weeks time (Fig. 2B). The putatively transformed shoots were directly rooted on soil (soilrite) watered with sterile water containing 100 mg/L kanamycin (Fig. 2C) and after hardening in the green house established in the field (Fig. 2D).

Vegetative (such as root, stem and leaves) and floral parts (anther, ovary, stigma, calyx, petals and pollen grains) were analyzed for the GUS gene expression (Jefferson *et al.*, 1987) and was found to be intense in leaves (Fig. 2E). There was however, variation in the intensity of blue coloration as the plant matured. Out of 124 plants assayed for GUS, 84 plants were positive in showing the characteristic deep blue coloration. Control leaves taken from plants regenerated normally did not show any blue coloration. Since GUS intron gene was used, we can safely conclude that the blue color is due to actual integration of the gene after splicing and not due to *Agrobacterium* contamination. In transgenic plants containing this chimeric gene the intron splices efficiently, giving rise to GUS enzymatic activity. No GUS activity can be detected in *Agrobacteria* containing this construct due to the lack of an eukaryotic splicing apparatus in prokaryotes (Vancaneyt *et al.*, 1990). Fari *et al.* (1995) found a strong cell-type preference of the GUS expression in the phloem tissues and in the veins, whereas the mesophyll and epidermis

Figure 2 : Different developmental stages of transgenic brinjal. A: Differentiation of shoot buds from transformed calli on selection medium. B: Shoot elongation. C: Shoots rooted on soil rite. D: Plants established in the field. E: Leaf showing GUS activity.

cells showed a little activity. They also found that in the vegetative tissues, the GUS-activity gradually decreased from the base toward the top of the plants and almost no activity could be detected in the leaves of the newest nodes. On the other hand, a surprisingly high level of GUS activity could be localized in anthers and pollen grains. But, in the present investigation we observed GUS activity in complete plant even at field conditions. The variation in the expression may be due to the usage of different promoters in the present (CAMV 35S) and previous study (T_R2^1*mas 2*).

Hundred and twenty four randomly selected shoots, rooted on kanamycin sulfate were analyzed for PCR amplification. The genomic DNA from young leaves of those plants was isolated by CTAB method and used for PCR amplification by using NPTII specific primers. One hundred and twenty two plants were found to be positive for NPTII sequence in their genome. Expected 750 bp of NPTII was amplified (Fig. 3)

Various parameters like the type of *Agrobacterium* strain, preculture period, nature of wounding, kanamycin concentration seems to have a major effect on transformation efficiency. It was observed that *Agrobacterium* strain LBA 4404 was superior to others in the present study. Whereas in capsicum, Manoharan *et al.* (1998) found EHA 105 found to be superior for transformation studies. One-day preculture period

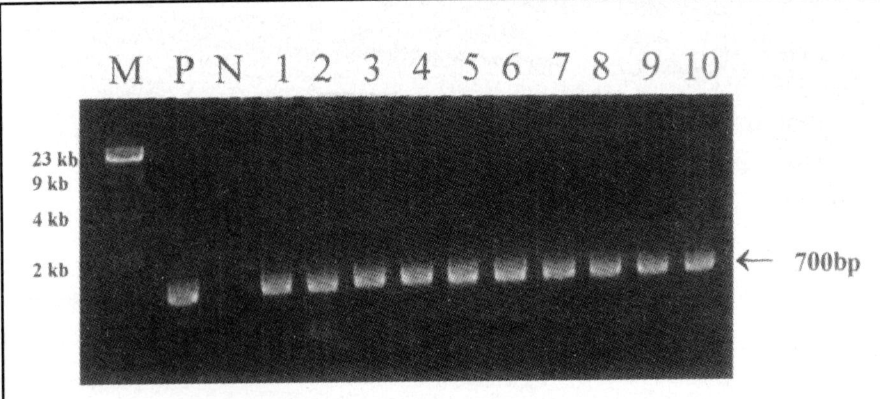

Figure 3 : PCR analysis of kanamycin resistant transgenic eggplants showing the presence of an expected 700 bp DNA fragment of *npt II* gene; M: λ *Hin*dIII Marker; P: Positive control (pBAL 2); N: Negative control (untransformed plant); Lanes 1-10: Transgenic plants showing amplification of the expected 700 bp sequence of *npt II*.

was more than sufficient for eggplant. Explants precultured and co-cultivated on hormone-free medium showed higher transformation efficiency in *S. integrifolium* (Rotino *et al.*, 1992). Controversially, our experiments with eggplant showed less efficient transformation when the explants were precultured and co-cultivated on hormone free medium. In our experiments wounding of the explants significantly increased the frequency of transgenic calli formation in eggplant, but, unwounded one showed very few or nil transgenic calli. Whereas Curtis *et al.* (1999) found that less number of transformed shoots produced from wounded explants than unwounded one. Differences in the kanamycin sensitivity (10-250 mg/L) of different explants co-cultivated with the same strain have also been recorded. Untransformed leaf and cotyledonary explants did not callus or regenerate shoots at 100 mg/L kanamycin. On the other hand both callus induction and shoot regeneration were achieved at same kanamycin level in the transformed explants in our studies.

Elongated shoots showed a little or no response in root formation by customary methods of transferring them on culture media containing auxins and antibiotics. However, in the present study we achieved 100% of rooting by a simple method of transferring shoots excised from TDZ containing selection media directly to soil rite irrigated with antibiotics. The number of roots induced per shoot was also significantly higher than all other rooting media tested. Such rooted plants upon transplantation to field showed 100% survival.

4. CONCLUSIONS AND FUTURE PROSPECTS

In view of the recent developments in biotechnology in general and genetic transformation in particular many of the crops including vegetables are going to be improved in quality to meet the requirements of growing population. Allover the world lot of emphasis is now given to crops other than rice and wheat. Among these tomato, potato, capsicum, brinjal occupy an important role among vegetable crops. Results presented demonstrate the potential for commercialization of capsicum and brinjal against viral resistance as well as insect resistance.

REFERENCES

Agrawal S, Chandra N and Kothari SL (1989) Plant regeneration in tissue cultures of pepper (*Capsicum annuum* L. cv. mathania). *Plant Cell Tiss. Org. Cult.*, **16** : 47-55.

Arroyo R and Revilla R (1991) *In vitro* plant regeneration from cotyledon and hypocotyl segments in two bell pepper cultivars. *Plant Cell Rep.*, **10** : 414-416.

Chakraborty S, Chakraborty N and Datta A (2000) Increased nutritional value of transgenic potato by expressing a nonallergenic seed albumin gene from *Amaranthus hypochondriacus*, *Proc. Natl. Acad. Sci. USA*, **97** : 3724-3729.

Christopher T and Rajam MV (1994) *In vitro* clonal propagation of *Capsicum* spp. *Plant Cell Tiss. Org. Cult.*, **38** : 25-27.

Curtis IS, Power JB, Hedden P, Ward DA, Philips A, Lowe KC and Davey MR (1999) A stable transformation system for the ornamental plant, *Datura meteloides* D.C., *Plant Cell Rep.*, **18** : 554-560.

Ebida AIA and Hu CY (1993) *In vitro* morphogenetic responses and plant regeneration from pepper (*Capsicum annuum* L. cv. Early california wonder) seedling explants. *Plant Cell Rep.*, **13** : 107-110.

Fari M and Czako M (1981) Relationship between position and morphogenetic response of pepper hypocotyl explants cultured *in vitro*. *Scientia Hort.*, **15** : 207-213.

Fari M, Nagy I, Csanyi M, Mityko J and Andrasfalvy A (1995) *Agrobacterium* mediated genetic transformation and plant regeneration *via* organogenesis and somatic embryogenesis from cotyledon leaves in egg plant (*Solanum melongena* L. cv. kecskemet; lila). *Plant Cell Rep.*, **15** : 82-86.

Gunay AL and Rao PS (1978) *In vitro* plant regeneration from hypocotyl and cotyledon explants of red pepper (*Capsicum*). *Plant Sci. Lett.*, **11** : 365-372.

Hanyu H, Murata A, Park EY, Okabe M, Billings S, Jelenkovic G, Pedersen H and Chin C-K (1999) Stability of luciferase gene expression in a long term period in transgenic eggplant, *Solanum melongena*, *Plant Biotechnol.*, **16** : 403-407.

Harini I and Lakshmi Sita G (1993) Direct somatic embryogenesis and plant regeneration from immature embryos of chilli (*Capsicum annuum* L.). *Plant Sci.*, **89** : 107-112.

Haq TA, Mason HS, Clements JD and Arntzen CJ (1995) Oral immunization with a recombinant bacterial antigen produced in transgenic plant. *Science*, **268** : 714-716.

Hyde CL and Phillips GC (1996) Silver nitrate promotes shoot development and plant regeneration of chile pepper (*Capsicum annuum* L.) *via* organogenesis. *In vitro Cell. Dev. Biol. Plant.*, **32** : 72-80.

Jefferson RA, Kavanagh TA and Bevan MW (1987) GUS fusions: b-glucronidase as a sensitive and versatile gene fusion marker in higher plants. *EMBO J.*, **6** : 3901-3907.

Kumar PA, Mandaokar A, Sreenivasu K and Chakrabarti SK (1998) Insect-resistant transgenic brinjal plants, *Mol. Breed.*, **4** : 33-37.

Lakshmi Sita G and Manoharan M (2001) *Agrobacterium* mediated genetic transformation in crop plants. In: *Advances in Plant Biotechnology*. (Ed. PC Trivedi), Oxford IBH, pp 49-68.

Magioli C, Rocha APM, de Oliveira DE and Mansur E (1998) Efficient shoot organogenesis of eggplant (*Solanum melongena* L.) induced by thidiazuron, *Plant Cell Rep.*, **17** : 661-663.

Manoharan M, Sree vidya CS and Lakshmi Sita G (1998) *Agrobacterium*-mediated genetic transformation in hot chilli. *Plant Sci.*, **131** : 77-83.

McCormick S, Niedermeyer J, Fry J, Barnason A, Horsch R and Fraley R, (1986) Leaf disc transformation of cultivated tomato (*L. esculentum*) using *Agrobacterium tumefaciens*. *Plant Cell Rep.*, **5** : 81-84.

Metz TD, Dixit R and Earle ED (1995) *Agrobacterium tumefaciens*-mediated transformation of brocoli (*Brassica oleracea* var. italica) and cabbage (*B. oleracea* var. capitata). *Plant Cell Rep.*, **15** : 287-292.

Murashige T and Skoog F (1962) A revised medium for rapid growth and bioassays with tobacco tissue cultures. *Physiol Plant.*, **15** : 473-497.

Peter KV (1995) Spices research and development: An updated overview. *Plant. Chron.*, **90 :** 471-477.

Phillips GC and Hubstenberger JF (1985) Organogenesis in pepper tissue cultures. *Plant Cell Tiss. Org. Cult.*, **4** : 261-269.

Rotino GL and Gleddie S (1990) Transformation of eggplant (*Solanum melongena* L.) using binary *Agrobacterium tumefaciens* vector. *Plant Cell Rep.*, **9** : 26-29.

Rotino GL, Perrone D, Ajmone-Marsan P and Lupotto E (1992) Transformation of *Solanum integrifolium* Poir *via Agrobacterium tumefaciens*: Plant regeneration and progeny analysis. *Plant Cell Rep.*, **11** : 11-15.

Sripichitt P, Nawata E and Shigenaga S (1987) *In vitro* shoot forming capacity of cotyledon explants in red pepper (*Capsicum annuum* L. cv. yatsufusa). *Jpn. J. Breed.*, **37** : 133-142.

Szasz A, Nervo G and Fari M (1995) Screening for *in vitro* shoot forming capacity of seedling explants in bell pepper (*Capsicum annuum* L) genotypes and efficient plant regeneration using thidiazuron. *Plant Cell Rep.*, **14** : 666-669.

Tsafataris A (1996) The development of herbicide tolerant transgenic crops. *Field Crops Res.*, **45** : 115-124.

Valera-Montero LL and Ochoa-Alejo N (1992) A novel approach for chili pepper (*Capsicum annuum* L.) plant regeneration: shoot induction in rooted hypocotyls. *Plant Sci.*, **84** : 215-219.

Vancaneyt R, Schnidt R, O' conner-Sanchez A, Willmitzer L and Rocha-Sosa M (1990) Construction of an intron-containing marker gene: Splicing of the intron in transgenic plants and its use in monitoring early events of *Agrobacterium*-mediated plant transformations. *Mol. Gen. Gen.*, **220** : 245-250.

Zhu K, Wen-Jun, Ou-Yang, Zhang YF and Chen ZL (1996) Transgenic sweet pepper plants from *Agrobacterium* mediated transformation. *Plant Cell Rep.*, **16** : 71-75.

Chapter 7

ADVANCES TOWARDS THE PRODUCTION OF TRANSGENIC GARLIC PLANTS

Barandiaran Xabier

Universidad de Córdoba, Departamento de Microbiología, C/. San Alberto Magno s/n, Facultad de Ciencias - 14004 Córdoba, Spain.

Summary

Cultivated garlic (Allium sativum) is a sterile plant reproduced exclusively through the formation of bulbs or floral topsets. This characteristic has seriously limited the breeding of this crop to the selection of the existing genetic variability. Transgenic technology is especially important in this crop because it opens the possibility of gene transfer in the species. Although garlic, as other Alliums, is considered recalcitrant both for regeneration and transformation, recent advances in the development of in vitro culture protocols and the demonstration that exogenous DNA can be introduced and expressed in different garlic tissues offers the possibility to obtain whole transgenic plants. Most important applications of this technology would be the introduction of disease resistance genes, especially for virus resistance, and the modification of the sulphur metabolism pathway in order to increase the pharmaceutical value of garlic bulbs.

Keywords : *Allium sativum*, alliinase, disease resistance, organogenesis, particle bombardment, somatic embryogenesis, transgenic plants

Abbreviations : 2,4D - 2,4 dichlorophenoxiacetic acid; 2iP - isopentenyl adenine. BA - 6-benzylaminopurine, NAA - α-Naphthaleneacetic acid, Picloram - (4-amino-3, 5,6-

*Corresponding author : E-mail : xbarandiaran@wanadoo.es

trichloropicolinic acid), K - 6-furfurylaminopurine, GUS - β-glucuronidase; ATA - aurintricarboxylic acid.

1. INTRODUCTION

Garlic (*Allium sativum* L.) is an important and widely cultivated crop used for food and therapeutic purposes. This crop is present in the five continents being Asia, by far, the region presenting the highest productions (>80% world production). China is the mean garlic producer followed by India, Republic of Korea, and the United States (FAOSTAT, 1999).

Cultivated garlic (*Allium sativum* L.) is an obligated vegetatively propagated plant due to its inability to produce seeds of sexual origin. This characteristic has limited breeding of this crop to clonal selection, including the production of healthy stocks via meristem culture. In spite of the efforts conducted by private companies and public institutions to obtain sexual seeds, results to date no registered variety exists that is produced via sexual breeding.

The inability of garlic to produce fertile flowers extremely restricts the breeding possibilities of this crop, which are practically limited to the selection of the existing genetic variability. Although, this variability is important and considerable efforts are being conducted to conserve and characterise it in a number of germplasm banks (Maggioni *et al.*, 1997), the impossibility of a gene flux among the different accessions prevents the unification of the desired traits in a single variety.

Gene isolation and cloning techniques, together with those of genetic transformation are producing remarkable results in the directed breeding of many species. Although, the use of these techniques can be of great utility in all crops, it is especially interesting in garlic because it constitutes the only way to transfer genes in the strictly vegetatively propagated cultivated clones. However, garlic as the vast majority of monocotyledonous plants, have shown to be recalcitrant not only to genetic transformation but also to callus production and further shoot regeneration.

The development, in the last years, of efficient systems for *in vitro* regeneration of shoots, roots and embryos from axenic explants, together with the demonstration that exogenous DNA can be introduced and expressed in different garlic tissues has constituted an important step towards the obtention of whole transgenic garlic plants.

2. GARLIC TISSUE CULTURE

2.1. Introduction

Garlic (*Allium sativum*) belongs to the Liliaceae family of the monocots. The *in vitro* culture of this group of plants, especially cereals and grasses, presents common requirements different from those generally applied to the dicotyledonous plants.

The main characteristic of the monocot plants, concerning its *in vitro* culture, is the necessity to use meristematic tissues to start the culture. Another general characteristic of the *in vitro* culture of monocot plants is the requirement of strong auxins such as 2,4D (2,4 dichlorophenoxiacetic acid) or picloram (4-amino-3, 5,6-trichloropicolinic acid). Explants of these species show little or no reaction to growth regulators such as NAA or IAA that are extensively used in dicots tissue culture. An additional difference among the *in vitro* culture of both groups of plants is that, in general, monocot plants do not need an exogenous cytokinin source, neither for the starting of the culture nor for further regeneration. Growth and differentiation of the culture are mainly manipulated by modifying exogenous auxin levels and not via the relative balance among auxins and cytokinins, as it occurs in the vast majority of the organogenic processes in dicotyledonous plants.

These general requirements frequently present in the *in vitro* cultures of important monocots such as rice, maize and wheat have notably influenced the *in vitro* culture research of other monocots such as *Alliums*. Sometimes, as is the case of the use of high doses of strong auxins, this influence has negatively affected the results obtained.

2.2. Callus formation and organogenesis

The most frequent use of the organogenic techniques is as part of the protocols for production of transgenic plants. Other uses include the induction of somaclonal variants or the mere vegetative propagation. Whatever the final goal, it is always convenient to obtain high levels of callus production from the chosen explants, and a further regeneration rate as high as possible. In spite of the relatively high number of published papers on garlic callus culture and organogenesis, there is a lack of precise data about these two key parameters, that are essential for evaluating the efficiency of any *in vitro* regeneration protocol.

The first studies on garlic *in vitro* organogenesis date from 1973. In this work, the regeneration of adventitious shoots from calli induced from young leaf explants, cultivated in presence of 2,4D was described (Havránek and Novák, 1973). Subsequently, a number of protocols have been published using different explants and media.

Usually, the organogenic process starts with the establishment of a decontaminated explant on a medium designed for callus induction and growth. Calli can be sub-cultivated indefinitely in the same medium, or they can be transferred to an appropriate medium for organogenic induction. When friable calli are transferred to a liquid medium, cell suspensions are generated, sometimes maintaining its organogenic potential (Barrueto *et al.*, 1994). Most frequently used explants have been those including meristematic tissue such as: apical buds (Kehr and Schaeffer, 1976; Nagasawa and Finer, 1988), basal parts of leaves (Havránek and Novák, 1973; Novák, 1974; Dolezel and Novák, 1984; Nagasawa and Finer, 1998; Rauber and Grunewaldt, 1988; Barrueto *et al.*, 1994), basal plates (Koch *et al.*, 1995) and root tips (Haque *et al.*, 1997; Myers and Simon, 1998; Barandiaran *et al.*, 1999).

Explant treatment protocols include the typical decontamination of the plant material with different combinations of ethanol and hypochlorite. Other decontamination protocols involve flaming of the bulbs (Bhojwani, 1980) or the use of fungicides like benomyl (Mohamed-Yassen *et al.*, 1994). However, these protocols are usually inefficient when the explants are taken from plant parts that have been in permanent contact with the soil, such as bulbs. Rauber and Grunewaldt (1988) described contamination rates up to 60% when using leaf explants treated with a solution containing 2.6% of active chlorine. In order to avoid contamination problems, it is highly recommendable to use axenic explants excised from micropropagated material.

Environmental conditions used for callus induction are diverse according to different authors. Temperature varies among 25°C (Koch *et al.*, 1995; Barandiaran *et al.*, 1999) and 28°C (Nagasawa and Finer, 1988; Haque *et al.*, 1997). Concerning light requirements, some authors describe callus induction and growth in the dark (Havránek and Novák, 1973; Dolezel and Novák, 1984) while others use different combinations of light intensity and photoperiods. Organogenic tests have always been conducted in the presence of light, usually maintaining the same regime of temperatures. However, light is apparently not essential for the organogenic process since we have been able to regenerate roots and etiolated shoots in the dark (unpublished results).

Culture media used in garlic tissue culture are mainly based on MS (Murashige and Skoog, 1962). However, some researches have used others like AZ (Abo El-Nil, 1977), LS (Rauber and Grunewaldt, 1988), BDS (Koch *et al*., 1995) or B5 (Barandiaran *et al*., 1999). In order to induce callus, the presence of growth regulators in the medium is essential. Strong auxins like 2,4D have been used in 90% of the works published. Besides auxins, the inclusion in the medium of cytokinins such as kinetin (6-Furfurylaminopurine) or BA (6-Benzylaminopurine) has been frequent. According to Havránek and Novak (1973), the combined use of 2,4D and K improves callus growth. On the other hand, Rauber and Grunewaldt (1988) found a higher callus production frequency when the cytokinin BA was used instead of K. In order to regenerate shoots, calli are generally transferred to a medium without strong auxins, containing or not cytokinins. Havránek and Novak (1973) showed that increasing the concentration of K in the medium, a higher number of proliferating shoots were obtained. Once shoots have been obtained, these are micropropagated following any of the numerous protocols available, where is possible to multiply and induce bulbing in the regenerated material (Table I).

In the last years, significant advances in the *in vitro* organogenesis of garlic have been published. These reports have in common the use of root tips as explants, and show high rates of callus formation, organogenesis and shoots per explant or gram of callus tissue. Myers and Simon (1998), using a medium with a high dose of 2,4D (4.5 μM), reported a 29.1% calli formation rate, a 85.3% regeneration rate and a maximum number of shoots per explant of 5.4. Barandiaran *et al*. (1999), using a medium with a low level of 2,4D (0.14 μM) reported callus formation levels closc to 100%, a mean number of shoots per gram of callus of 16.2. The same authors, using a medium without strong auxins reported a callus formation rate mean close to 80%, a mean percentage of regeneration of 62.5% and a mean shoot production of 10.7 shoots/g.

On the other hand, Haque *et al*. (1997) reported a remarkable system of direct regeneration of shoots from root tips, showing high levels both of regenerating explants and number of shoots regenerated per explant. In this system, cells competent for regeneration are formed within 8 days culture in the epidermis (Haque *et al*. 1999). In case these cells are also competent for accepting and expressing foreign DNA, the protocol described would be of a great interest for the biolistic transformation of garlic.

Table 1. Summary of some of the reports on garlic micropropagation showing explant and medium used and the results obtained for the multiplication rate (Mult.) and bulb formation (Bulb.). B: Bulb; b: Topset bulbil.

Authors	Explant	Culture Medium*	Mult.	Bulb.
Bhojwani, 1980	Apical bud (B)	B5; 2ip, 0.5; NAA, 0.1	x 6.6/month	————
Peña-iglesias, 1982	Apical bud (B)	MS; 2ip, 0.5; NAA, 0.01	————	1.5 cm
Bhojwani, 1982	Apical bud (B)	B5; 2ip, 0.5; NAA, 0.1	x 4/month	————
Walkey, 1987	Apical bud (B)	B5; K, 10	————	————
Matsubara, 1989	Apical bud (B)	MS; BAP, 0.01; NAA, 0.1	————	5 mm
Moriconi, 1990	Apical bud (B)	MS; 2ip, 3; NAA, 0.3	x 6.5/month	89%
Conci, 1991	Apical bud (B)	MS; 2ip, 3; NAA, 0.3	————	————
Ravnikar, 1993	Apical bud (B)	B5; 2ip, 1; AJ, 2	x 2/month	56% 4 mm
Seabrook, 1993	Apical bud (B)	MS; BAP, 2; NAA, 0.1	x 9.5/month	6 mm
Nagakubo, 1993	Apical bud (B)	LS; BAP, 2; NAA, 1	x 4/month	74%
Yasseen, 1994	Apical bud (B)	MS; BAP, 2; NAA, 0.02	x 8.2/month	————
Ma, 1994	Apical bud (b)	MS; K, 0.5; NAA, 0.5	————	6 mm
Verbeek, 1995	Apical bud (b)	MS; BAP, 1; NAA, 0.5	x 6/month	————
Barandiaran, 1999	Axillary buds	B5; 2ip, 0.5; NAA, 0.1	x 5/month	0.5 cm

*concentration of PGRs; mg/L

Apart from the promising results published in the last years on callus formation and shoot regeneration in specific garlic cultivars (White Roppen, Morado de Cuenca, Piacenza), organogenic studies including many additional clones have been conducted. Barandiaran *et al.* (1999) evaluated callus formation and regeneration in 20 garlic genotypes, identifying two highly regenerable clones. Myers and Simon (1999) evaluated the regeneration potential of five clones using calli of different ages. They obtained high regeneration levels in some clones using young calli.

The use of root tips as explants in genotypes selected for their high regeneration capacity seems a crucial step for success in the production of transgenic garlic plants.

Protocol 1 : Callus culture and shoot regeneration

- *Take axenic root tips (about 1 cm long) obtained from micropropagated material (see protocol 2).*
- *Culture root sections in sterile 9 cm Petri dishes containing 25ml of B5 medium supplemented with BA 13.3 µM, 2,4D 0.14 µM and NAA 10.7 µM. Add to the media 3% sucrose and solidify it with 0.8% Bacto-Agar. Adjust medium to pH 5.7 before autoclaving.*
- *Incubate cultures at 25°C. in the dark.*
- *After three months, more than 80% explants will present callus formation depending on the genotype used.*
- *Transfer calli obtained to 9 cm Petri dishes containing 25ml B5 medium containing BA 13.3 µM, supplemented with 3% sucrose and solidified with 0.8% Bacto-Agar. Before autoclaving, adjust medium to pH 5.7.*
- *Incubate cultures in a growth chamber at a temperature of 25° ± 2°C, under 16 h daily illumination with Grolux fluorescent light (55 µmol m^{-2} s^{-1}).*
- *After 4 months cultured calli have produced a variable number of shoots (0 to 20) depending on the genotype and explant used.*

2.3. Somatic Embryogenesis

Somatic embryogenesis is probably the most adequate *in vitro* regeneration system to be used in a genetic transformation program, especially in the case of embryos with unicellular origin. A number of studies on somatic embryogenesis have been conducted in different *Allium* species: *Allium sativum* (Abo El-Nil, 1977; Suh and Park, 1988; Xue *et al.*, 1991; Koul *et al.*, 1994; Bockish *et al.*, 1997; Haque *et al.*, 1998), *Allium cepa* (Dunstan and Short, 1977; Van der Valk *et al.*, 1992; Saker, 1997; Eady *et al.*, 1998), *Allium porrum* (Van der Valk *et al.*, 1992; Hong and Debergh, 1995), *Allium ampeloprasum* (Buiteveld *et al.*, 1994; Silvertand *et al.*, 1996), *Allium tuberosum* (Matsuda and Adachi, 1996; Matsuda *et al.*, 1998), *Allium fistulosum* (Shahin and Kaneko, 1986; Van der Valk *et al.*, 1992) and *Allium carinatum* (Havel and Nóvak, 1988). However, despite the high number of reports mentioned, the number of available protocols ready to be used in transgenic plant production are quite limited. The vast majority of reports lack some of the four important requirements to confirm somatic embryogenesis: an appropriate histogical analysis, the development of independent synchronic bipolar structures, the presence of only one cotyledon and the competence to germinate. However, there are outstanding reports in some *Allium* species like leek (Buiteveld *et al.*, 1994) and onion (Eady *et al.*, 1998) where practically all the mentioned requirements are accomplished.

In spite of the fact that garlic was the first *Allium* plant in which somatic embryogenesis was reported (Abo El-Nil, 1977) to date there is no a reliable protocol available to be used in a garlic transgenic plant production program. None of the reports meets all four requirements mentioned above, and none of them present a histological evidence of somatic embryogenesis. However, the morphological description of garlic embryogenesis and the identification of somatic embryos by the simultaneous development of the root and shoot poles reported by Haque *et al.* (1998) are promising.

In our lab we have shown that calli induced from different explants and cultured in presence of 2,4D produce nodular structures when transferred to a basal medium (Fig. 1). These structures resemble somatic embryos in the globular stage, similar to those described in many reports. However, these apparent pro-embryos rarely evolve adequately to bipolar embryos. In the vast majority of the cases, many roots and some shoots are produced without an apparent connection. Occasionally, a cluster of embryo- resembling structures is formed (Fig. 2) but they are not able to germinate.

The vast majority of somatic embryogenesis reports in other *Allium*

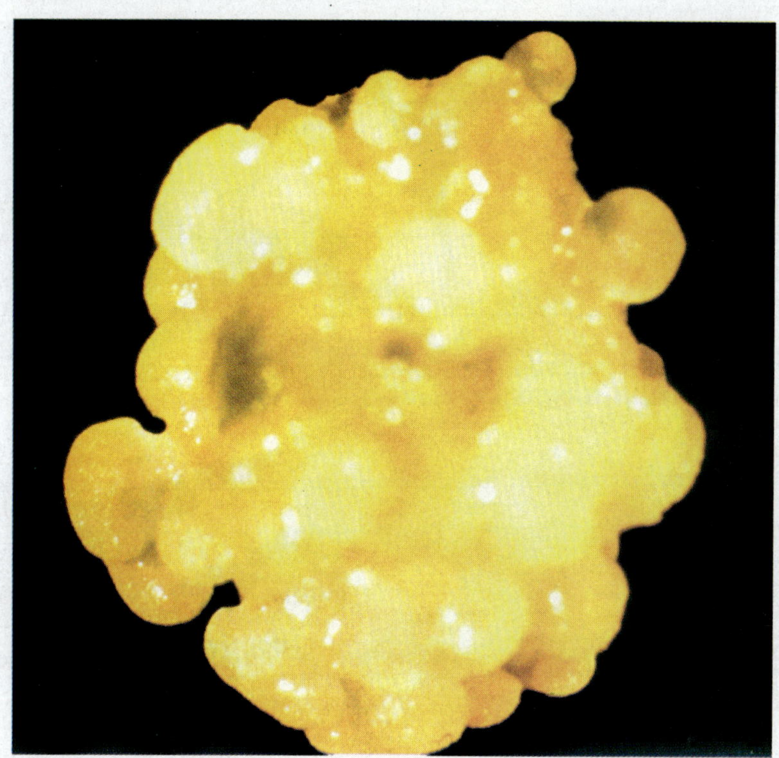

Figure 1 : Garlic callus obtained from a basal plate cultured in presence of 2, 4-D, showing globular structures.

species have been conducted using seeds as a source of explants. The best results have been obtained culturing immature embryos in media containing strong auxins like 2,4D or picloram. Unfortunately, cultivated garlic is sterile therefore, it is not possible to use this kind of highly regenerative explants. In order to establish an efficient somatic embryogenesis system adequate to be used in the production of transgenic garlic plants, it would be interesting to use seeds from fertile garlic accessions as initial explants.

2.4. Micropropagation

Clonal multiplication of the shoots regenerated in a selection medium is convenient in any protocol of plant transformation in order to have enough available material to conduct expression and integration analyses. However, in plants like garlic where the analysis of sexual offspring populations is impossible, it is essential to have a protocol to

Figure 2 : Cluster of embryo-like structures produced on a garlic callus.

micropropagate the putative transformants. On the other hand, the *in vitro* establishment of garlic plants can be used as a continuous source of axenic explants for organogenic studies.

Fortunately, garlic micropropagation has been used frequently for the production of healthy stocks and many protocols are available. Main factors to be considered in a garlic *in vitro* multiplication program include multiplication rate and the degree and quality of bulb formation. Multiplication rate is the number of propagula produced by initial propagula in a defined period of time. This rate limits the speed at which it is possible to multiply a determined material and consequently, the number of plants it is possible to obtain in each cycle. On the other hand, "microbulbing" or *in vitro* bulbing determines the hardening off easiness of the plants produced and therefore, their extra-vitrum survival rate. Moreover, on the size of the micro-bulbs obtained depends the time necessary to obtain commercial-sized bulbs. Table 1 shows a summary of some of the papers on garlic micropropagation, including the achievements on multiplication rate and *in vitro* bulbing.

The micropropagation process starts with the *in vitro* establishment of a small bud that in the vast majority of the published papers corresponds to the apical bud present in the shoot existing inside the mature cloves. Occasionally, other kinds of buds have been used like those present in the aerial bulbils (Table 1). Barandiaran *et al.* (1999) reported for the first time the use of buds excised from immature bulbs in a micropropagation protocol.

Decontamination protocols are similar to those described in the previous section. However, the fact that buds are protected within an axenic environment facilitates the successful establishment of in *vitro* cultures.

Environmental conditions routinely used in garlic micropropagation include: temperatures around 25° C, 16/8 h photoperiods and light intensities around 55 mmol m^{-2} s^{-1}.

Nearly all the papers concerning garlic micropropagation describe protocols where different media are used in each one of the successive stages of the culture: explant establishment, multiplication and bulb formation. Table 1 shows the media producing the best results concerning multiplication rate. As it can be observed, basal media more frequently used were MS (Murashige and Skoog, 1962) and B5 (Gamborg *et al.*, 1968). Growth regulators used in the multiplication media include a combination of the auxin NAA with a cytokinin, that usually 2iP (isopentenyl adenine) or BA.

In vitro bulbing of garlic has been studied by numerous researchers. Commonly, three strategies have been followed in order to obtain a satisfactory quantity and quality of micro-bulbs: the change of the culture physical environment (light and temperature), the change in the kind and concentration of the carbon source used, and the inclusion of specific chemicals in the medium. Ravnikar *et al.* (1993) using jasmonic acid obtained a 56% of bulb formation, producing microbulbs up to 4 mm (Table 1). On the other hand, Seabrook (1993) obtained *in vitro* bulbs ranging from 3.6 mm up to 6 mm according the accession cultured by reducing to the half the sucrose concentration typically used, and by supplementing the medium with 8 g/l mannitol. Likewise, Nagakubo *et al.* (1993) obtained multiplication rates up to 74% submitting the cultures to a 5°C temperature treatment during 6 months. Finally, Mohamed-

Yassen *et al.* (1994) made use of an 18/6 photoperiod and a 25°C temperature together with a medium containing 12% sucrose and 5 g/l active charcoal in order to induce *in vitro* bulb formation. On the other hand, Barandiaran *et al.* (2000) reported 100% *in vitro* bulb formation, in many different accessions, just leaving the explants in the same medium during time enough (up to 16 weeks).

Protocol 2. Garlic micropropagation

- *Harvest immature bulbs approximately two months before the crop cycle is completed and check for presence of axillary buds.*
- *Clean bulbs containing* axillary buds *in soapy water and dip them in 96% ethanol. Rub the base of each bulb with a scouring pad and again dip them in 96% ethanol. Finally, decontaminate the bulbs for 20 minutes in a sodium hypochlorite solution (1% of active chlorine) followed by three washes of 5 minutes each with autoclaved, double-distilled water.*
- *Excise aseptically, from the bulbs, axillary buds of approximately 0.5 mm, with the aid of a binocular microscope.*
- *Culture the excised axillary buds individually in sterile 10 ml plastic tubes containing 5 ml B5 medium with 0.5 mgL^{-1} of 2iP and 0.1 mgL^{-1} of NAA. This medium should be supplemented with 3% sucrose and solidified with 0.8% Bacto-Agar. Before autoclaving, the medium is adjusted to a pH of 5.7.*
- *6 weeks after the explants are established, transfer the plantlets obtained from the surviving buds to baby-food jars (175 ml, Magenta B-caps as closures) containing 50 ml of the same medium.*
- *Incubate tubes and baby food jars in a growth chamber at a temperature of 25° ± 2°C, under a 16 h photoperiod with Sylvania Grolux fluorescent lamps (55 µmol $m^{-2}s^{-1}$).*
- *20 weeks later, separate the clusters of plants formed into individual plants when possible, excise out the roots and transfer shoots to fresh medium. After 8 weeks culture in the growth chamber, incubate the jars in a cold room at 4°C in the dark during 4 weeks.*

- *Take out the plants from jars and remove all medium remaining with tap water. Sow plants showing micro-bulbs in 5L pots containing sand, clay, compost and humus (20:10:4:1). Place pots outdoors protected with an anti-aphid screen. Sow bulbs from November to June and harvest them from April to June, depending on the genotype and the weather conditions.*

3. GARLIC TRANSFORMATION

3.1. Introduction

Monocots have been much more challenging than dicots regarding the development of efficient systems for transgenic plant production. Genetic transformation systems more commonly used *Agrobacterium*, particle bombardment and protoplast transformation have shown particular problems in this group of plants. On principle, monocotyledonous plants are out of the *Agrobacterium* natural host range. This limitation has moved researchers to develop specific protocols, using a number of strategies such as the use of more virulent *Agrobacterium* strains or the application of infection inducing chemicals (Graves and Goldman, 1986; Raineri *et al.*, 1990; Deng *et al.*, 1988; Gould *et al.*, 1991; Chan *et al.*, 1992). On the other hand, biolistic transformation requires the use of cells both with organogenic capacity and competent to accept and express exogenous DNA. This kind of cells, that are relatively easy to obtain in dicotyledonous plants, are rare in monocots. Moreover, the use of protoplasts in monocot transformation is limited to some specific cases, due to the low regeneration level shown by these systems.

Besides the specific problems associated to transformation systems, monocot plants have shown other peculiarities regarding the genetic sequences used in the process as the promoters or the selectable genes. This is the case of the CaMV 35S promoter that renders high gene expression levels in the majority of dicotyledonous plants, while in monocots levels are usually 10 to 100 times lower (Fromm *et al.*, 1985; Hauptmann *et al.*, 1988). Furthermore, the low expression levels obtained with the marker genes limits the development of efficient selection systems. Consequently, it has been necessary to design specific vectors, containing promoters obtained from monocotyledonous plants, besides the use of gene expression enhancers (Vasil, 1993).

An additional problem, frequently mentioned in the development of monocot transformation protocols, appears in the transgenic cells selection procedure. Antibiotics like kanamycin, commonly used with dicotyledonous plants, are inadequate in monocots due to the natural resistance of this kind of plants and its interference with the regenerative processes. Due to these inconvenient, the use of the broad range herbicide phosphinothricin has become common as a selection agent (Vasil, 1993).

3.2. *Allium* transformation

Although a relatively low number of investigations directed to the production of transgenic plants in *Allium* species, have been developed, this genus has been included in the group of "recalcitrants". This categorization is due to two main reasons: on one side, other mococots including some pertaining to the *Liliaceae* family are among the agronomic species most difficult to be transformed (Eady, 1995). On the other hand, *Allium* species are characterised by containing large genomes where the probability to obtain a proper expression of the introduced genes is reduced due to the risk of insertion into non-translatable regions (Eady, 1995).

In spite of the difficulties, important advances have been obtained in the last years in the stable transformation of some Liliaceae: *Gladiolus* (Kamo *et al.*, 1995), *Lilium* (Watad *et al.*, 1997), *Asparagus* (Cabrera-Ponce *et al.*, 1997). The first report on stable transformation of an *Allium* species (onion) has been reported recently (Eady *et al.*, 2000) using *Agrobacterium* and *gfp* (green fluorescent protein) assisted selection. A number of investigations are being conducted in other *Allium* species: leek, shallot and garlic. Recent advances in *Allium* transformation have been mainly due to: the development of very efficient regeneration systems (see previous sections), the identification of appropriate selection agents (Eady and Lister, 1998), and the finding that most of the monocot promoters used with cereals are inadequate for a correct gene expression in *Liliaceae*. (Wilmink *et al.*, 1995)

Barandiaran *et al.* (1998) reported for the first time the introduction and expression of a reporter gene (*uidA*) in different garlic tissues (bulb, leaf and callus), using the biolistic particle delivery system. Figure 3 shows the expression of the *uidA* reporter gene in garlic epidermis cells. In this report a very high nuclease activity, in all the tissues tested, is described. This nuclease activity prevented transient expression of the exogenous DNA until it was blocked with an appropriate nuclease inhibitor (aurintricarboxylic acid). Additional research on garlic nuclease activity in a garlic core collection has shown that this activity is widely extended in the species, independently of the genotype tested (Fig. 4).

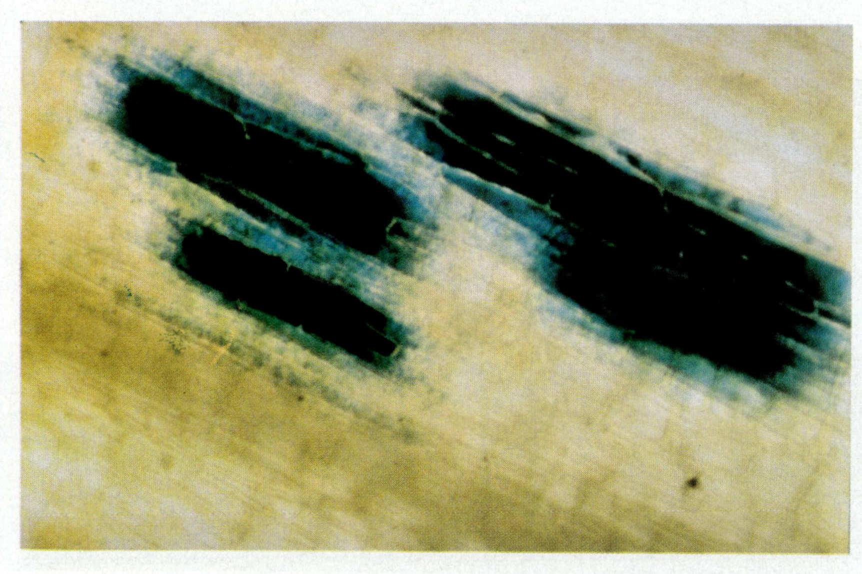

Figure 3 : Expression of the *uidA* reporter gene in garlic epidermic cells.

The identification of tissues both competent for regeneration and DNA integration and expression is a prerequisite in any transformation program. DNA expression has been reported in different garlic tissues including regenerable callus (Barandiaran *et al.*, 1999). Additional research conducted in our laboratory has shown clear differences in the competence of different tissues, treated at the same time with the biolistic device, to express exogenous DNA. Clove sections containing a central shoot, showed that young leaf tissue is much more amenable to express the *uidA* gene than the storage tissue (Fig. 5).

Rapid progress in the recovery of transgenic cereals has been the result of the utilization of organized and regenerable tissues such as immature embryos, in combination with particle bombardment technology (Christou, 1995). In garlic, we have used axillary buds (Ø 0.5 mm) as a source of organized and regenerable tissue in combination with the particle delivery system. Six weeks after bombardment of groups of excised

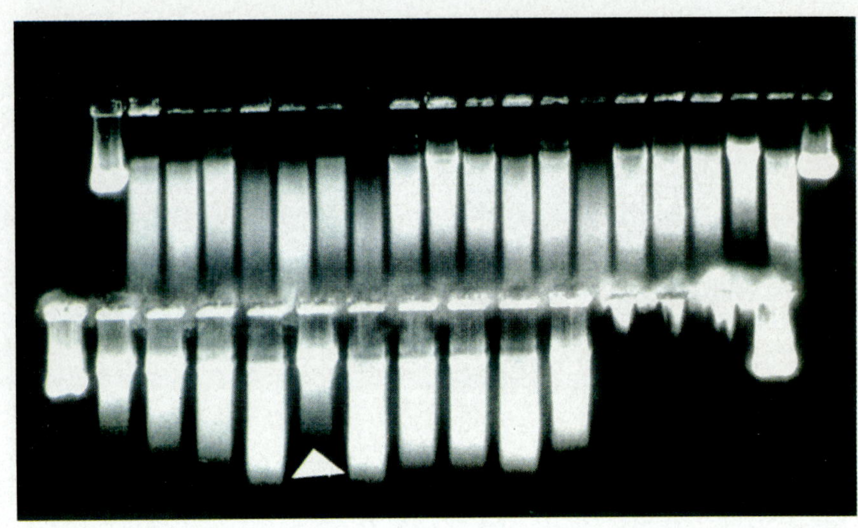

Figure 4 : Plasmid DNA degradation after 10 min. incubation with 20 mg tissue of 31 different accessions of garlic. Arrow show the degradation pattern produced by a Brazilian cultivar showing a slightly less nuclease activity. The four extreme lanes show the control plasmid DNA without garlic tissue.

buds precultured in a B5 medium (Gamborg *et al.*, 1968) containing 2 mM aurintricarboxylic acid, chimeric plants (⌀ 1 cm) stably expressing the *uidA* gene were produced (Fig. 6). In order for this system to be useful in the production of whole transgenic garlic plants it will be necessary to establish an efficient selection protocol of the transgenic cells. This selection is usually difficult in organized systems like buds, where transgenic cells can protect the non-transgenic ones to be killed.

Protocol 3. Transient expression of a reporter gene in garlic leaves

- *Bombardments will be carried out using a PDS 1000/He microprojectile gun (Bio-Rad).*

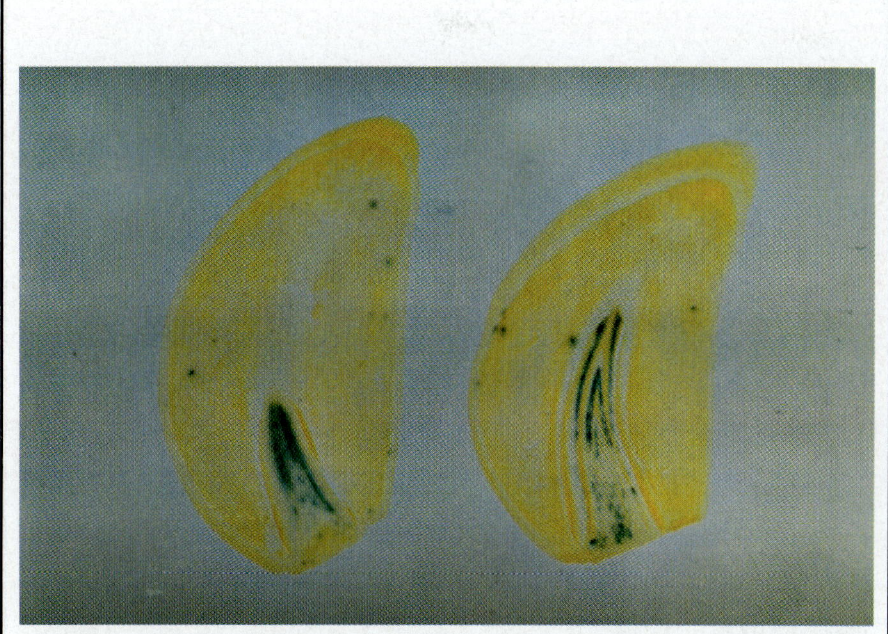

Figure 5 : Clove section showing a differential expression of the *uidA* gene in two different garlic tissues : shoot and storage leaf.

- *Harvest immature bulbs approximately two months before the crop cycle is completed. Decontaminate bulbs as described in protocol 2. Separate the storage leaves and with the aid of a cork-borer make leaf circles of about 1cm diameter.*

- *Centre the leaf circles on 9 cm Petri dishes containing 25ml B5 medium supplemented with 0.2 M Sorbitol, 0.2 M Mannitol and 2 mM ATA. Preculture the leaf explants in a growth chamber at a temperature of 25° ± 2°C in the dark, during a week.*

- *Precipitate plasmid DNA containing a 35S + uidA cassette onto gold particles (Bio-Rad 1.6 ❍m). The precipitation mixture should contain 2 mg gold particles suspended in 50 ❍ filter-sterilised water, 10 ❍ plasmid DNA, 50 ❍ 2.5 M $CaCl_2$ and 20 ❍ 0.1 M spermidine free base. (In order to reduce gold*

Figure 6 : Chimeric garlic plant (≅1 cm) stably expressing the *uidA* gene, produced after treating garlic buds (≅ 0.5 mm) with a particle delivery system.

clumping add $CaCl_2$ and spermidine at the same time by introducing two pipette tips containing each chemical into the tube containing the gold suspension, and by pipetting up and down several times with both pipettes simultaneously).

- *Discard the supernatant after microprojectile-DNA precipitation. Wash the particles in 150 µl absolute ethanol and resuspend them in 60 µl absolute ethanol.*
- *For each bombardment place 5 µl of microprojectile-DNA solution onto each macrocarrier (160 µg of particles and 0.83 µl of DNA per shot).*
- *Use 1550 psi rupture disks. Set macrocarrier travel distance constant in all experiments by fixing the stopping screen*

support above both spacer rings. Set chamber vacuum to 0.06 atm. (28 in Hg) before shooting.

- *Transient gene expression should be assayed 24 h after bombardment by incubating the whole bombarded tissues in X-Gluc buffer (McCabe et al., 1988). After overnight incubation at 37°C, count blue cells per explant using and inverted microscope.*

4. CONCLUSIONS AND FUTURE PROSPECTS

Garlic, as all *Allium* crops, has been considered recalcitrant for both regeneration and transformation. However, considering the important advances made in the last years in the *in vitro* regeneration systems of garlic, and the demonstration that exogenous DNA can be introduced and stably expressed in this crop, it is reasonable to expect that a protocol to efficiently produce whole transgenic garlic plants will be developed soon.

The availability of a system to produce transgenic garlic plants will have a great impact in the breeding of this sterile crop. The main applications will probably be breeding for disease resistance, especially virus resistance and breeding for quality.

The vegetative propagation of garlic throughout the years has resulted in a widespread infection with various virus mixtures all over the world. Besides nematodes (*Dytilenchus*), virus infections constitute the most important garlic diseases, in terms of yield losses. Nowadays, the only solution available is the use of "certified seed" produced via meristem culture. However, garlic plants are usually re-contaminated in the same year of planting, forcing farmers to buy new seed every year.

Virus resistant transgenic plants would constitute a permanent solution to one of the most important problems in garlic cultivation. The first example of genetically engineered virus resistance in plants was published in 1986 (Powell-Abel *et al.*, 1986). Since then, dozens of reports concerning virus resistant transgenic plants have been produced. In practically all the cases reported, the resistance genes were derived from the genome of the virus including: coat protein genes, replicase genes, movement protein genes, antisense sequences and sense defective RNAs and satellite and defective interfering RNAs (Beachy, 1997).

Fortunately, many of the viruses infecting garlic have been

characterized at the molecular level, being available a number of genes of the main families: Potyvirus (Tsuneyoshi *et al.*, 1997; Tsuneyoshi *et al.*, 1998), Potexvirus (Song *et al.*, 1995), Allexvirus (Sumi *et al.*, 1993; Yamashita *et al.*, 1996; Sumi *et al.*, 1999), Carlavirus (Tsuneyoshi *et al.*, 1998), Tobamovirus (Yamanaka *et al.*, 1998). The appropriate expression of one or several of these cloned genes in garlic will probably conduct to the production of virus resistant transgenic plants.

Although less important than virus, fungal diseases can cause important losses in garlic, depending on the state of development of the crop and the weather conditions. Main fungal diseases in garlic are caused by: *Slerotium cepivorum*, *Stemphylium vesicarium* and *Puccinia alli*.

While important progress has been made in the control of plant fungal diseases, via genetic engineering, the availability of suitable genes is limited. Some genes tested in transgenic plants that could be used for garlic breeding are those encoding for: osmotin (Liu *et al.,* 1994), chitinase (Lin *et al.*, 1995), thionin (Epple *et al.*, 1997) cationic peptides (Hancock and Lehrer, 1998) and the so-called killing proteins (Clausen *et al.*, 2000).

Another important aspect of garlic breeding that could be tackled by means of transgenic technology is quality. Garlic is used for food and has a long historical record for its use in medicine. The therapeutic effect of garlic is directly linked to its sulphur metabolism, characterised by the synthesis and storage of large amounts of S-alk(en)yl cysteine sulphoxides (CSOs) (Agarwal, 1996). The main cysteine sulphoxide found in garlic is alliin, the precursor to the main bioactive compound, allicin. In order for extraction of this bioactive compound to be economically viable, a minimum level of 1% alliin (by fresh weight) is usually requested by the pharmaceutical companies. The alteration of this metabolism in order to increase the CSOs level would produce a garlic more suitable for pharmaceutical applications. This alteration could be achieved through the introduction of an engineered copy of the alliinase gene (the enzyme which catalyses the conversion of CSOs) that has been cloned and sequenced in garlic (Rabinkov *et al.*, 1994)

REFERENCES

Abo El-Nil M (1977) Organogenesis and embryogenesis in callus of garlic (*Allium sativum* L.) *Plant Sci. Lett.,* **9** : 259-264.

Agarwal KC (1996). Therapeutic actions of garlic constituents. *Medicinal Res. Rev.,* **6** : 111-124.

Ayuso P and Pena Iglesias A (1982). The elimination of garlic viruses by thermotherapy and/or tissue culture. *Cell Biology International Rep.,* **5** : 835.

Barandiaran X, Di Pietro A and Martin J (1998) Biolistic transfer and expression of a *uidA* reporter gene in different tissues of *Allium sativum* L. *Plant Cell Rep.,* **17** : 737-741.

Barandiaran X, Martin N, Rodriguez-Conde MF, Di Pietro A and Martin J (1999). Genetic variability in callus formation and regeneration of garlic (*Allium sativum* L.). *Plant Cell Rep.,* **18** : 434-437.

Barandiaran X, Martin N, Rodriguez-Conde MF, Di Pietro A and Martin J (1999). An efficient method for callus culture and shoot regeneration of garlic (*Allium sativum* L.). *HortSci.,* **34** : 348-349.

Barandiaran X, Martin N, Rodriguez-Conde MF, Di Pietro A and Martin J (1999). An efficient method for the *in vitro* management of multiple garlic accessions. *In Vitro Cell. Dev. Biol. Plant.,* **35** : 166-169

Barrueto LP, Dieter R and Piedrabuena A (1994). Regeneration of garlic plants (*Allium sativum* L., cv. "Chonan") via cell culture in liquid medium. *In vitro Cell. Dev. Biol.,* **30** : 150-155.

Beachy RN (1997). Mechanisms and applications of pathogen-derived resistance in transgenic plants. Curr. Opin. Biotech., **8** : 215-220.

Bhojwani SS (1980) *In vitro* propagation of garlic by shoot proliferation. *Scientia Horticulturae,* **13** : 47-52.

Bhojwani SS, Cohen D and Fry PR (1982/83) Production of virus-free garlic and field performance of micropropagated plants. *Sci. Hortic.,* **18** : 39-43.

Bockish T, Saranga Y and Altman A (1997). Garlic micropropagation by somatic embryogenesis. *Hort. Biotech. In vitro Cult. Breed.* (Eds. Altman A and Ziv M) *Acta Hort.,* **447**, ISHS.

Buiteveld J, Fransz J and Creemers-Molenaar J (1994). Induction and charazterization of embryogenic callus types for the initiation of suspension cultures of leek (*Allium ampeloprasum* L.). *Plant Sci.,* **100** : 195-202.

Cabrera-Ponce JL, Lopez L, Assad-Garcia N, Medina-Arevalo C, Bailey AM and Herrera-Estrella L (1997). An efficient particle bombardment system for the genetic transformation of asparagus (*Asparagus officinalis* L.). *Plant Cell Rep.,* **16** : 255-260.

Chan MT, Lee TM and Chang HH (1992). Transformation of indica rice *(Oryza sativa* L.) mediated by *Agrobacterium tumefaciens. Plant Cell Physiol.,* **33** : 577-583.

Christou P (1995). Strategies for variety-independent genetic transformation of important cereals, legumes and woody species utilizing particle bombardment. *Euphytica,* **85** : 13-27.

Clausen M, Krauter R, Schachermayr G, Potrykus I and Sautter C (2000). Antifungal activity of a virally encoded gene in transgenic wheat. *Nature Biotech.,* **18** : 446-449

Conci VC and Nome SF (1991) Virus free garlic (*Allium sativum* L.) plants obtained by thermotherapy and meristem tip culture. *J. Phytopathol.,* **132** : 186-192.

Deng W, Lin X and Shao Q (1988). Transformation of some cereal crops with *Agrobacterium tumefaciens. Gene. Manipul. Crops Newsl.,* **4** : 1-2.

Dolezel J and Novak F (1984). Karyological and cytophotometric study of callus induction in *Allium sativum* L. *J. Plant Physiol.,* **118** : 421-429.

Dunstan DI and Short KC (1977). Improved growth of tissue cultures of the onion (*Allium cepa* L.). *Physiol Plant.,* **41** : 70-72.

Eady CC (1995). Towards the transformation of onions (*Allium cepa*). *NZJ. of Crop Hort. Sci.,* **23** : 239-250.

Eady CC, Butler RC and Suo Y (1998). Somatic embryogenesis and plant regeneration from immature embryo cultures of onion (*Allium cepa* L.). *Plant Cell Rep.,* **18** : 111-116.

Eady CC and Lister CE (1998). A comparison of four selective agents for use with *Allium cepa* L. immature embryos and immature embryo-derived cultures. *Plant Cell Rep.,* **18** : 117-121

Eady CC, Weld RJ and Lister CE (2000). *Agrobacterium tumefaciens*-mediated transformation and transgenic plant regeneration of onion (*Allium cepa* L.). *Plant Cell Rep.,* **19** : 376-381

Epple P, Apel K and Bohlmann H (1997). Overexpression of an endogenous thionin enhances resistance of *Arabidopsis* against *Fusarium oxysporum. Plant Cell,* : 509-520

FAOSTAT. (1999). http://apps.fao.org/

Fromm ME, Taylor LP and Walbot V (1985). Expression of genes transferred into monocot and dicot plant cells by electroporation. *Proc. Natl. Acad. Sci. USA,* **82** : 5824-5828.

Gamborg OL, Miller RA and Ojima K (1968). Nutrient requirements of suspension cultures of soybean root cells, *Exp. Cell. Res.,* **50** : 151.

Graves ACF and Goldman SL (1986). The transformation of *Zea mays* seedlings with *Agrobacterium tumefaciens. Plant Mol. Biol.,* **7** : 43-50.

Gould J, Devey M, Hasegawa O, Ulian EC, Peterson G and Smith RH (1991). Transformation of *Zea mays* L. using *Agrobacterium tumefaciens* and the shoot tip. *Plant Physiol.,* **95** : 426-434.

Haque MS, Wada T and Hattori K (1997). High frequency shoot regeneration and plantlet formation from root tip of garlic. *Plant Cell Tiss. Org. Cult.,* **50** : 83-89.

Haque MS, Wada T and Hattori K (1998). Efficient plant regeneration in garlic through somatic embryogenesis from root tip explants. *Plant Prod. Sci.,* **1** : 216-222.

Haque MS, Wada T and Hattori K (1999). Anatomical changes during *in vitro* direct formation of shoot bud from root tips in garlic (*Allium sativum* L.). *Plant Prod. Sci.,* **2** : 146-153.

Hancock REW and Lehrer R (1998). Cationic peptides, a new source of antibiotics. *TIBTECH.,* **16** : 82-88.

Hauptmann RM, Vasil V, Ozias-Akins P, Tabaeizadeh Z, Rogers SG, Fraley RT, Horsch RB and Vasil IK (1988). Evaluation of selectable markers for obtaining stable transformants in the graminae. *Plant Physiol.,* **86** : 602-606.

Havel L and Novak FJ (1988). Regulation of somatic embryogenesis and organogenesis in *Allium carinatum* L. *J. Plant Physiol.,* **132** : 373-377.

Havranek P and Novak FJ (1973). The bud formation in callus cultures of *Allium sativum* L. *Z. Pflanzenphysiol. Bd.,* **68** : 308-318.

Hong and Debergh (1995). Somatic embryogenesis and plant regeneration in garden leek. *Plant Cell Tiss. Org. Cult.,* **43** : 21-28.

Kamo K, Blowers A, Smith F, Van Eck J and Lawson R (1995). Stable transformation of *Gladiolus* using suspension cells and callus. *J. Amer. Soc. Hort. Sci.,* **120** : 347-352.

Kamo K, Blowers A, Smith F and Van Eck J (1995). Stable transformation of *Gladiolus* by particle gun bombardment of cormels. *Plant Sci.,* **110** : 105-111.

Kehr AE and Schaeffer GW (1976). Tissue culture and differentiation of garlic. *HortSci.,* **11** : 422-423.

Koch M, Tanami Z and Salomon R (1995). Improved regeneration of shoots from garlic callus. *HortSci.,* **30** : 378.

Koul S, Sambyal M and Grewal S (1994). Embryogenesis and plantlet formation in garlic (*Allium sativum* L.) *Spices Aromatic Crops,* **3** : 43-47.

Lin W, Anuratha CS, Datta K, Potrykus I, Muthukrishnan S and Datta SK (1995). Genetic-Engineering of Rice for Resistance to Sheath Blight. *Bio/Technology,* **13** : 686- 691.

Liu D, Kashchandra G, Raghothama G, Hasegawa PM and Bressan RA (1994). Osmotin overexpression in potato delays development of disease symptoms. *Proc. Natl. Acad. Sci. USA.,* **91** : 1888-1892

Ma Y, Wang HL, Zhang CJ and Kang YQ (1994) High rate of virus-free plantlet regeneration via garlic scape-tip culture. *Plant Cell Rep.,* **14** : 65-68.

Maggioni L, Astley D, Rabinowitch H, Keller J and Lipman E (1997). Report of a working group of *Allium. European Cooperative Program for Crop Genetic Resources Networks* (ECP/GF). Sixth meeting 23-25 October 1997. Plovdiv, Bulgaria.

Matsubara S and Chen D (1989) *In vitro* production of garlic plants and field acclimatization. *HortSci.,* **24** : 677-679.

Matsuda Y and Adachi T (1996). Plant regeneration via embryogenesis in commercial cultivars of Chinese chive (*Allium tuberosum* Rottl.). *Plant Sci.,* **119** : 149-156.

Matsuda Y, Hoffmann F and Adachi T (1998). Embryogenic callus formation from protoplasts of Chinese chive (*Allium tuberosum* Rottl.). *J. Breed. Genet.,* **30** : 91-96.

Mohamed-Yasseen Y, Splittstoesser WE and Litz RE (1994) *In vitro* shoot proliferation and production of sets from garlic and shallot. *Plant Cell Tiss. Org. Cult.,* **36** : 243-247.

Moriconi DN, Conci V and Nome SF (1990) Rapid multiplication of garlic (*Allium sativum* L.) *in vitro. Phyton.,* **51** :145-151.

Murashige T and Skoog F (1962). A revised medium for rapid growth and bioassays with tobacco tissue cultures. *Physiol. Plant.,* **15** : 473-479.

Myers M and Simon P (1998). Continuous callus production and regeneration of garlic (*Allium sativum* L) using root segments from shoot tip-derived plantlets. *Plant Cell Rep.,* **17** : 726-730

Myers M and Simon P (1999). Regeneration of garlic callus as affected by clonal variation plant growth regulators and culture conditions over time. *Plant Cell Rep.,* **19** : 32-36.

Nagakubo T, Nagasawa A and Ohkawa H (1993) Micropropagation of garlic through *in vitro* bulblet formation. *Plant Cell Tiss. Org. Cult.,* **32** : 175-183.

Nagasawa A and Finer J (1988a). Induction of morphogenic callus cultures from leaf tissue of garlic. *HortSci.,* **23** : 1068-1070.

Nagasawa A and Finer J (1988b). Development of morphogenic suspension cultures of garlic (*Allium sativum* L.) *Plant Cell Tiss. Org. Cult.,* **15** : 183-187.

Powell-Abel P, Nelson RS, De B, Hoffman N, Rogers SG, Fraley RT and Beachy RN (1986). Dealay of disease development in transgenic plants that express the tobacco mosaic virus coat protein gene. *Science,* **232** : 738-743.

Rabinkov A, Zhu X, Grafi G, Galili G and Mirelman D (1994). Allin lyase (Allinase) from garlic (*Allium sativum*). Biochemical characterization and cDNA cloning. *Appl. Biochem. Biotech.,* **48** : 149-170

Raineri DM, Bottino P, Gordon MP and Nester EW (1990). *Agrobacterium*-mediated transformation of rice (*Oryza sativa* L.). *Bio/Technology,* **8** : 33-38.

Rauber M and Grunewaldt J (1988). *In vitro* regeneration in *Allium* species. *Plant Cell Rep.,* **7** : 426-429.

Ravnikar M, Zel J, Plaper I and Spacapan A (1993) Jasmonic acid stimulates shoot and bulb formation of garlic *in vitro. Plant Growth Regul.,* **12** : 73-77.

Saker MM (1997). *In vitro* regeneration of onion through repetitive somatic embryogenesis. *Biol. Plant.,* **40** : 499-506.

Seabrook JEA (1993) *In vitro* propagation and bulb formation of garlic. *Can. J. Plant Sci.,* 155-158.

Shain EA and Kaneko K (1986). Somatic embryogenesis and plantlet regeneration from callus cultures of non bulbing onions. *HortSci.,* **21** : 294.

Silvertand B, Van Rooyen A, Lavrijsen P, Van Harteb AM and Jacobsen E (1996). Plant regeneration via organogenesis and somatic embryogenesis in callus cultures derived from mature zygotic embryos of leek (*Allium ampeloprasum* L.). *Euphytica,* **91** : 261-270.

Song JT, Choi JN, Song SI, Lee JS and Choi YD (1995). Identification of a potexvirus in Korean garlic plants. *J. Korean Agric. Chem. Soc.,* **38** : 55-62.

Suh S and Park H (1988). Somatic embryogenesis and plant regeneration from flower organ culture of garlic (*Allium sativum* L.). *Korean J. Plant Tissue Cult.,* **15** : 121-132.

Sumi SI, Tsuneyoshi T and Furutani H (1993). Novel rod-shaped viruses isolated from garlic, *Allium sativum*, possessing a unique genome organization. *J. Gen Virol.,* **74** : 1879-1885.

Sumi SI, Matsumi T and Tsuneyoshi T (1999). Complete nucleotide sequences of garlic viruses A and C, members of the newly ratified genus Allexvirus. *Arch. Virol.,* **144** : 1819-1826.

Tsuneyoshi T, Ikeda Y and Sumi SI (1997). Nucleotide sequences of the 3' terminal region of onion yellow dwarf virus isolates from *Allium* plants in Japan. *Virus Genes.,* **15** : 73-77.

Tsuneyoshi T, Matsumi T, Deng TC, Sako I and Sumi SI (1998). Differentiation of *Allium* carlaviruses isolated from different parts of the world based on the viral coat protein sequence. *Arch. Virol.,* **143** : 1093-1107.

Tsuneyoshi T, Matsumi T, Natsuaki KT and Sumi SI (1998). Nucleotide sequences analysis of virus isolates indicates the presence of three potyvirus species in *Allium* plants. *Arch. Virol.,* **143** : 97-113.

Van Der Valk P, Scholten OE, Verstappen F, Jansen RC and Dons JJM (1992). High frequency somatic embryogenesis and plant regeneration from zygotic embryo-derived callus cultures of three *Allium* species. *Plant Cell Tiss. Org. Cult.,* **30** : 181-191

Vasil IK (1993). Molecular genetic improvement of cereal and grass crops. IAPTC Newsletter. December 1993.

Verbeek M, Van Dijk P and Van Well PMA (1995) Efficiency of eradication of four viruses from garlic (*Allium sativum* L.) by meristem-tip culture. *Eur. J. Plant Pathol.,* **101** : 231-239.

Watad AA, Yun DJ, Matsumoto T, Niu X, Wu Y, Kononowicz AK, Bressan RA and Hasegawa PM (1997). Microprojectile bombardment-mediated transformation of *Lilium longiflorum. Plant Cell Rep.,* **17** : 262-267.

Wilmink A, Van De Ven BCE and Dons JJM (1995). Activity of Constitutive promoters in various species from, the Liliaceae. *Plant Mol. Biol.,* **28** : 949-955

Xue H, Araki H, Shi L and Yakuma T (1991). Somatic embryogenesis and plant regeneration in basal plate and receptacle derived-callus cultures of garlic (*Allium sativum* L.) *J. Jpn. Soc. Hort. Sci.,* **60** : 627-634

Yamanaka T, Komatani H, Meshi T, Naito S, Ishikawa M and Ohno T (1999). Complete nucleotide sequence of the genomic RNA of tobacco mosaic virus strain Cg. *Virus Genes,* **16** : 173-176.

Yamashita K, Sakai J and Hanada K (1996). Characterization of a new virus from garlic (*Allium sativum* L.), garlic mite-borne mosaic virus. *Ann. Phytopath. Soc. Japan,* **62** : 483-489.

Chapter 8

TRANSGENIC CUCUMBER PLANT

Yutaka Tabei*

National Institute of Agrobiological Sciences, Kan-nondai, Tsukuba, Japan 305-8602

Summary

Transgenic cucumber were produced by Agrobacterium-mediated and microprojected –mediation. Since regeneration system is one of crucial point for efficient production of transgenic cucumber plants introduced, it focused on regeneration condition for production of transgenic cucumber plants in this chapter. Under the present condition, building public acceptance is important for utilization of genetically modified organisms, brief history and world trend for risk assessment are introduced.

Keywords : Cucumber, embryogenesis, organogenesis, risk assessment, transformation

Abbreviations : NAA - naphthaleneacetic acid, BA - benzyladenine, 2,4-D - 2,4-dichlorophenozy-acetic acid, Z - zeatin, ABA - abscisic acid.

1. INTRODUCTION

Cucumber (*Cucumis sativus* L.) is one of the most important vegetables in many countries. The total amount of production was about 10,757 tons per year in the world, and the cultivation area was about 817 thousands ha per year in the world (Esquinas-Alcazar and Gulick, 1983). The most limiting factor for the field cultivation of cucumber is yield instability. The yield instability is caused by biotic pathogens such as

*Corresponding author : E-mail : tabei@affrc.go.jp

Fusarium wilt, angular leaf spot, gray mold, cucumber mosaic virus (CMV) and Zucchini yellow mosaic virus (ZYMV) as well as by abiotic stress such as low temperature and drought. Although some resistant breeding materials against these stress factors can be found in wild *Cucumis*, application of conventional plant breeding method (*e.g.* cross breeding) is strongly limited due to interspecific incompatibility among *Cucumis sativus* L. (2n = 14) and most wild *Cucumis* (2n = 24) except *C. hardwickii* (2n = 14) (Deakin 1971). Somatic hybridization seems to be used to settle this problem. However, Trusulon and Shahin (1986) and Orczyk *et al.* (1988) reported the regeneration of whole cucumber plants from cucumber protoplasts, application of somatic hybridization might be prevented due to inefficient protoplast culture system and great morpholic changes of hybrid between cucumber and wild *Cucumis*. The recent advances in plant biotechnology would especially be welcomed in improvement of this economically important crop.

Cucumber had been thought that one of the most recalcitrant plants for tissue culture two decades ago. And the difficulty of plant regeneration in tissue culture has been the major limiting factor for the advancement of genetic engineering. However, from the late 1980's, several groups have reported the regeneration of whole plants from cucumber tissue and cell culture via adventitious shoots formation (Chee and Tricoli, 1988; Tabei *et al.*, 1989; Orczyk *et al.*, 1988; Cade *et al.*, 1990; Chee, 1990a; Gambley and Dodd, 1990; Punja *et al.*, 1990; Raharjo and Punja, 1994; Bruza and Malepszy, 1995) or via somatic embryogenesis (review, Debeaiujon and Branchard, 1993). These efficient regeneration systems from tissue and cell culture have been contributed to the development of producing transgenic cucumber. Transformation technology could be used to protect cucumber against biotic and abiotic stresses. All of these improvements depend upon the existence of a genetic transformation system. This technology is now developing and the purpose of this chapter is to review published transformation systems as well as to present our data on the production of transgenic cucumber.

2. GENERAL TRANSFORMATION STRATEGIES OF CUCUMBER

The importance and enormous potential applications of cucumber transformation led to active research both at academic and industrial laboratories worldwide.

2.1. General review

Production of transgenic plants depend on three major factors which are 1) efficient introduction of foreign gene, 2) establishment of efficient regeneration system, 3) reliable and efficient selection of transformed callus or regenerated plants. Since *Cucurbit* including cucumber is infected with *Agrobacterium* (Smarrelli *et al.*, 1986), many researchers introduced foreign genes by *Agrobacterium*-mediated method (Table 1). Moreover, in this transformation system, an intact DNA segment can be integrated in to the plant genome at low copy numbers (Hansen *et al.*, 1994). On the other hand, two research group used microprojectile-mediated transformation method (Chee and Slightom, 1992; Schulze *et al.*, 1995). Recently, efficient regeneration systems for shoot organogenesis and embryogenesis were reported as mentioned above. For selection of transgenic callus or plants, kanamycin and hygromycin resistant genes have been used as selection markers (Table 1). Each research group has used different selection conditions by traits of tissues and culture conditions.

2.2. *Agrobacterium*-mediated transformation using petiole, hypocotyl and leaf via somatic embryogenesis and/or shoot organogenesis

Recently, several research groups reported plant regeneration from several tissues in cucumber (Chee, 1990a; Punja *et al.*, 1990; Burza and Malepszy, 1995). Based on these regeneration systems via somatic embryogenesis and adventitious organogenesis, some research groups reported production of transgenic cucumber plants from petiole and leaf (Sarmento *et al.*, 1992), petiole (Raharjo *et al.*, 1996), hypocotyls (Nishibayashi *et al.*, 1996) and leaf (Szwacka *et al.*, 1996).

Sarmento *et al.* (1992) developed transgenic cucumber plants using regeneration system reported by Punja *et al.* (1990). Raharjo *et al.* (1996) also used the protocol reported by Punja *et al.* (1990) as well as by Raharjo and Punja (1994). Punja *et al.* (1990) investigated several factors for regeneration efficiency. They investigated several combinations of phytophormones and their concentrations, regeneration ability of various tissues (petiole, cotyledon and leaf) and three varieties. They concluded that the highest frequency of plant formation occurred with petiole cultured on NAA and BA, NAA and Z or 2,4-D and BA. The range of regeneration frequency for cotyledon, leaf and petiole were 0-38, 0-75 and 14-96%, respectively.

Table 1. Summary of various studies for production of transgenic cucumber

Explant	Introduction method	Integrated gene	Selection (concentration)	Reference
hypocotyl	*A. rhizogenus*	*nptII*	kanamycin (25 mg/l)	Trulson and Shahin (1986)
cotyledon	*A. tumefaciens* (C58Z707)	*nptII*	kanamycin (100 mg/l)	Chee *et al.* (1990)
cotyledon	*A. tumefaciens* (C58Z707)	*nptII*, *uidA* CP of CMV	kanamycin (100 mg/l)	Chee and Slightom (1991)
embryogenic cell	particle bombardment	*nptII*	kanamycin (50~100 mg/l)	Chee *et al.* (1992)
petiole, leaf	*A. tumefaciens* (LBA4404)	*nptII*	kanamycin (75 mg/l)	Sarmento *et al.* (1992)
cotyledon of mature seed	*A. tumefaciens* (GV3101)	*hpt*	hygromycin (20 mg/l)	Tabei *et al.* (1994)
embryogenic cell	particle bombardment	*uidA*, *nptII*	kanamycin (50~100 mg/l)	Schulze *et al.* (1995)
petiole	*A. tumefaciens* (LBA4404)	Chitinase gene from petunia, tobacco and bean *nptII*	kanamycin (50 mg/l)	Raharjo *et al.* (1996)
hypocotyl	*A. tumefaciens* (EHA101)	*hpt*, *nptII*, *uidA*	kanamycin (50~100 mg/l) hygromycin (20, 25 and 30 mg/l)	Nishibayashi *et al.* (1996)

Table 1. Continued

Explant	Introduction method	Integrated gene	Selection (concentration)	Reference
leaf	*A. tumefaciens* (LBA4404)	thaumatin II cDNA, *nptII*	kanamycin (100~200 mg/l)	Szwacka *et al.* (1996)
cotyledon of mature seed	*A. tumefaciens* (LBA4404)	Chitinase gene from rice, *nptII*	kanamycin (25~100 mg/l)	Tabei *et al.* (1998)
cotyledon of mature seed	*A. tumefaciens* (LBA4404)	Coat protein of ZYMV, *nptII*	kanamycin (25~100 mg/l)	Wako *et al.* (2001)

uidA (GUS) : ß-glucronidase, *nptII* : neomycin phosphotransferase, *hpt* : hygromycin phosphotransferase, CP : coat protein gene

Sarmento *et al.* (1992) used petiole and leaf as explants cultured on medium supplemented with NAA/BA, NAA/Z or 2,4-D/BA according to the procedure by Punja (1990). This group investigated the influence of acetosyringone, utilization of tobacco feeder layer, size of explants, exposure period of *Agrobacterium,* concentration of *Agrobacterium* and co-cultivation period on production efficiency of transgenic cucumber. They proved the optimal procedure involved exposing segments of petiole (4-6 mm) or leaf (0.5 cm^2) segments to a bacterial suspension (10^8 cells/ml) containing 20 μM acetosyringone for 5 min, followed by a 48h co-cultivation period on a tobacco feeder layer. Since addition of acetosyringone slightly increased the frequency of antibiotic resistant callus formation, they recommended utilization of acetosyringone. But the difference of effect between with and without acetosyringone was not significant. Simultaneously, under these conditions, they recovered 23 individual plants and 21 out of 23 plant exhibited *nptII* positive. Raharjo *et al.* (1996) investigated the effect of different *Agrobacterium* strain on transformation frequency and introduced chitinase genes from petunia, tobacco and bean into cucumber. The frequency of embryogenic calli that developed further and grew on kanamycin-containing medium ranged from 0 to 12%. The highest frequency (12%) was achieved when suprevirulant strain (EHA105) was used. Other strains yielded a maximum frequency of around 5 %. They pointed out that suspension culture was useful for multiplication of embryogenic aggregate and for production of cell clumps capable of shoot formation. Protein extracts from the transgenic plants showed varying but enhanced level of chitinase activity in leaves compared to the untransformed cucumber plant. From results for response to inoculation with *Alternaria cucumerina*, *Botrytis cinerea*, *Colletotrichum lagenarium* and *Rhizoctonia solani*, no difference in disease development were detected between transgenic and nontransgenic cucumber plants (Punja and Raharjo 1996). Nishibayashi *et al.* (1996) compared the effects of kanamycin and hygromycin on the selection of transgenic cucumber. In all the reports for transformation of cucumber except Tabei *et al.* (1994), kanamycin is used as selectable antibiotic. Although kanamycin allows to grow callus of explants uninfected *Agrobacterium*, hygromycin is more effective to suppress callus formation from uninfected explants (Nishibayashi *et al.* 1996). In this report, acetosyringone treatment (100 μM) during co-cultivation was effective in enhancing the efficiency of transformation at the cut surface cells of hypocotyls. Szwacka *et al.* (1996) introduced thaumatin II cDNA and *nptII* genes into cucumber using regeneration system reported by Burza and Malepszy (1995).

Selection was carried out by two ways; one is to culture explants infected on medium with 100mg/l kanamycin after co-cultivation and the other is to culture explants in liquid medium with 100 mg/l kanamycin overnight and then transferred them on the same solid medium. This report described that *nptII* gene could be detected by PCR, although the presence of thaumatin II gene was detected.

2.3. *Agrobacterium*-mediated transformation using plant regeneration system via somatic embryogenesis

Chee (1990b) reported that the production of transgenic cucumber by *Agrobacterium*-mediated transformation and the recovery of fertile plants via somatic embryogenesis. Cotyledons of young seedling were used as explants in this regeneration system (Chee, 1990a). Cucumber cotyledon pieces were cultured on the initiation medium supplemented with various concentrations (25–200 mg/L) of kanamycin. One-hundred mg/l kanamycin made possible the selection of putative transformed cells and transgenic cucumber plants. Transgenic plantlets were visually identified by their ability to form adventitious roots in MS medium (Murashige and Skoog, 1962) with 50 mg/L kanamycin. Finally, more than 100 kanamycin resistant cucumber plants were obtained and all of them exhibited *nptII* activity. Seventy-five percent of the progenies obtained from four randomly selected R_0 cucumber lines indicated expressing *nptII* activities. These data proved that *nptII* gene was integrated in cucumber genome (Chee, 1990b). This transformation system was applied to introduction of coat protein gene of CMV (Chee and Slightom, 1991), and resistance to CMV was evaluated in test filed from 1989 to 1991 (Gonsalves *et al.*, 1992). Four to 65% of transgenic cucumber plants were infected with CMV when 84% of non-transgenic Poinsett 97 was infected with CMV.

2.4. *Agrobacterium*-mediated transformation using cotyledon of mature seed and selection of transgenic callus in suspension culture

Various tissues have been used as *in vitro* manipulation in cucumber. Although plant regeneration frequency from cotyledon of seedling was not high, petiole was good matcrial for tissue culture for plant recover (Punja *et al.*, 1990). In melon, which belongs to the same *Cucumis* genus as cucumber, petiole also exhibited comparatively higher ability of shoot organogenesis and somatic embryogenesis, and cotyledon of mature seed had much higher competence for plant regeneration via shoot organogenesis and embryogenesis than cotyledon of seedling, hypocotyls, petiole and leaf (Tabei *et al.*, 1991). The growth of regenerated shoots was often inhibited gradually during subculture on solid media containing kanamycin. However introduced neomycin phosphotransferase II (*nptII*) gene were detected by polymerase chain reaction (PCR) from explants of those regenerated shoots. From this result, it was speculated that regenerated shoots consisted of transformed cells and non-transformed cell (chimera). To establish reliable selection conditions and transformation system using cotyledon of mature seed, the effects of three antibiotics and culture methods were investigated (Tabei *et al.*, 1994). When uninfected cotyledons of mature seeds were cultured on solid medium or in liquid medium with different concentrations of three antibiotics (100-300 mg/l kanamycin, 20-80 mg/L geneticin and 20-80 mg/L hygromycin). Kanamycin could not completely suppress callus proliferation from explants on solid and in liquid medium. This result suggested cucumber has partially resistant to kanamycin. On the other hand, callus proliferation and greening of explants were suppressed well in liquid medium with geneticin and hygromycin. Liquid culture is better procedure than solid medium culture for selection of transgenic callus because antibiotic can contact whole surface of explants. Although geneticin excessively suppressed callus formation and inhibited callus formation from *Agrobacterium*-infected explants, combination of liquid culture and hygromycin selection seemed to be reliable selection method for transgenic callus regeneration from cotyledon of mature seeds. Many hygromycin resistant calluses were induced by this procedure and six putative transgenic cucumber plants regenerated via somatic embryogenesis. PCR-Southern hybridization confirmed the presence of transgenes in all the regenerated plants, and this procedure did not permit the regeneration of escaped plants. This transformation procedure made possible reliable selection. However, frequency of plant regeneration should be improved though plant recovery tends to be very difficult from completely dedifferentiated callus in cucumber.

2.5. *Agrobacterium*-mediated transformation using cotyledon of mature seed and plant regeneration via shoot organogenesis

Novel regeneration system by direct shoot organogenesis was developed and this regeneration system was applied to production of transformant. To improve the efficiency of plant regeneration itself and production of transgenic cucumber, effect of abscisic acid on induction of multiple shoots was also examined. Addition of 1 mg/L ABA into regeneration medium enhanced regeneration frequency and induction of multiple shoots (Tabei *et al*., 1998). Based on novel regeneration system, Chitinase gene derived from rice and coat protein gene from ZYMV were transferred into cucumber by Tabei *et al*. (1998) and Wako *et al*. (2001), respectively. Rice chitinase gene was cloned (Nishizawa and Hibi 1991) and vector, pBI121-RCC2, were constructed. This vector integrated in *Agrobacterium* LBA4404 by tri-parental mating. After co-cultivation, both explants infected and uninfected by *Agrobacterium* cultured on regeneration medium with 25 mg/L kanamycin and 200 mg/L Claforan. Multiple shoots transferred on MS medium with higher concentration (100 mg/L) of kanamycin. The selective antibiotic gradually slowed the growth of adventitious shoots from uninfected explants and escaped shoots from infected explants. They were also gradually bleached. Crucial point of this transformation system was to culture lateral branch at several times to reduce the degree of chimera in regenerated shoots. After confirmation of existence of introduced gene, leaves of transgenic cucumber were placed on 1.2% agar disk containing conidia of *Botrytis cinerea* and incubated at 20°C. Three of 60 transgenic cucumber plants tested exhibited high level of resistance against *B. cineria* (Fig. 1).

Wako *et al*. (2001) introduced coat protein gene of ZYMV and developed ZYMV resistant cucumber using transformation procedure by Tabei *et al*. (1998). As a result of transformation, total 26 transgenic cucumber plants were regenerated from 4619 explants infected. Some transformants showed a delay of 2-6 day in symptom development and the disease severity were significantly lower than non-transgenic cucumber plant. Especially, one of 26 transformants exhibited only mild mosaic symptom and showing high level of ZYMV resistance (Fig. 2).

2.6. Microprojectile-mediated transformation using embryogenic cell

The majority of transgenic plants including cucumber have been obtained using the *Agrobacterium*-mediated DNA transfer system (Uchimiya *et al*., 1988). However, this transfer system can drastically reduce the efficiency of plant regeneration due to toxicities associated with the *Agrobacteria* or the high concentrations of antibiotics (Weising *et al*., 1988). Consequently, it seemed that production of transgenic cucumber by *Agrobacterium*-mediated was not efficient. For these reasons, direct DNA transfer technique was applied to transformation of cucumber (Chee and Slightom, 1992). And there are only a few reports in transformation of cucumber to date (Schulze *et al*., 1995). In both reports, embryogenic callus in suspension culture was used. Chee and

Slightom (1992) described that embryogenic callus was not only the preferred target tissue in cucumber but other species such as papaya (Fitch *et al.* 1990) and cotton (Finer and McCullen, 1990). A total of

Figure 1 : Comparison of resistance to gray mold (*Botrytis cinerea*) by artificial infection onto leaf in transgenic cucumber. C1 and C2 are non-transgenic cucumber plants of which varieties are 'Sagamihanjiro' and 'Shimoshirazu', respectively. 33, 34 and 37 indicate transgenic cucumber with foreign chitinase gene, 33 (high resistance), 34 (middle resistance) and 37 (susceptible).

107 independently regenerated cucumber, and 17 (16%) of them were transformed with a portion of the Nos-*nptII* coding region. However, only four of 17 transgenic cucumber plants expressed introduced NPT-II gene. The four plants contained a single copy of Nos-*nptII*, though most of remaining 13 plants contained multiple copies. If multiple copies of introduced gene are present, the gene is inactivated by methylation of the introduced gene (Matzke *et al.*, 1989). In case of the suppression of the introduced gene, they speculated that Nos promoter was subject to methylation. Thus, even though microprojectile-mediated DNA transfer method is very effective and does not adversely affect plant regeneration, improvements are needed to reduce the integration of multiple copies (Chee and Slightom, 1992).

On the contrary, Schulze *et al.* (1995) also transferred *nptII* and *uidA* genes into embryogenic suspension callus by microprojectile-mediation and produced 28 vigorous transgenic cucumber plants. All selected plants were proved to be *nptII* positive and no escapes could

Figure 2 : Comparison of Zucchini yellow mosaic virus (ZYMV) resistance. Left : T_1 transgenic cucumber plant, Right: Non-transgenic cucumber plant

be detected. Co-integration efficiency of the linked unselectable marker (*uidA*) gene was 67% and all transplanted cucumber plants showed normal fruit development. Although transplanted cucumber plant showed rich flower development and possessed vital pollen, the harvested fruits did not contain any seed while plants regenerated from non-bombarded suspensions were fully fertile. With regard to the microprojectile-mediation, sterile plants have been described in maize (Fromm *et al.*, 1990) and wheat (Vasil *et al.*, 1992). Moreover, in regeneration procedure, the relatively long *in vitro* phase, stress by shooting and using high selection pressure by kanamycin seem to be obstacles in getting fertile plants. Although transformation by microprojectile-mediated can get high number of transgenic cucumber plant (four per bombardment), shortening the culture period is necessary to establish embryogenic suspension culture for getting fully fertile plants.

3. RISK ASSESSMENTS OF GENETICALLY MODIFIED ORGANISMS (GMO)

First success of transformation of *E. coli* was reported in 1973 (Cohen *et al.*, 1973). In 1975, Ashiroma conference was held at Ashiroma in California to get consensus for control of GMO. Conclusion of the conference recommended for matching types of containment with types of experiments under self-imposed control of GMO (Berg *et al.*, 1975). Then Organization for Economic Co-operation and Development (OECD) started discussion for risk assessment of GMO in industrial use. OECD worked out basic concepts 'substantial equivalence' as food safety evaluation (OECD, 1994) and 'familiarity' as environmental risk assessment (OECD, 1993). Environmental risk is evaluated on a 'step by step' and 'case by case' basis. 'Cartagena protocol on biosafety' in biodiversity was adopted on January 2000 and the protocol is opened for signature and ratifications. This protocol provides environmental risk assessment to each country when exporting and importing GMO (http://www.biodiiv.org/biosafety/iccp.asp.). On the other hand, Ad Hoc Intergovernmental Task Force on Foods Derived from Biotechnology in Codex (Food Associated Organization (FAO)/World Health Organization (WHO)) started discussion to make two major texts which are a set of broad general principle for risk analysis of foods derived from biotechnology, and specific guidance on the risk assessment of foods derived from biotechnology (Codex, http://www.codexalimentarious.net/Reorts.htm). Under the present condition, as importance for risk assessment will be steadily gaining ground among people, there is a growing demand for reliable risk assessment methods.

Regarding to risk assessment of transgenic cucumber plants, virus resistant cucumber (Gonsalves *et al.*, 1992) had been evaluated their resistance in the field trial. Gray mold resistant cucumber (Tabei *et al.*, 1998) was assessed environmental risk. The cucumber got official approval from Minister of Agriculture, Forestry and Fisheries, and was allowed to cultivate in open field in 1999 (http://www.s.affrc.go.jp/docs/sentan/eguide/edevelp.htm). However, any transgenic cucumber plants had not gotten to all non-regulated status required for commercialization (OECD biotrack online).

4. CONCLUSIONS AND FUTURE PROSPECTS

Regarding to development of transformation system, investigation of regeneration system, introduction method of foreign gene and selection condition of transformants are essential.

Fortunately, several groups in the world started to develop efficient regeneration system via shoot organogenesis and embryogenesis from various tissues. Especially shoot organogenesis from petiole (Punja *et al.*, 1990), shoot organogenesis from cotyledon of mature seed (Tabei *et al.*, 1998) and embryogenesis from cotyledon and young leaf (Chee, 1990b; Burza and Malepszy, 1995) were often used for transformation of cucumber.

Since cucumber is sensitive to *Agrobacterium*-infection, foreign genes were introduced into cucumber by *Agrobacterium*-mediated in most reports. But Chee (1990) and Schulze *et al.* (1995) produced transgenic cucumber by microprojectile transformation method. Although it was reported that *Agrobacterium*-mediated transformation often reduces the efficiency of plant regeneration due to high concentrations of antibiotics (Weising, 1988), the efficiency seemed to be improved by the investigation of the regeneration and selection conditions. On the contrary microprojectile transformation did not adversely affect plant regeneration (Chee, 1990; Schulze *et al.*, 1995), but it was serious problem that multiple copies of gene transferred by microprojectile transformation suppress the expression of the target gene (Christou *et al.*, 1989; Chee, 1990).

Kanamycin selection is the most frequently used in cucumber transformation and the range of kanamycin concentration is mainly 50 – 100 mg/L. Nishibayashi *et al.* (1996) compared effect of kanamycin and hygromycin, and they conclude that hygromycin was more effective to eliminate escapes. Tabei *et al.* (1994) also reported that hygromycin was preferable antibiotic for selection if escape must be completely suppressed from cotyledon of mature seed. However, the total number of transgenic cucumber plants depend on such conditions as selective agents (concentrations), type of explants and regeneration culture systems. Since high level of selection pressure by antibiotics may reduce regeneration frequency, and sensitivity to antibiotics vary from the type of tissues, it should be considered to maintain balance between regeneration frequency and generation of escapes.

Until now, several research groups tried to introduce useful genes into cucumber. Coat protein gene of CMV was transferred (Chee and Slightom, 1991) and Gonsalves *et al.* (1992) evaluated CMV resistance of transgenic cucumber plants harboring CMV coat protein gene in the field. Wako *et al.* (2001) also introduced coat protein gene of ZYMV and confirmed one transgenic cucumber plants exhibited obviously high resistance to ZYMV. Chitinase genes were also used to transform cucumber (Raharja *et al.*, 1996 and Tabei *et al.*, 1998). Raharja *et al.* (1996) confirmed introduced chitinase genes and their expression in cucumber plants. However, there is no difference for development of three fungal diseases between transgenic and nontransgenic cucumber plants (Punja and Raharjo, 1996). On the contrary, three transgenic cucumber plants harboring rice chitinase gene exhibited high level of resistance to gray mold disease (Tabei *et al.*, 1998). Since gray mold resistant cucumber varieties have not been found in genetic resources, this cucumber may be useful for disease resistant breeding. Szwacka *et al.* (1996) transferred thaumatin II cDNA and just confirmed the existence of the gene in cucumber plants.

For last decade, many results of researches have reported not only transformation of cucumber but also regeneration system in cucumber. Many factors influence success of transformation, and one of crucial points for transformation is the establishment of reliable and efficient regeneration system.

As transformation technology can manipulate gene itself and add novel characteristics into target crops, it means that broad range of living species can be utilized as breeding material. We should make the best use of this technology to improve crops, and also consider food and environmental safety of GMO to promote building public acceptance.

REFERENCES

Berg P, Baltimore D, Brenner S, Roblin RO and Singer MF (1975) Summary Statement of the Asilomar conference on recombinant DNA molecules. *Proc. Natl. Acad. Sci. USA,* **72** : 1981-1984.

Burza W and Malepszy S (1995) Direct plant regeneration from leaf explants in cucumber (*Cucumis sativus* L.) is free of stable genetic variation. *Plant Breed.,* **114** : 341-345

Cade RM, Wehner TC and Blazich FA (1990) Somatic embryos derived from cotyledons of cucumber. *J. Amer. Soc. Hort. Sci.,* **115** : 691-696

Chee PP and Tricoli DM (1988) Somatic embryogenesis and plant regeneration from cell suspension cultures of *Cucumis sativus* L. *Plant Cell Rep.,* **7** : 274-277

Chee PP (1990a) High frequency of somatic embryogenesis and recovery of fertile cucumber plants. *HortSci.,* **25** : 792-793.

Chee PP (1990b) Transformation of *Cucumis sativus* tissue by *Agrobacterium tumefaciens* and the regeneration of transformed plants. *Plant Cell Rep.,* **9** : 245-248.

Chee PP and Slightom JL (1991) Transfer and expression of cucumber mosaic virus coat protein gene in the genome of *Cucumis-sativus. J. Am. Soc. Hortic. Sci.,* **116** : 794-802.

Chee PP and Slightom JL (1992) Transformation of cucumber tissues by microprojectile bombardment: identification of plants containing functional and non-functional transferred genes. *Gene,* **118** : 255-260

Christou P, Swain WF, Yang YS and McCabe DE (1989) Inheritance and expression of foreign genes in transgenic soybean plants. *Proc. Natl. Acad. Sci. USA,* **86** : 7500-7504

Cohen SN, Chang ACY, Boyer HW and Helling RB (1973) Construction of Biologically Functional Bacterial Plasmids *In Vitro. Proc. Natl. Acad. Sci. USA,* **70** : 3240-3244.

Deakin JR, Bohn GW and Whitaker TW (1971) Interspecific hybridization in *Cucumis. Econ. Bot.,* **25** : 195-211

Debeaujon I and Branchard M (1993) Somatic embryogenesis in *Cucurbitaceae. Plant Cell Tiss. Org. Cult.,* **34** : 91-100

Esquinas-Alcazar JT and Gulick PJ (1983) Genetic resources of *Cucurbitaceae. International Board for Plant Genetic Resources Secretariat,* Roma, Italy, pp. 1-101.

Finer JJ and McCullen MD (1990) Transformation of cotton (*Gossypium hirsutum* L.) via particle bombardment. *Plant Cell Rep.,* **8** : 586-589

Fitch MMM, Manshardt RM, Gonsalves D, Slightom JL and Sanford JC (1990) Stable transformation of papaya via microprojectile bombardment. *Plant Cell Rep.,* **9** : 189-194

Fromm ME, Morrish F, Armstrong C, Williams R, Thomas J and Klein TM (1990) Inheritance and expression of chimeric genes in the progeny of transgenic maize plants. *Bio/Technology,* **8** : 833-839

Gambley RL and Dodd WA (1990) An *in vitro* technique for the production *de novo* of multiple shoots in cotyledon explants of cucumber (*Cucumis sativus* L.). *Plant Cell Tiss. Org. Cult.,* **20** : 177-183

Gonsalves D, Chee P, Provvidenti R, Seem R and Slightom JL (1992) Comparison of coat protein-mediated and genetically-derived resistance in cucumbers to infection by cucumber mosaic virus under field conditions with natural challenge inoculations by vectors. *Biotechnology,* **10** : 1562-1570

Hansen G, Das A and Chilton MD (1994) Constitutive expression of the virulence genes improves the efficiency of plant transformation by *Agrobacterium. Proc. Natl. Acad. Sci. USA,* **91** : 7603-7607

Matzke MA, Primig M, Trnovsky J. and Matzke AJM (1989) Reversible methylation and inactivation of marker genes I sequentially transformed tobacco plants. *EMBO J.*, **8** : 643-649.

Murashige T and Skoog F (1962) A revised medium for rapid growth and bioassays with tobacco tissue culture. *Physiol. Plant.*, **15** : 473-497.

Nishibayashi S, Kaneko H and Hayakawa T (1996) Transformation of cucumber (*Cucumis sativus* L.) plants using *Agrobacterium tumefaciens* and regeneration from hypocotyl explants. *Plant Cell Rep.*, **15** : 809-814.

Nishizawa Y and Hibi T (1991) Rice chitinase gene: cDNA cloning and stress-induced expression. *Plant Sci.*, **76** : 211-218

OECD (1993) *Safety considerations for biotechnology: Scale-up of crop plants.* OECD publication, Paris, France, pp. 1-40.

OECD (1994) *Safety evaluation of foods derived by modern biotechnology: concepts and principles.* OECD publication, Paris, France, pp. 1-79.

OECD biotrack online http://www.oecd.org/ehs/cd.htm

Orczyk W, Nadolska-Orczyk A and Malepszy S (1998) Plant regeneration of wild *Cucumis* species. *Acta Soc. Bot. Pol.*, **57** : 195-200

Punja ZK, Abbas N, Sarmento GG and Tang FA (1990) Regeneration of *Cucumis sativus* var. *sativus* and *C. sativus* var. *hardwickii*, *C. melo*, and *C. meturiferus* from explants through somatic embryogenesis and organogenesis. *Plant Cell, Tiss. Org. Cult.*, **21** : 93-102

Punja ZK and Raharjo SHT, (1996) Response of transgenic cucumber and carrot plants expressing different chitinase enzymes to inoculation with fungal pathogens. *Plant Dis.*, **8** : 999-1005.

Raharjo SHT and Punja ZK (1994) Regeneration of plantlet from embryogenic suspension cultures of pickling cucumber (*Cucumis sativus* L. cv. Endeavor). *In Vitro Cell. Dev. Biol.*, **30** : 16-20

Raharjo SHT, Hernandez MO, Zhang Y and Punja Z K (1996) Transformation of pickling cucumber with chitinase-encoding genes using *Agrobacterium tumefaciens*. *Plant Cell Rep.*, **15** : 591-596.

Sarmento GG, Alpert K, Tang FA and Punja ZK (1992) Factors influencing *Agrobacterium tumefaciens*-mediated transformation and expression of kanamycin resistance in pickling cucumber. *Plant Cell Tiss. Org. Cult.*, **31** : 185-193.

Schulze J, Balko C, Zellner B, Koprek T, Hansch R, Nerlich A and Mendel RR (1995) Biolistic transformation of cucumber using embryogenic suspension cultures: long-term expression of reporter genes. *Plant Sci.*, **112** : 197-206.

Smarrelli JJ, Watters MT and Diba LH (1986) Response of various cucurbits to infection by plasmid-harboring strains of *Agrobacterium*. *Plant Physiol.*, **82** : 622-624.

Somers DA, Rines H W, Gu W, Kaeppler HF and Bushnell WR (1992) Fertile, transgenic oat plants. *Bio/Technology,* **10** : 1589-1593

Szwacka M, Morawski M and Wurza W (1996) *Agrobacterium tumefaciens*-mediated cucumber transformation with thaumatin II cDNA. *Genet. Pol.*, **37A** : 126-129.

Tabei Y and Kanno T (1989) Effect of three kinds of auxins on the regeneration of cucumber (*Cucumis sativus* L.). *Bull. Natl. Res. Inst. Veg., Ornam. Plants and Tea, Japan* **3** : 97-105.

Tabei Y, Kanno T and Nishio T, (1991) Regulation of organogenesis and somatic embryogenesis by auxin in melon, *Cucumis melo* L. *Plant Cell Rep.*, **10** : 225-229

Tabei Y, Nishio T, Kurihara K and Kanno T (1994) Selection of transformed callus in a liquid medium and regeneration of transgenic plants in cucumber (*Cucumis sativus* L.). *Breed Sci.,* **44** : 47-51.

Tabei Y, Kitade S, Nishizawa Y, Kikuchi N, Kayano T, Hibi T and Akutsu K (1998) Transgenic cucumber plants harboring a rice chitinase gene exhibit enhanced resistance to gray mold (*Botrytis cinerea*). *Plant Cell Rep.,* **17** : 159-164

Trulson A J, Simpson RB and Shahin EA (1986) Transformation of cucumber *Cucumis sativus* L. plants with *Agrobacterium. Theor. Appl. Genet.,* **73** : 11-15.

Uchimiya H, Handa T and Brar DS (1989) Transgenic plants. *J. Biotechnol.* **12** : 1-20

Vasil V, Castillo AM, Fromm ME and Vasil IK (1992) Herbicide resistant fertile transgenic wheat plants obtained by microprojectile bombardment of regenerable embryogenic callus. *Bio/Technology,* **10** : 667-674

Wako T, Terami F, Hanada K and Tabei Y (2001) Resistance to Zucchini yellow mosaic virus (ZYMV) in transgenic cucumber plants (*Cucumis sativus* L.) harboring the coat protein gene of ZYMV. *Bull. Natl. Res. Inst. Veg., Ornam. Plants and Tea, Japan* **16** : 175-186.

Weissing K, Schell J and Kahl G (1988) Foreign genes in plants : transfer, structure, expression, and applications. *Annu. Rev. Genet.,* **22** : 421-477.

Chapter 9

IN-VITRO CULTURE AND GENETIC TRANSFORMATION OF PEA

C Nirmala[1]★ and MLH Kaul[2]

[1]*Department of Botany, Panjab University, Chandigarh - 160 014, India*
[2]*Department of Biotechnology, Kurukshetra University, Kurukshetra - 132 119, India*

Summary

Pea, is an important legume vegetable crop. Its grain yield is poor and efforts to increase the yield through mutation breeding has been unsuccessful. Genetic engineering techniques combined with conventional breeding programs could enhance both the yield and nutritive value. However, genetic engineering in pea is still in the ab-initio stage owing mainly due to the difficulties with regeneration in tissue culture. This necessitates the development of efficient regeneration and transformation protocol for production of fertile transgenic plants with the desired traits. In-vitro culture has been done using various explants like immature embryos, epicotyl, hypocotyl, nodal and internodal segments, meristems, leaves and roots and also protoplasts. Though fertile plants have been produced via organogenesis or somatic embryogenesis, limitations exist like poor regenerability, long term callus phase, phenotypic abnormalities, altered ploidy and reduced plant fertility. Also strong in-vitro genetic differences exist in the morphogenetic capacity of different pea genotypes and explants. Whereas only a few reports exist for direct gene transfer into protoplasts, successful genetic transformation has been achieved mostly by using Agrobacterium vectors in pea in Australia, Europe and USA. Efforts are being made by the present authors to develop protocols which are efficient, economical, reproducible and with a broad action

★Corresponding author : E-mail : cnirmala@mantramail.com

spectrum applicable to the Indian pea genotypes. Preliminary efforts have been successful in the most commonly cultivated pea variety Bonneville. The work already done and being done is reviewed in this paper.

Keywords : Pea, organogenesis, somatic embryogenesis, protoplast culture, genetic transformation

Abbreviations : BAP - 6-Benzylaminopurine, GUS - β-glucuronidase, HPT - Hygromycin Phosphotransferase, IAA - Indole-3-acetic acid, IBA - Indole-3-butyric acid, NAA - Naphthalene acetic acid, NPTII - Neomycinphosphotransferase II, PEG - polyethylene glycol, 2,4D - 2,4-Dichlorophenoxyacetic acid, 4-CPA - 4-chlorophenoxyacetic acid, TDZ - Thidiazuron, 2,4,5-T - Trichlorophenoxyacetic acid.

1. INTRODUCTION

Though in number of genera and species, legumes follow composites and orchids, the 600 genera and 18,000 species of legumes are distributed throughout the world occupying diverse climatic and geographic areas including temperate zones, savannas and lowlands. Legume seeds also known as grain legumes and the ones whose dried seeds are consumed by humans as the pulses are richer in proteins than cereals. In addition, seeds of many grain legumes are rich in oils. Their seeds also serve as an excellent source of vitamins, thiamine and niacin as well as minerals, calcium, iron and carbohydrates. Many legumes are exploited for human consumption being used in several forms as immature seeds, dry grain, roasted grain in flour production, as condiments and as fermented products. Of the thousands of legume species, those included for consumption are peanut, soybean, pea, lentil, pigeon pea, chickpea, mungbean, kidney bean, cowpea, alfalfa, clover and vetch. In all the developing countries, increased cultivation of legumes is the best way to combat shortages in food and protein supplies.

Amongst legumes, peas, the fourth most important legume crop in the world, rose to fame and honour as they were used by Mendel in his seminal experiments over 135 years ago. Of all the various peas, the garden and vegetable pea, *Pisum sativum*, represents an important model plant for biochemists, cytogeneticists, molecular biologists and physiologists. Currently, it has become one of the world's most important grain legume serving a variety of roles as a source of both human food and animal feed (Casey and Davies, 1993). The seed storage protein content varies from 15.5% to 39.7% (Monti and Grillo, 1983). This wide range is due to both genetic differences among varieties and interactions of genotype and environment. Peas are limited in the sulphur containing aminoacids cysteine and methionine but are rich in lysine and other essential amino acids. Other constituents include lecithin, vitamins A, B1, B2, C in seeds and vitamin E in the radicle. Peas also contain 5%-6% fibre and 2%-4% ash. Potassium, phosphorous, magnesium, sulphur, calcium and iron are present in legume seeds (Wright, 1985). Pea seeds are harvested either in dry mature form or in an immature state - the precise stage of harvest depends upon the end use. While the dry mature seed is used extensively as animal feed, as human food and as an ingredient of a variety of prepared foods, those harvested in immature form are used directly as vegetable or are canned, dehydrated or frozen. Pea, like other legumes has the ability to undergo symbiosis with *Rhizobia* and hence, increasing the acreage of pea is desirable not only for economical but also for ecological reasons especially for India, so that they are able to grow and yield in different ecological zones of our country as well as enrich soil in these areas. Their wide cultivation will facilitate their increased-availability, price-reduction and consumer-purchase. However, pea is a poor grain yielder with relatively low yield stability. Improvement in it has not been possible as intervarietal and interspecific hybrid sterility barrier obviate wide hybridizations and thus limits its gene pool. Moreover, induced *in-vivo* mutations have not produced any marked yield improvements in this vegetable crop; *in-vitro* mutagenesis in it has not been tried (Kaul and Nirmala, 1999). The productivity and nutritional value of pea needs to be greatly increased. Being a multifactorial problem, strategies for improvement are manifold and include :

1. Improvement of harvesting characteristics viz. improved standing ability, concentration of pods at the top of the plant, more pods per node, more seeds per pod and reduced pod abortion due to stress.

2. Introduction of or selection for resistancc against diseases like pea wilt (*Fusarium oxysporum* and *F.solani*), powdery mildew (*Erysiphe polygoni*), seedling and foot rot (*Aschocyta ssp*), bacterial blight (*Psuedomonas syringae pv pisi*), common root rot (*Aphanomyces eutiches*), and certain viral pathogens (PSbMV, PEV, PSV).

3. Improvement of storage protein composition by increasing the amount of essential amino acids.

4. Increase the percentage of amylopectin in the starch.

Diseases caused by fungi, viruses and pests are the serious constraints of pea pod production. Breeding resistant peas by conventional methods has not been possible so far. Hence, the conventional breeding programs need to be complemented with *in-vitro* techniques to develop transgenic peas resistant to viruses, pathogens and pests. A major prerequisite for the *in-vitro* introduction of foreign genes into a plant and selection of desirable traits is the ability to regenerate fertile plants from explants or from isolated protoplasts (Nirmala and Kaul, 1997a; Trigiano and Gray, 2000). The totipotency of plant cells have made it possible to culture plant cells and tissues and have provided the opportunity to micropropagate elite plant clones by stimulation of axillary buds and somatic embryos and also facilitated the adventitious regeneration of plants from whole organs, explants, callus and isolated protoplasts. The ability to regenerate whole plants from cells and tissues has paved the way for genetic transformation which involves inserting a gene of interest into an organism that expresses the transferred gene. This results in the development of the gene modified plants known as transgenic plants. Of the various dicots, legumes are the most recalcitrant. Also, for each species and cultivar, the protocol for *in-vitro* culture, regeneration and genetic transformation has to be standarized. This had led to the enormity of transgenic work as legumes constitute a major group of edible plants, each species having numerous cultivars specific to the area of cultivation. The present paper reports about *in-vitro* regeneration and genetic transformation in *Pisum sativum*, the garden pea.

2. *IN-VITRO* REGENERATION

2.1. Explant culture

Regeneration of whole plants from cells or tissues cultured *in-vitro* occurs either by induced embryogenesis or stepwise organogenesis. Legume tissues though tough to culture *in-vitro*, regeneration in pea by organogenesis or somatic embryogenesis have been reported in the exotic pea (Ozcan *et al.*, 1992). Though reports on shoot regeneration from pea explants exist (Griga and Novak, 1990) in hardly a few of these, true de novo regeneration is demonstrated (Puonti-Kaerlas, 1993) the main reason being plant regeneration of pea is time consuming, requires numerous media transfers involving different combinations of plant growth regulators, the regeneration frequency is too low and final plant survival still lower. Various explants have been tried for *in-vitro* morphogenesis induction (Table 1). Culture medium used is mostly MS salts (Murashige and Skooge, 1962) supplemented with 3% sucrose,

Table 1. Explant culture and regeneration in *Pisum sativum.*

Explant type	Regeneration stage	Regen-eration	Reference
Immature embryos from callus	Organogenesis and somatic embryos	Plants	Atanassov and Mehandjiev (1979); Natali and Cavallini (1987); Kysely *et al.*, (1987)
Embryonic Axes	Direct and indirect organogenesis	Plants	Kosturkova *et al.* (1997)
Root segments	-do-	Shoots	Kunakh *et al.* (1984)
Hypocotyl	Direct organogenesis	Plants	Sanago *et al.* (1986)
Epicotyl	-do-	Shoot Plants	Kunakh *et al.* (1984); Sladky and Jandova (1984)
Shoot apex	Somatic embryos from callus	Plants	Kysely *et al.* (1987)
Apical meristem	Organogenesis from callus	Plants Shoots	Gamborg *et al.* (1974); Preobrazenskaja (1983)
Internodal and nodal segments	Organogenesis from callus	Shoot Plants	Hildebrandt *et al.* (1963); Gostimskij *et al.* (1985); Present report
Leaf segments	Somatic embryos in cell suspension	Torpedo embryonic stage with scant root formation	Jacobsen and Kysely (1984); Kysely and Jacobsen (1984)
Whole leaves	-do-	Shoot Plants	Kunakh *et al.* (1984); Mroginski and Kartha (1981); Rubluo *et al.* (1984)

0.9% or 0.3% gelrite an alternative for agar, B5 vitamins and different auxin-cytokinin combinations. First *in-vitro* regeneration of pea was reported by Hildebrandt *et al.* (1963) who used stem segments as explants. Shoots regenerated after a callus phase in a medium supplemented with coconut milk. Gamborg *et al.* (1974) first reported fertile pea plant regeneration via organogenesis from a macerated cell mass derived from apical meristems. Instead of macerated apical meristem that distorted differentiation and destroyed many potential calli, Kartha *et al.* (1974) used the complete shoot apical meristems and obtained a higher regeneration frequency. Further success of regeneration was obtained by Malmberg (1979) who regenerated pea plants via organogenesis from calli derived from epicotyl explants. An effective and generalized protocol for *in-vitro* pea regeneration was developed by Mroginski and Kartha, 1981 and Rubluo *et al.* (1984). They utilized immature leaflets (0.9-1.8mm) which proved the best explants for shoot regeneration. Of the broad spectrum of growth regulators (BAP, Kinetin, 2iP, Zeatin, NAA, ABA, 2,4-D, Picloram) tested, combination of BAP (10 µM) and NAA (10 µM) proved the best for callus induction and subsequent regeneration. Optimal environment was an alternating temperature of 20^0C/15^0C with a 16hr photoperiod at 2000 Lx. But by using this system for Indian pea, it was found by the present authors that though immature leaf explant is best for regeneration, frequency of genetic transformation in this tissue is too low (present report). Instead nodal and internodal explants excised from 7 day old etiolated seedlings proved best for the two Indian pea cultivars, Bonneville and Arkel. After inducing callus growth on MS medium supplemented with 0.1 mg/L BAP and 0.1 mg/L picloram (Fig. 1, 2) multiple shoot induction occured Subsequently calli were placed on MS medium with 4.5 mg/L BAP and 0.02 mg/l NAA (Fig. 3). The fast developing shoots (Fig. 4) were transferred on root induction medium and maintained on it for 6 weeks. The rooted shoots were planted in a sterile 1:1 soil-vermiculite mixture in cotton capped jars. The caps were removed gradually to harden the plantlets against low moisture conditions. In an attempt to improve the *in-vitro* regeneration frequency for genetic transformation, another *in-vitro* culture protocol using thidiazuron (TDZ), a substituted phenylurea with cytokinin like activity was developed by Sanago *et al.* (1996). In this protocol, 0.1, 0.8, 1.0, 2.0, 5.0 and 10 µM of TDZ were added to the MS basal media. After a week in culture, multiple shoots appeared. Shoot formation was significantly influenced by TDZ concentrations. Whereas, 0.5 µM and 1.0 µM was

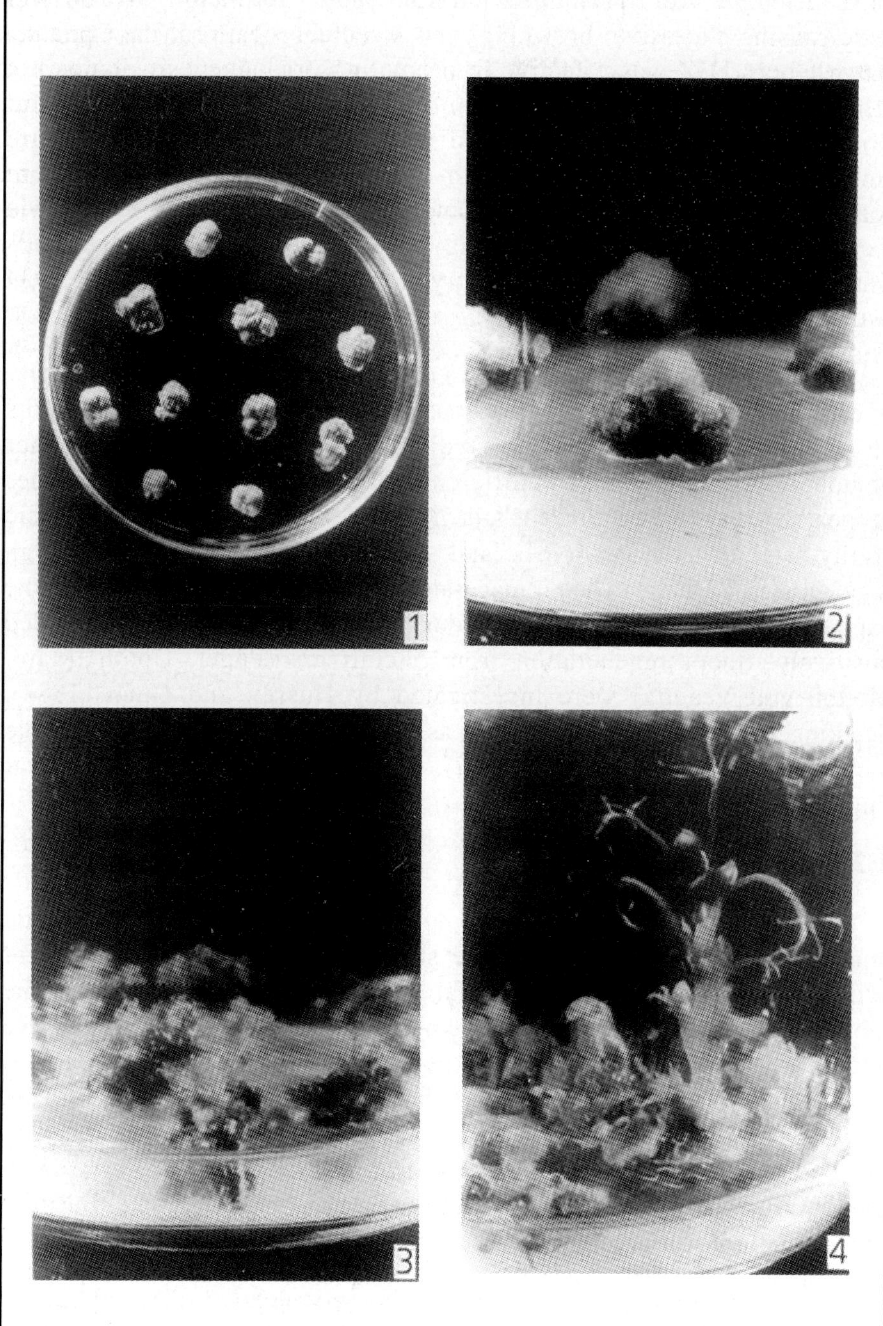

Figures 1-2 : Growing callus; *Figures 3-4* : Regenerated shoots

most effective for promoting multiple shoot formation, 10 μM was toxic causing excessive browning and eventual death of the explants. But whereas TDZ was effective in promoting organogenesis, it was not effective in inducing somatic embryogenesis. On the other hand Kosturkova *et al.* (1997) cultured embryonic axes from immature embryos of ten pea genotypes in a medium containing different concentrations of 2,4-D, α-NAA and BA. Whereas shoots formed via a callus phase in medium containing 0.2 μM 2,4-D and 5 μM BA, buds were directly formed from the embryonic axis on medium with 10 μM BA and 1 μM NAA.

Strong *in-vitro* genetic differences exist in the morphogenetic capacity of different explants and pea genotypes. This is evidenced by the results of Jacobsen *et al.* (1980) who studied morphogenetic responses after phytohormone applications in different mutant and non-mutant genotypes of pea. Likewise, Preobrezenskaja, (1983) studied the organogenetic ability of eight pea genotypes and only three genotypes showed high regeneration capacity when macerated shoot tips were used as the explant. Kunakh *et al.* (1984) studied 6 genotypes and only two responded positively. Shoot regeneration from 5 cultivars (Puget, Upton, Maro, Melton and Vedetta) were investigated by Hussey and Gunn (1984). Seedling epicotyl bases were used as the initial explant and out of the 5 cultivars, efficiently growing calli could be obtained only from cultivar Puget and Upton.

2.2. Meristem culture

Meristem culture refers to the culture of the apical meristematic dome situated at the tip of the shoot and being the centre of activity of various developmental programs of a plant, has the inbuilt totipotent capacity. Thus, the meristem in its juvenile vegetative phase is the most appropriate explant to initiate meristem culture. Accordingly, both the meristem and shoot tip cultures have been utilized in diverse areas such as rapid clonal multiplication of vegetatively propagated crop plants, for virus elimination and for germplasm preservation of both vegetatively and seed propagated plants. Because of the inherent totipotency potential, the meristems have now been used as explants for genetic transformation protocols.

In pea tissue culture, Kartha *et al.* (1974) used meristems of selected lines for rapid propagation. Later Kartha and Gamborg (1978), Griga *et al.* (1986) used axillary meristems and axillary buds for plant regeneration.

The effect of various gamma radiation doses of seeds in the growth of a long term pea meristem was studied by Novak *et al.* (1984). A dose of 10GY was found to stimulate growth *in-vitro* when compared with the untreated control whereas a higher dose inhibited growth. Though suitable for *in-vitro* culture, only a low frequency of transformation following *Agrobacterium tumefaciens* inoculation of shoot tip was found (Kathen and Jacobsen, 1995; Jacobsen, 1992).

2.3. Protoplast culture

Wall less naked plant cells are termed as protoplasts. These barrier less cells represent the only source of single-cell system obtainable in higher plants. These cells have the capability to divide to form millions of individual cells in culture and their totipotency allows regeneration of fully mature flowering plants. Constabel *et al.* (1973) and Landgen and Torrey (1973) first used pea protoplasts to raise cultures upto the callus stage but could not regenerate plants. Puonti-Kaerlas and Eriksson (1988), Lehminger-Mertens and Jacobsen (1989), Bohmer *et al.* (1995) have successfully regenerated plants from protoplasts. They isolated the protoplasts from the lateral shoots of embryonal axis and used the cytokinin analogue, thidiazuron (TDZ) as an effective shoot inducing growth regulator in a medium consisting of MS salt enriched with B5 vitamins, 2% sucrose, 3% mannitol and 0.05% MES (2-N-morpholino ethane acid). TDZ induced shoot formation with the optimal being at 10 μM. In all these cases, the protoplast division rates and regenerative competence are totally genotype dependent. Also, various cultivars exhibit different hormone requirements specific for each.

2.4. Somatic embryogenesis

Somatic tissues, the products of mitotic division, retain the ability to regenerate via somatic embryogenesis when grown under appropriate culture conditions. This phenomena known as somatic embryogenesis occurs both in diploid and haploid cells without the intervention of sexual fusion. Production of somatic embryos have been obtained in pea through callus mediated regeneration from leaves, immature embryos, shoot apices, nodal segments and protoplasts (Jacobsen and Kysely, 1984; Kysely *et al.,* 1987; Kysely and Jacobsen, 1990; Griga, 1990, 1998; Stejskal and Griga, 1992; Van Doorne *et al.,* 1991, 1995; Lehminger-Mertens and Jacobsen, 1989a,b; Bohmer *et al.,* 1995; present report) and from immature zygotic embryos and shoot apices without callus intervention (Tetu *et al.,* 1990; Loiseau *et al.,* 1995; Griga,

1998). In all cases, exogenous auxins like NAA, 2,4-D; 2,4,5 T, Picloram,4-CPA were used for somatic embryos. But addition of cytokinin (BAP, zeatin, kinetin) to the auxin supplemented medium either inhibits or reduces somatic embryogenesis in pea (Kysely and Jacobsen, 1990; Tetu *et al.,* 1990; Loiseau *et al.,* 1995). In some pea genotypes, carbohydrate addition increases the somatic embryogenic efficiency and embryo development (Loiseau *et al.,* 1995). But in all the cases, in pea, not only the genotype and type of explant used but also the explant size its physiological state and the type and concentration of auxin used effects the kind of regeneration significantly, both the direct or indirect somatic embryogenesis. Whereas, the use of 3-5mm shoot apices as explants results in embryogenic callus formation followed by somatic embryo initiation, explants in the range of 0.5 to1mm with minimally differentiated structures yields direct somatic embryogenesis with negligible callus formation. Transfer of embryogenic callus cultures to MS medium with cytokinin stimulates development of young embryos and differentiation of plantlets. Regenerated plantlets grow well to maturity in soil and produce fertile viable seeds.

2.5. Anther culture

Compared to other food plants, in legumes, reports on anther cultures are too scanty. The first report is from White (1943) who obtained haploid callus from pollen of *Pisum sativum* cultivar B22 cultured on White's medium supplemented with 15% v/v coconut milk and 0.5 mg/l 2,4-D. Subsequent trials by Gupta (1975) Gosal and Bajaj (1988) also did not yield stable viable haploids. Nearly complete euploids were obtained by Davey *et al.* (1994). Extensive emperical research is still required to develop efficient and reliable techniques that can be used to generate large number of haploid plants from cultures anthers of legumes.

3. GENETIC TRANSFORMATION

Precisely cut foreign genes can be delivered into plant cells either through a vector or by direct gene introduction into the nucleus; the former comprises the most popular and successful *Agrobacterium* mediated gene transfer whereas, the latter includes vectorless methods which comprise both chemical and physical induction of DNA uptake (Nirmala and Kaul, 1997; Hansen and Chilton, 2000; Finer *et al.*, 2000). The most popularly used vectorless methods are electroporation, microinjection and particle bombardment. For the latter two methods, protoplasts are needed as the gene receipts. This requirement has confined their applicability to those species only where efficient, economical and reproducible protoplast culture and regeneration systems are established. In pea, these systems though known are strictly and strongly genotype specific.

3.1. Vector mediated

3.1.1. Explant transformation

Of the various *in-vitro* gene transfer methods, *Agrobacterium* mediated gene transfer is preferred to others being economical, efficient and efficacious (Table 2). In pea, though *in-vitro* culture of tissues have been done, its recalcitrancy in *de-novo* regeneration of whole plants due to slow response in tissue culture, a low frequency of cellular transformation, poor number of cells competent for regeneration and inefficient selection of transformed cells have become serious impediments of pea genetic transformations (Puonti-Kaerlas, 1993). However, *Agrobacterium* mediated transformation of pea explants like (i) stem explants (Lulsdorf *et al.,* 1991) (ii) embryonic axis and epicotyl segments (Fillipone and Lurquin, 1989; Puonti-Kaerlas *et al.,* 1989; Schroeder *et al.,* 1993) (iii) nodal explants (Kathen and Jacobsen, 1990; Nauerby *et al.,* 1991) (iv) root explants and protoplasts (Schaerer and Pillet, 1991, (v) cotyledonary nodes (Jordan *et al.,* 1992) and (vi) cotyledonary meristems (Davies *et al.,* 1993; Bean *et al.,* 1997) has been documented. But the stable transformation of pea and production of mature flowering transgenic plants was first reported by Puonti-Kaerlas *et al.* (1990) followed by Zubko *et al.* (1990); Davies *et al.* (1993), Schroeder *et al.* (1993) and Bean *et al.* (1997). Puonti-Kaerlas took explants from axenic shoot cultures and seedling epicotyls of two cultivars Stivo and Puget and co-cultivated them with *Agrobacterium.* The transformed calli were selected on medium containing either hygromycin or kanamycin. The response to different pea cultivars in callus induction studies on selective medium differed between the cultivars used and the types of selection applied. Whereas. transgenic shoots were selected for hygromycin resistance, no regenerants could be obtained from kanamycin resistant clones. Transmission of the introduced gene was stable over two generations. Schroeder *et al.* (1993) and Davies *et al.* (1993) reported recovery of transgenic pea expressing *nptII* and *uidA* respectively. In the first report, explant sections from the embryonic axis of immature seeds were used whereas in the later, lateral cotyledonary meristems of germinating seeds were inoculated with *Agrobacterium.* In both the cases full plants were regenerated. The protocols used

Table 2. Agrobacterium mediated genetic transformation in pea.

Explant marker	Selectable assay	Marker analysis	Status	DNA	Reference
Embryonal axis of immature seeds	Phosphino-thricine (15mg/l)	*nptII*	Fertile plants	Yes	Schroeder *et al.* (1993)
Cotyledonary meristems	Kanamycin (100 mg/l)	*nptII*	Fertile plants	Yes	Davies *et al.* (1993)
Cotyledonary meristems	Phosphino-thricine (5mg/l)	*pat*	Fertile	Yes	Bean *et al.* (1997)
Roots	Kanamycin (100mg/l)	Opine, *nptII*	Callus	-	Schaerer and Pilet (1991)
Epicotyl	Kanamycin (75mg/l)	Opine	Callus	Yes	Puonti-Kaerlas *et al.* (1989)
	Hygromycin (15mg/l)	-	Fertile plants	Yes	Puonti-Kaerlas *et al.* (1992)
Nodal segments	Kanamycin (50mg/l) Hygromycin (5-10mg/l)	Opine, *nptII*	Plants	-	Kathen and Jacobsen (1990)
Internodal and nodal segments	Kanamycin (50mg/l) Hygromycin (25mg/l)	Opine, *nptII*	Callus	Yes	Lulsdorf *et al.* (1991)
Shoot and leaf segments	Kanamycin (75mg/l) Hygromycin (15mg/l)	Opine, *nptII*	Callus Fertile plants	Yes	Puonti-Kaerlas *et al.* (1990)

suffered from some drawbacks like poor regenerability, complicated and long term regeneration via callus phase, reduced plant fertility, phenotype abnormalities, altered ploidy and loss of transgene activity in subsequent generations. Though Schroeder's method obviated these problems by *in-vitro* management and rigorous regenerant selection, their approach had disadvantages like requiring the production of donor pea plants under specific growth conditions and the subsequent accurate identification of material at the highly sensitive and strongly specific correct developmental stage. By targeting highly regenerable lateral cotyledonary meristems capable of rapidly producing transformants independently of a callus phase, reducing thereby the tissue culture phase and minimising effective selection coupled with rapid grafting, Bean *et al.* (1997) rendered possible the reproducible production of phenotypically normal, fertile transgenic plants which transmit functional transgenes to the progeny. Increase in competence of plant cells by application of auxins and abscissic acid during cocultivation was reported by Kathen and Jacobsen 1994, 1995. The cells competent for transformation were mainly restricted to the dedifferentiating cells neighbouring the vascular system of cotyledon and epicotyl explants. In the etiolated seedlings, competence for transformation decreased as the distance of the epicotyl explant from the shoot apex increased. The competence could be induced by the exogenous application of auxins. Factors contributing to the competent state of the tissue include : availability of specific plant cell attachment sites for *Agrobacterium*, the production of phenolic compounds inducing excision and processing of the T-DNA protein complex, targeting of the plant cell nucleus and finally its integration into the plant genome as well as the expression of the introduced genes.

3.1.2. Protoplast transformation

Amongst legumes, transgenic plants have been produced by cocultivating protoplasts with *Agrobacterium* only in mothbean and pea (Eapen *et al.,* 1987; Puonti-Kaerlas *et al.,* 1989) . Unlike in mothbean, in pea, for cocultivation with *Agrobacterium*, freshly isolated pea protoplast are not suitable as they are very sensitive and hence small protoplast derived colonies, the microcalli, are used. After co-cultivation of microcalli, kanamycin resistant transgenic calli were selected. Gene insertion in them was confirmed by DNA analysis. Pea protoplasts have also been transformed using direct gene transfer methods. Hobbs *et al.* (1990) studied transient GUS expression in electroporated pea protoplasts but stable transformants were not developed. Stably transformed hygromycin resistant clones were obtained after integration of pea protoplasts in an electric field of 400V/cm for 20 mins using the HPT gene as a selectable marker (Puonti-Kaerlas *et al.,* 1992). The transformation frequencies varied depending on the genotype, tissue source, DNA concentration and other physical parameters. On the basis of dividing protoplasts, the transformation efficiency was 1.4×10^{-3} but no regeneration was obtained from resistant calli. Failure to obtain transformed calli after selection on kanamycin indicates that the selectable marker has a significant influence on the transformation frequency. Genotypic differences in transformation frequencies have been reported in soybean (Christou *et al.,* 1987) and mothbean (Kohler *et al.,* 1987). The transformed nature of the calli was confirmed by GUS assay and DNA analysis but no transgenic plants could be regenerated. So far no protocols have been published for PEG mediated transformation in pea. A protocol for PEG mediated transformation has been developed in which a pUC9 derived plasmid was used containing the GUS and the pat gene both under the control of the 35S promoter. Two days after transformation,GUS positive protoplasts were detected. Resistant calli expressing the GUS and/or the pat gene were obtained.

4. CONCLUSIONS AND FUTURE PROSPECTS

Successful *in-vitro* culture coupled with genetic transformation is a means of introducing useful foreign genes of immediate use for pea breeding improvement. Targets for pea gene engineering include introduction of genes for resistance to herbicides, disease and pests, improving grain protein content and nutritional quality, introducing genes to improve nitrogen fixation by *Rhizobium*-legume symbiotic relationships. Compared to the major cereals where enormous resources have been expended, in legumes, only soybeans and peanuts have been exploited for genetic engineering. Though some methods for regenerating plants from various explants and protoplasts of pea are known, an efficient, economical, rapid and reproducible protocol is needed for *in-vitro* regeneration and conservation of rare genotypes and for genetic transformation. Present limitations include low reproducibility between individual *in-vitro* culture procedures, a prolonged culture period requirement for regeneration, difficult, scanty rooting of transgenic shoots, somaclonal variation and low stability of transgenics in field grown populations. Though explant transformation systems are more successful than the protoplast systems, they need to be standarised. Screening for high performance genotypes and best responding explant type is needed to develop rapid regeneration system compatible with transformation. In view of rapid escalation of prices of animal protein and the risks associated with its consumption and cheap availability and easy storage of legumes seed vegetable protein, the importance of legumes in world agriculture is increasing fast especially in the developing countries; there is an urgent need to develop efficient methods for rapid production of improved legume varieties for the increasing demand to ever expanding population. Detection of molecular markers for high seed protein content and quality would contribute to legume seed improvement (Nirmala and Kaul, 1999).

ACKNOWLEDGEMENT

The authors are grateful to Prof. HJ Jacobsen, Head, Department of Molecular Genetics, University of Hannover, Germany for providing lab. facilities during which a part of the work was done

REFERENCES

Atanassov AI and Mehandjiev AD (1979) *In-vivo* induced morphogenesis in pea embryos. *R Acad Bulg Sci.,* **32** : 115-118

Bean SJ, Gooding PS, Mullineaux PM and Davies DR (1997) A simple system for pea transformation. *Plant Cell Rep.,* **16** : 513-519.

Bohmer P, Meyer B and Jacobsen HJ (1995) Thidiazuron induced high frequency of shoot induction and plant regeneration in protoplast derived pea callus. *Plant Cell Rep.,* **15** : 26-29.

Casey R and Davies DR (1993) *Peas: genetics, molecular biology and biotechnology.* CAB International, Wallingford, UK.

Christou P, Murphy JE and Swain WF (1987) Stable transformation of soybean by electroporation and root formation from transformed callus. Proc. Natl. Acad. Sci., **84** : 3962-3966.

Constabel F, Kirkpatrick JW and Gamborg OL (1973) Callus formation from mesophyll protoplasts of *Pisum sativum. Can J Bot.,* **51** : 2105-2106.

Davey MR, Kumar V and Hammatt N (1994) *In-vitro* culture of legumes. In: Plant Cell and Tissue Culture (eds) Vasil IK, Thorpe TA. Kluwer Acad Publ. The Netherlands.

Davies DR, Hamilton J and Mullineaux P (1993) Transformation of peas. *Plant Cell Rep.,* **12** : 180-183.

Eapen S, Kohler F, Gerdmann M and Schieder O (1987) Cultivar dependence of transformation rates in moth bean after cocultivation of protoplasts with *Agrobacterium tumefaciens. Theor. Appl. Genet.,* **75** : 207-210.

Filippone E and Lurquin PF (1989) Stable transformation of pea tissues after cocultivation with two *Agrobacterium tumefaciens* strains. *Pisum Newslett.,* **21** : 16-18.

Finer JJ, Finer KR and Ponappa T (2000) Particle bombardment mediated transformation. In: (Eds Hammond J, Mc Garvey P, Yusibov V). Plant Biotechnology. Sringer Verlag, Germany pp. 59-82.

Gamborg OL, Constabel F and Shyluk JP (1974) Organogenesis in callus from shoot apices of *Pisum sativum. Physiol. Plant.,* **30** : 125-128.

Gosal SS and Bajaj YPS (1988) Pollen embryogenesis and chromosomal variation in anther culture of three food legumes - *Cicer arietinum, Pisum sativum* and *Vigna mungo. Sabrao J.,* **20** : 114-119.

Gostimskij SA, Bagrova AM and Ezhova TA (1985) Discovery and cytogenetic analysis of variation originated during plant regeneration in pea tissue culture. *Dokl. Akad. Nauk. (USSR),* **283** : 1003-1011.

Griga M (1990) The study of *in-vitro* regeneration systems in pea, horsebean and soybean. *Ph.D. Thesis,* Masaryk Univ. Brno.

Griga M (1998) Direct somatic embryogenesis from shoot apical meristems of pea and thidiazuran induced high conversion rate of somatic embryos. *Biol. Plant.*, **41** : 481-495.

Griga M and Novak FJ (1990) Pea (*Pisum sativum* L). In: *Biotechnology in Agriculture and Forestry. Vol.10 Legumes and Oilseed Crops 1.* (Ed Bajaj YPS) Springer-Verlag, Berlin. pp. 65-99.

Griga M, Tejklowa E, Novak FJ and Kubalakova M (1986) *In-vitro* clonal propagation of *Pisum sativum* L. *Plant Cell Tiss. Org. Cult.*, **6** : 95-104.

Gupta S (1975) Morphogenetic response of haploid callus tissue of *Pisum sativum. Indian Agric.*, **19** : 11-21.

Hansen G and Chilton MD (2000) Lessons in gene transfer to plants by a gifted microbe. In: *Plant Biotechnology.* (Eds Hammond J, McGarvey P, Yusibov V). Springer Verlag, Germany, pp 21-58.

Hildebrandt AL, Wilmar JC, Johns H and Riker AJ (1963) Growth of edible chlorophyllous plant tissue *in-vitro. Amer. J. Bot.*, **50** : 248-254.

Hobbs SLA, Jacksson JA, Balisky DS, Delong CMO and Mahon JD (1990) Genotype and promoter induced variability in transient α-glucuronidase expression in pea protoplasts. *Plant Cell Rep.*, **9** : 17-20.

Hussey G and Gunn HV (1984) Plant production of pea in long term callus with superficial meristems. *Plant Sci. Lett.*, **37** : 143-148.

Jacobsen HJ (1992) Biotechnology applied to grain legumes - current state and prospects. Proceedings of the 1st European conference on grain legumes. Angers, pp 99-103.

Jacobsen HJ and Kysely W (1984) Induction of somatic embryos in pea *Pisum sativum* L. *Plant Cell Tiss. Org. Cult.*, **3** : 319-324.

Jacobsen HJ, Ingensiep HW, Herlt M and Kaul MLH (1980) Tissue culture studies in *Pisum sativum.* In: *Plant Cell Cultures: Results and Perspectives : Developments in Plant Biology.* (Eds Sala F, Pasisi B, Cella R, Ciferri O) Vol 5. Elsevier, Amsterdam pp 319-324.

Jordan M, Rempel H and Hobbs SLA (1992) Genetic transformation of *Pisum sativum* L via *Agrobacterium tumefaciens* or particle bombardment Proceedings of the 1st European conference on grain legumes.Angers pp 115-116.

Kathen de A and Jacobsen HJ (1990) *Agrobacterium tumefaciens* - mediated transformation of *Pisum sativum* using binary and cointegrate vectors. *Plant Cell Rep.*, **9** : 276-279.

Kathen de A and Jacobsen HJ (1994) Transformation of pea (*Pisum sativum* L) In: *Biotechnology in Agriculture and Forestry Vol. 23* (Ed Bajaj YPS) Springer-Verlag, Germany, pp. 331-347.

Kathen de A and Jacobsen HJ (1995) Cell competence for *Agrobacterium* mediated DNA transfer in *Pisum sativum* L. *Trans Res.*, **4** : 184-191.

Kartha KK and Gamborg OL (1978) Meristem culture techniques in the production of disease free plants and free preservation of germplasm of tropical tuber crops and grain legumes. In: *Diseases of tropical food crops.* (Eds Maraite H, Meyer JA) Univ. Catholique de Louvain, Belgium pp 267-283.

Kartha KK, Gamborg OL and Constabel F (1974) Regeneration of pea (*Pisum sativum* L) plants from shoot apical meristems. *Z. Planzenphysiol.*, **72** : 172-176.

Kosturkova G, Mehandjiev A, Dobreva I and Tzvetkova V (1997) Regeneration systems from immature embryos of Bulgarian pea genotypes. *Plant Cell Tiss. Org. Cult.*, **48** : 139-142.

Kaul MLH and Nirmala C (1999) Biotechnology: Miracle or Mirage IV. In-vivo and *in-vitro* mutagenesis. In: *Breeding Crop Plants* (Eds Siddique B, Khan S) Kalyani Publ. N. Delhi. pp. 80-110.

Kohler F, Golz C, Eapen S, Kohn H and Schieder O (1987) Stable transformation of moth bean *Vigna aconitifolia* via direct gene transfer. *Plant Cell Rep.*, **6** : 313-317.

Kunakh VA, Voitluk LI, Alkhimova EG and Alpatova LK (1984) Callus tissue formation and induction of organogenesis in *Pisum sativum* L. *Fiziol. Rast.* (Moscow) **31** : 542-548.

Kysely W and Jacobsen HJ (1984) Induction of somatic embryos in pea and soybean. In: (Eds Novak FJ, Havel L, Dobzel J). *Proc. Int. Symp. Plant Tissue Cell Cult. - appl crop improv, Olomouc* pp 131-132.

Kysely W and Jacobsen HJ (1990) Somatic embryogenesis from pea embryos and shoot apices. *Plant Cell Tiss. Org. Cult.,* **20** : 7-14.

Kysely W, Myers JR, Lazzari PA, Collins GB and Jacobsen HJ (1987) Plant regeneration via somatic embryogenesis in pea (*Pisum sativum* L) *Plant Cell Reg.,* **6** : 305-308.

Landgren CR and Torrey GJ (1973) The culture of protoplast derived from explants of seedling pea roots. Colloq Int C NRS. p. 212.

Lehminger-Mertens R and Jacobsen HJ (1989a) Protoplast regeneration and organogenesis from pea protoplasts. *In Vitro cell. Dev. Biol.,* **25** : 571-574.

Lehminger-Mertens R and Jacobsen HJ (1989b) Plant regeneration from pea protoplasts via somatic embryogenesis. *Plant Cell Rep.,* **8** : 379-382.

Loiseau J, Marche C and Deunff Le Y (1995) Effects of auxins, cytokinins, carbohydrates and amino acids on somatic embryogenesis induction from shoot apices of pea. *Plant Cell Tiss. Org. Cult.,* **41** : 267-275.

Lulsdorf MM, Rempel H, Jackson JA, Baliski DS, Hobbs SLA (1991) Optimizing the production of transformed pea (*Pisum sativum* L callus using disarmed *Agrobacterium tumefaciens* strains. *Plant Cell Rep.,* **9** : 479-583.

Malmberg RL (1979) Regeneration of whole plants from callus culture of diverse genetic lines of *Pisum sativum* L *Planta,* **146** : 243-244.

Monti LM and Grillo S (1983) Legume seed improvement for protein content and quality. Qual. *Plant Food Human Nutr.,* **32** : 253-266.

Mroginski LA and Kartha KK (1981) Regeneration of pe (*Pisum sativum* cv Century) plants by *in-vitro* culture of immature leaflets. *Plant Cell Rep.,* **1** : 64-66.

Murashige T and Skooge F (1962) A revised medium for rapid growth and bioassays with tobacco tissue cultures. *Physiol Plant.,* **15** : 473-497

Natali L and Cavallini A (1987) Regeneration of pea (*Pisum sativum* L) plantlets by *in-vitro* culture of somatic embryos. *Plant Breed.,* **99** : 172-176.

Nauerby B, Madsen J, Christiansen J and Wyndaele R (1991) A rapid and efficient regeneration system for pea (*Pisum sativum* L), suitable. *Plant Cell Rep.,* **9** : 676-679.

Novak FJ, Lucreth S, Hermelin T, Donini B, Afza R, Daskalo S, Kubalakova M and Griga M (1984) Gamma ray radiation effects on multiple shoot cultures of pea (*Pisum sativum* L). In: *Proc. Int Sym Plant Tissue Cell Cult - Appl Crop Imrov.* (Eds Novak FJ, Havel L, Dolezel J) Olomouc pp 453-454.

Nirmala C and Kaul MLH (1997a) Biotechnology in Plant Breeding I. Foreign gene transfer: Aims, avenues and accomplishments.pp 208-249. In: *Plant Breeding Advances and In-vitro culture.* (Eds Siddique BA, Khan S) CBS Publ. N. Delhi.

Nirmala C and Kaul MLH (1997b) Biotechnology: Miracle or Mirage II. Remarkable genetic system of *Agrobacterium*. In: *Trends in Plant Tissue Culture and Biotechnology* (Eds Pareek LK and Swarnkar PL) Agro Botanical Publ. New Delhi pp. 31-46.

Nirmala C and Kaul MLH (1999) Biotechnology: Miracle or Mirage IV. Molecular markers in crop improvement. In: *Breeding Crop Plants.* (Eds Siddique B, Khan S) Kalyani Publ. N. Delhi. pp. 80-110.

Ozcan S, Barghichi M, Firek S and Draper J (1992) High frequency adventitious shoot regeneration from immature cotyledons of pea. *Plant Cell Rep.,* **11** : 44-47

Puonti-Kaerlas J (1993) Genetic improvement in pea crop improvement. *Acta Agric Scand Sect B Soil and Plant Sci.,* **43** : 65-73.

Puonti-Kaerlas J and Eriksson T (1988) Improved protoplast culture and regeneration of shoots in pea (*Pisum sativum* L) *Plant Cell Rep.,* **7** : 242-245.

Puonti-Kaerlas J, Stabel P and Eriksson T (1989) Transformation of pea by *Agrobacterium tumefaciens. Plant Cell Rep.,* **8** : 321-324

Puonti-Kaerlas, Eriksson T and Engstrom P (1990) Production of transgenic pea (*Pisum sativum* L) plants by *Agrobacterium tumefaciens* mediated gene transfer. *Theor. Appl. Genet.,* **80** : 246-252.

Puonti-Kaerlas J, Ottosson A and Eriksson T (1992) Survial and growth of pea protoplasts after transformation by electroporation. *Plant Cell Tiss. Org. Cult.,* **30** : 141-148.

Preobrazenskaja EV (1983) The comparision of organogenetic ability of different pea (*Pisum sativum* L) lines and cultivars in callus culture. Tr 14 Konf Mol Ucenych Biol Fak MGU, Moscow, pp. 134-137.

Rubluo A, Kartha KK, Mroginski LA and Dyck J (1984) Plant regeneration from pea leaflets cultured *in-vitro* and genetic stability of regenerants. *J. Plant Physiol.,* **117** : 119-130.

Sanago MHM, Shattuck VI and Strommer J (1996) Rapid plant regeneration. *Plant Cell Tiss. Org Cult.,* **45** : 165-168.

Schaerer S and Pilet PE (1991) Roots, explants and protoplasts from pea transformed with strains of *Agrobacterium tumefaciens* and *rhizogenes. Plant Sci.,* **78** : 247-258.

Schroeder HE, Schotz AH, Wardley-Richardson T, Spencer D and Higgins TJV (1993) Transformation and regeneration of two cultivars of pea (*Pisum sativum* L). *Plant Physiol.,* **101** : 751-757.

Sladky Z and Jandova B (1984) Micropropagation of pea, cucumber and potato. In: Proc Int Symp Plant Tissue Cult Appl Crop Improv. (Eds Novak FJ, Havel L, Dolezel J) Olomouc pp. 509-510.

Stejskal J and Griga M (1992) Somatic embryogenesis and plant regeneration in *Pisum sativum* L. *Biol. Plant.,* **34** : 15-22.

Tetu T, Sangwan RS and Sangwan-Noreel BS (1990) Direct somatic embryogenesis and organogenesis in cultured immature zygotic embryos of *Pisum sativum* L. *J. Plant Physiol.,* **137** : 102-109.

Trigiano RN and Gray DJ (2001) Plant Tissue Culture Concepts and Laboratory Excercises. CRC Press. USA pp. 454.

Van Doorne LE, Marshall G and Kirkwood RC (1991) Development of herbicide tolerance in peas II. Regeneration via somatic embryogenesis. In: *Aspects of Applied Biology 27, Production and protection of legumes.* (Eds Froud-Williams RJ, Gladders P, Heath MC, Jenkyn JF, Knott CM, Lane A, Pink D) Assoc Appl Biol, Wellesbourne. pp 271-274.

Van Doorne LE, Marshall G and Kirkwood RC (1995) Somatic embryogenesis in pea (*Pisum sativum* L)- effect of explant, genotype and culture conditions. *Ann. Appl. Biol.,* **126** : 169-179.

Wright PR (1985) Combining peas for human consumption. In: *The pea crop.* A basis for improvement. (Eds Hebblewaite PD, Heath MC, Dawkins TCK) Butterworths, london pp 441-451.

White PR (1943) *A handbook of plant tissue culture.* Cattel J, Lancaster. UK.

Zubko EI, Kuchul NV, Tumanova LG, Vikonskaya NA and Gleba YY (1990) Genetic transformation of pea plants mediated by *Agrobacterium tumefaciens. Biopolymers Cells,* **3** : 80-84.

Chapter 10

NUTRITIONAL IMPROVEMENT IN VEGETABLE CROPS USING BIOTECHNOLOGICAL APPROACHES

Jagdish Singh★, Sanjeev Kumar and G Kalloo

Indian Institute of Vegetable Research, 1, Gandhi Nagar (Naria) P.O. Box No 5002, P.O. B.H.U., Varanasi - 221 005, U.P., India

Summary

Biotechnology has emerged as one of the most innovative achievements in the life sciences of this century and is influencing almost every aspect of human life. The development of plant tissue culture techniques target somatic phase for genetic manipulation. These technologies help the breeder to wider the genetic base by generating variability in vitro, by bringing together genomes for recombination, which cannot be brought together by sexual crosses and by direct introduction of genes into cells by transformation. Similarly techniques of molecular biology permit us to locate, clone and sequence individual genes from a complex misture of DNA sequences. Using molecular methods plant breeders can follow genes of interest in segregating populations and improve selection efficiency.

This chapter describes the progresses that have been made in genetically manipulating vegetables to alter colour, carotene content, as well as their post harvest properties. Most of the studies have concentrated on tomato, which has been the subject of intensive study and genetically modified tomato varieties are now in the market with extended shelf life. Many ripening genes have now been cloned including polygalacturonase and pectinase involved in cell wall

★Corresponding author : E-mail : singh.jagdish@lycos.com; pdveg@up.nic.in

softening; phytoene synthase, required for carotene synthesis; and ACC synthase and ACC oxidase, which catalyze the production of ethylene. The techniques of gene silencing, either using antisense genes or sense-suppression have been used successfully to reduce or inactivate the expression of specific genes. Tomatoes genetically modified to have low polygalacturonase activity have specific advantages for fresh market or processing applications, whereas reduction in ethylene synthesis has been shown to improve quality and storage life of tomato and melon. Studies to alter the solids content in tomato are also in progress. Field trials are being conducted in EU and US for broccoli, cabbage, carrot, cauliflower, eggplant, onion, pea, pepper, sweet potato and watermelon. Brief accounts of genetic modifications of carbohydrate metabolism in potato, nitrogen assimilation in leafy vegetables and nutritional quality of seed proteins in leguminous vegetables have also been discussed.

Keywords : Anti oxidants, antisense suppression, carotenoids, post-harvest, quality traits, QTLs, transgenics, vegetables.

1. INTRODUCTION

Vegetables are an important component of the daily diet and fairly large number of vegetable crops serve to meet the need of nutritional requirement. Since the beginning of agriculture, the human race has selected and developed varieties of vegetables. Initially, better yield was the priority, but in the present day market, appearance and reproducibility of color and size as well as nutritional quality of vegetables are important factors. Earlier, color modification in vegetables has been achieved by selection for example; the orange carrot and yellow bell pepper were natural mutants. Pigmentation of flowers by anthocyanins is already being altered by genetic manipulation (Courtney-Gutterson, 1994), similar approaches to anthocyanins in fruits and vegetables could be taken. As the understanding of biosynthesis of carotenoids has improved to a great extent, similar approaches to produce novel colors in carotenoid pigmented vegetables could also be taken. Flavour and aroma compounds are also feasible for manipulation as are pigments. The pathways of some aromatic and terpenoid aromas are steadily being unraveled and some of the genes involved in such pathways have been isolated. The option to create novel flavours within fruits and vegetables lies in the near future. The intensity of the flavour compounds and their continued biosynthesis in the post harvest product could be enhanced. The processing characteristics of vegetables have received much attention. Considerable progress has been made in genetically manipulating vegetables to alter their post harvest properties. The tomato fruit has been the subject of intensive study and varieties ('Favr-Savr') are now in the market with extended shelf life as a result of decreasing polygalacturonase activity (Gray *et al.*, 1992; Sheehy *et al.*, 1988) or inhibiting ethylene production (Gray *et al.*, 1992; Oellar *et al.*, 1991). Tomatoes with altered solids content are also not far behind.

2. BREEDING FOR QUALITY ATTRIBUTES IN VEGETABLE CROPS

The classical approach for breeding cultivars is to select suitable phenotypes or mutants, which are then crossed, selfed, cloned or combined with populations. A major drawback of this approach is that for most agronomically important traits, the phenotypic variance is the base for selection, which, however, is not only composed of the genotypic variance but also comprises an environmental component as well as interactions between genotypes and environments. This tends to obscure selection progress and is one of the reasons why breeding for quantitative or polygenic traits is so tedious and time consuming. In addition, a combination of positively acting polygenes of one specific genetic background is difficult, if not impossible, because of recombinational dispersion of genes in each sexual generation.

2.1. Various approaches in plant breeding

Classical Breeding

- Selection of phenotypes
- Intercrossing, selfing, vegetative cloning of favorable phenotypes
- Inbred line or population or clonal varieties

Marker assisted selection

- Mapping of quantitative trait loci (QTL) (*e.g.,* solids content and pH of tomato pulp, chips quality of potato)
- Selection of genotypes instead of phenotypes

Genetic Engineering

- Isolation of gene of interest
- Gene suppression by antisense or co-supression
- Expression of foreign genes (constitutive or tissue specific)
- Over-expression of native genes

Biotechnological breakthroughs like *in vitro* fusion and regeneration of plant cells and marker assisted selection (MAS) of monogenic traits as well as the tracing of quantitative trait loci (QTL) by use of molecular genomic markers have gained increased importance for plant breeding. For example in tomato, QTL for quality traits such as soluble solids content, fruit mass, fruit pH, and fruit shape have been mapped (Grandillo *et al.*, 1996). In potato, traits, such as chip color, tuberization and tuber dormancy may be traced by molecular markers. For these characters, 5-13 QTLs have been identified. As a more recent achievement, transformation methods enabling the transfer of any isolated gene into virtually any important cultivated plant species have opened the tools of genetic engineering as a novel opportunity to the plant breeder, which may help to circumvent some of the obstacles in the breeding of complex traits. Biotechnology can be employed to improve the quality of vegetable crops and also to increase crop resistance to pathogens and insect pests besides increasing tolerance to abiotic stresses *viz.,* drought, flood or extreme temperatures. Nevertheless, at the present time, many vegetable crops are the targets of rapid biotechnological developments.

Biotechnology already has a strong impact on vegetable breeding by providing cultivars that are tolerant to herbicides, resistance to fungi, pests and viruses, or that have been rendered male sterile. From 1987 up to 1997, 4279 field trials at over 15,000 field sites were approved or acknowledged by the authorities. The most frequent vegetable crops that have been subject to genetic engineering in the U.S. and Europe are summarized in Fig.1. Potato is the major crop being field tested, followed by tomato, melons, squashes, cucumber and lettuce. Of the field releases in EU and the U.S.A., 26% correspond to fruits and vegetable crops, with potato (444) and tomato (428) leading the field far ahead of melons and squashes (132). Infrequent crops with five or less field-testings include broccoli, cabbage, carrot, cauliflower, eggplant, onion, pea, pepper, sweet potato and watermelon.

Amongst the field-testing in the U.S.A., herbicide tolerance is ranked first followed by insect resistance and product quality (Fig. 2). Product quality comprises one fifth of the field releases and may thus be considered as a significant objective of genetic engineering. Of the commercialized transgenic crops, which amounted to 12.8 million hectares worldwide, quality traits have a portion of less than 1% while herbicide tolerance, insect resistance and virus resistance are the dominant traits with a portion of 54, 31 and 14% respectively (James, 1997).

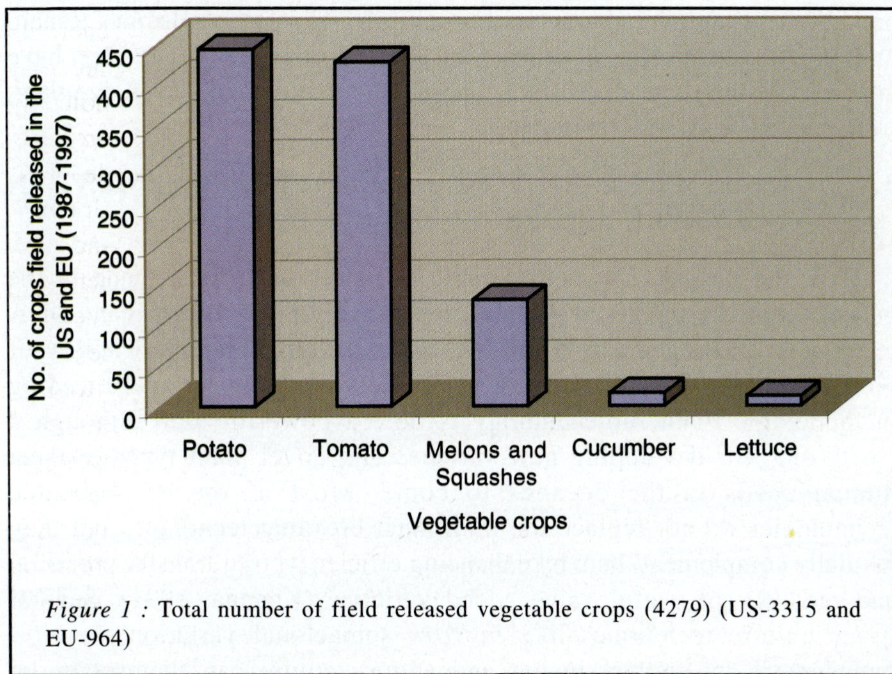

Figure 1 : Total number of field released vegetable crops (4279) (US-3315 and EU-964)

Among the fruit and vegetable crops, tomato has become the paradigm of 'high-tech' plant breeding and the breeding goals in tomato have to be grouped for the purpose of use. Besides general goals, tomatoes have to fulfill specific demands for either processing or fresh market use. Ripening characteristics, solids content, and fruit color have all been subject to genetic engineering. In the United States, the FDA and USDA have approved five lines or cultivars of tomato with delayed softening

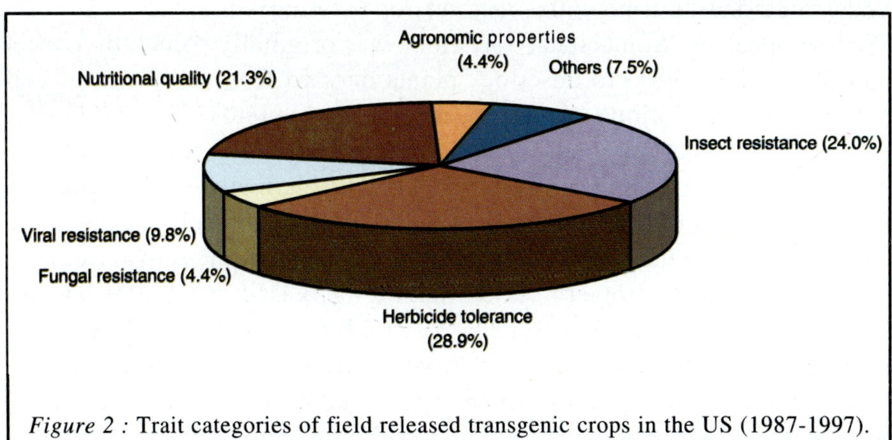

Figure 2 : Trait categories of field released transgenic crops in the US (1987-1997).

and ethylene ripening characters for commercial release. Besides tomato and potato, a variety of other transgenic fruits and vegetables have appeared on the stage since 1995. These include broccoli, carrot, eggplant, pea, pepper, and watermelon.

3. TISSUE CULTURE FOR QUALITY IMPROVEMENT IN VEGETABLE CROPS

Over the past decade considerable advances have been made in the genetic engineering of crop plants by utilization of recombinant DNA techniques and regeneration through tissue culture. The applicability of biotechnology to crop improvement involves non-conventional breeding methodology. Plant biotechnology is now a powerful tool, although a costly one, for developing new varieties and novel plant types to meet human needs in the decades to come. Most of the new/advance technologies do not replace the traditional breeding techniques but they mutually complement them by enhancing efficiency, trait transfer precision and recovery of useful value added varieties (Khanna, 1991). Several tissue culture techniques like *in vitro* somaclonal variation, *in vitro* mutagenesis, protoplast fusion and anther culture can themselves be utilized for quality improvement (Table 1). Applying selection pressure on isolated cells *in vitro* can also be a valuable tool to select desirable plants.

3.1. Somaclonal variation

For a considerable time, the frequently observed variability in plant population, regenerated from tissue culture was ignored. Now organogenic and embryogenic differentiation in callus and suspension cultures are being regarded as a potential method for creating variability. The term "Somaclones" or "Somaclonal variation" was originally coined by Larkin and Scowcraft (1981) to describe "plants derived from any form of cell culture" and variations displayed among such plants. The variations observed in tissue culture are usually due to changes induced by culture conditions. Somacloning can be used as a valuable method for production of strains with improved quality and yield. They also result in novel type suitable for different purposes (Karp, 1991). One such successful example is development of a tomato variety having increased dry matter content (Evans, 1989). Very recently, DNAP holding corporation, USA, has released a tomato variety Fresh World Farms® tomato. It is a fresh market tomato developed by somaclonal variation having superior color, taste, texture, dry matter content and also an increased shelf life.

Table 1. Vegetable crops modified for quality parameters using tissue culture techniques

Crop	Company	Method used	Trait	Reference
Lycopersicon esculentum	DNA plant technologies, USA	Somaclonal variation	High dry matter content	Evans, 1989
Lycopersicon esculentum	DNAP Holding corp., USA	Somaclonal variation	Superior taste	James, 1997
Capsicum annuum	DNAP Holding corp., USA	Anther culture	Increased sweetness	James, 1997
Solanum tuberosum	-	Mutant selection	High starch	Hovenkem *et al.*, 1987
Solanum tuberosum	-	Protoplast fusion	Improved yield and quality	Mattheij and Puite, 1992
Brassica oleracea var. *botrytis* + *B. napus*	-	Protoplast fusion	Low linolenic acid	Heath and Earle, 1995
Brassica oleracea var. *botrytis* + *B. napus*	-	Protoplast fusion	High Eurecic acid	Heath and Earle, 1997

3.2. Anther culture

Among the several biotechnological tools available to plant breeders, culture of anther and pollen, holds special promise because of its potential application in crop breeding. Anther culture has the potential to produce haploid plants that can be useful in generating homozygous diploids. These homozygous diploids have tremendous application in exploitation of hybrid vigour. Anther and pollen culture employs the same technique as that of tissue culture; only the explant used is either anther or microspore. The development of improved varieties by anther/microspore culture is already a reality in China. Most of the Chinese work is related to quality improvement in cereal crops like rice, barley etc. In cereal crops like rice, anther culture derived plants were identified with heavier seeds and high protein content (Yang and Fu, 1989). But in case of vegetables, less work has been done on this line. Table 2 summarizes some vegetable crops where haploids have been produced. This method involves production of F_1 hybrids between two parents of desired trait, and then microspores of F_1 are cultured to obtain plantlets, which are haploid and sterile. An application of mitosis inhibitors (like colchicine) are must to make these diploid, fertile and homozygous. Due to their inbred nature, the recessive deleterious as well as desirable quality traits are expressed in very first generation, which makes the selection very easy. The recent example of market release of sweet pepper confirms this strategy (James, 1997). This sweet pepper has increased sugar

Table 2. Anther/pollen culture in vegetable crops

Crop species	Mode of development
Asparagus officinalis	Direct/indirect androgenesis
Beta vulgaris	Direct/indirect androgenesis
Brassica oleracea	Direct/indirect androgenesis
Capsicum annuum	Direct/indirect androgenesis
Cucumis sativus	Indirect androgenesis
Lycopersicon esculentum	Direct/indirect androgenesis
Raphanus sativus	Direct androgenesis
Solanum tuberosum	Direct/indirect androgenesis
Solanum melongena	Direct/indirect androgenesis

content with deep red color and is nearly seedless. It was developed through this technique, which stabilizes preferred traits such as taste and quality.

3.3. Protoplast fusion and somatic hybrids

Protoplast (*i.e.* naked cells or cells without cell wall) can now be isolated and cultured at large scale. This protoplast is a very useful system for genetic manipulation. Protoplast culture employs a series of operations such as isolation of the protoplasts from cells, culturing them on a suitable medium, inducing them to divide and then regenerating plantlets from them. They are becoming an extremely valuable tool in biotechnology for improvement of complex characters such as quality and taste. Protoplasts can serve as a means of intensive *in vitro* multiplication and can lead to extensive somaclonal variation. Their usefulness lies in their ability for somatic hybridization of close or distant relatives. There are many reports in vegetables after the pioneering work of Cocking (1960), where substantial cell division has been observed in cultured protoplast. However, regeneration ability of fused protoplast is low. Nevertheless few successful reports opens up possibilities in other vegetables (Table 3). Mattheij and Puite (1992) were successful in fusing protoplast of *Solanum tubrosum* with *S. phureza* to obtain an improved potato variety. The same technique was used to resynthesize *B. napus* by fusing its protoplast with *B. oleracea botrytis* to get two different plants having low linolenic acid and high eurecic acid content (Heath and Earle, 1995, 1997).

At present the major studies on somatic hybridization are mainly concerned with *Brassica, Solanum* and *Lycopersicon* (Singh *et al.*, 1998). In case of potato, the technique is already being used on commercial scale. In India, several *Brassica* lines derived from somatic hybridization are being used in breeding programmes. Now it is likely that the approach of somatic hybridization will find greater application for quality improvement in sexually incompatible species.

4. GENETIC ENGINEERING FOR TRAITS AFFECTING QUALITY AND DELAYED RIPENING IN VEGETABLE CROPS

The major targets for genetic engineering of vegetable crops are for alteration of the parameters like fruit softening, fruit ripening, storage ability, total and soluble solids, fruit colour, carotenoid contents, sugar profile, parthenocarpy, starch branching ratio (amylose vs amylopectin)

Table 3. Protoplast fusion in vegetable crops

Crop	Method of fusion
Allium ameloprasum + *Allium cepa*	PEG mediated chemical fusion
Lycopersicon esculentum + *Lycopersicon peruvianum*	PEG mediated chemical fusion
Lycopersicon esculentum + *Lycopersicon pennelli*	Chemical method using DMSO
Lycopersicon esculentum × *Lycopersicon pennelli hybrid* + *Solanum melongena*	Chemical method using DMSO/PEG
Lycopersicon esculentum + *Solanum richii*	Chemical and electrical methods
Solanum melongena cv. Dourga + *Solanum khasianum*	Electrofusion
Solanum melongena + *S. nigrum*	Chemical and electrical fusion
Solanum melongena + *S. torvum*	Chemical and electrical fusion
Solanum melongena L + *Solanum symbriifolium*	PEG induced fusion

and bruising susceptibility etc. The advent of genetic engineering and recombinant DNA technology has opened up tremendous possibilities for transforming almost any plant by transferring any gene from any organism, across taxonomic barriers. Gene transfer has become routine and many genes of importance have been transferred in vegetable crops (Table 4). Achievements are so far restricted to one or two genes only, while, multigenic characters are still out of reach of the genetic engineers.

During the past few years, hundreds of transgenics have been field tested in several countries around the globe. Successful cases of transformations include major economic crop plants, which includes vegetable crops as well (Table 5).

4.1. Genetic engineering for inhibition of fruit tomato senescence

Ripening behavior is a compound trait and is the result of complex physiological pathways. Different approaches have been suggested to modify genes related to the ripening process. These may be grouped into

Table 4. Genes used in delaying ripening and prolonging shelf life in tomato and other vegetable crops.

Crop	Gene targeted	Organization/Company
Tomato	*acc* deaminase	Calgene
	acc oxidase	Petoseed
	acc synthase	Cornell University
	Acetolactate synthase	DNAP
	Ethylene receptor gene (*etr*-1-1)	DNAP
	Galactanase	University of Florida
	Galactanase	Zeneca
	ipt	Calgene
	Pectin esterase (*pe*)	Purdue University/ Peto seed
	Pectin methyl esterase (*pme*)	Purdue University
	Polygalacturonase (*pg*)	Calgene
	SAMase	Agritope
	sam transferase	Agritope
	sps	Calgene
	Transcriptional activator	DNAP
Cabbage	*acc* oxidase antisense	Seminis Vegetable
	acc synthase antisense	Seminis Vegetable
Melons	*acc* oxidase	INRA, France
	sam hydrolase	Agritope
	sam transferase	Agritope
Capsicum/Pepper	Acetolactate synthase	DNAP
	Galactanase	DNAP
	Hemicellulase	DNAP

Table 5. Genetically engineered vegetables modified for quality improvement, commercialized in USA/UK

Vegetables/Trade mark product	Altered trait	Technology used	Participant company
Tomato (Cherry)	Altered ripening, enhanced fresh market value	-	Agritope
Tomato (Flavr-Savr) (McGregor)	Delayed ripening, alteration in texture, enhanced fresh market value	Low PG trait produced by an antisense PG gene	Calgene
Tomato (Endless Summer)	Delayed ripening	-	DNA Plant technology
Tomato	Delayed ripening	-	Monsanto
Tomato	Thicker skin, altered pectin Enhanced processing value	Sense PG transgene to inhibit the expression of endogenous gene by sense supression	Zeneca/ Petoseed
Tomato (DNAP 9)	20% increase in soluble solids	Somaclonal variation	-
Broccoli	Slow ripening and the crop stays green longer	Insertion of a non-plant gene to control ethylene production	Agritope
Euromelon	Extended shelf life, high quality fruit and ripens on demand	Insertion of antisense gene (*pMEL.1*) which encodes the final enzyme for ethylene biosynthesis	-

efforts to reduce internal ethylene production as the physiological basis of ripening, and to delay fruit softening at the final steps of ripening. Tomato belongs to the group of climacteric fruits, which develop an autocatalytic release of ethylene at the beginning of maturation. The physiological effects of ethylene hormone are manifold and include seed germination, stimulation of ripening in fruits and vegetables, leaf abscission, fading in flowers, flower wilting, leaf yellowing and leaf epinasty. Ethylene is one of the simplest organic molecules with biological activity, which is considered to be the fruit-ripening hormone. Ethylene regulates fruit ripening by coordinating the expression of genes that are responsible for a variety of processes, including rate of respiration, autocatalytic ethylene production, chlorophyll degradation, carotenoid synthesis, conversion of starch to sugars and increased activity of cell wall degrading enzymes. It has always been a goal of plant biologists and post-harvest physiologists to prevent or delay fruit ripening in a reversible manner by controlling ethylene action or production. The cloning of fruit ripening genes and genes involved in ethylene biosynthesis opened the road to the development of ripening mutants in tomato using antisense RNA technology.

4.2. Ethylene biosynthesis genes and their manipulation for delayed ripening

Methionine is the biological precursor of ethylene in all-higher plants (Yang and Haffman, 1984). The biochemistry of ethylene has been intensively studied (Kende, 1993). The biosynthetic pathway for ethylene synthesis in higher plants has been elucidated in Fig. 3.

Briefly, there are two enzymatic steps dedicated to ethylene production. In the first reaction, S-adenosyl methionine is converted to cyclic amino acid, 1-aminocyclopropane-1-carboxylate (ACC) by ACC synthase (ACS). The ACC is then metabolized by ACC oxidase (ACO), which requires O_2, Fe^{2+} and is activated by CO_2, to produce ethylene. During ripening, the expression of *acs* and *aco* genes and the activity of their encoded enzymes govern the rate of ethylene production. Since a decrease in rate limiting enzymes ACC synthase or ACC oxidase should result in a concurrent decrease of ethylene production, these two enzymes have been the targets in different transgenic antisense or cosupression approaches. Each enzyme is encoded by a multigene family in tomato and their are 9 *acs* (Rottman *et al.*, 1991) and at least 3 *aco* (Barry *et al.*, 1996) genes. Inhibition of ethylene biosynthesis by production of antisense RNA to ACC synthase (Oeller *et al.*, 1991) and ACC oxidase

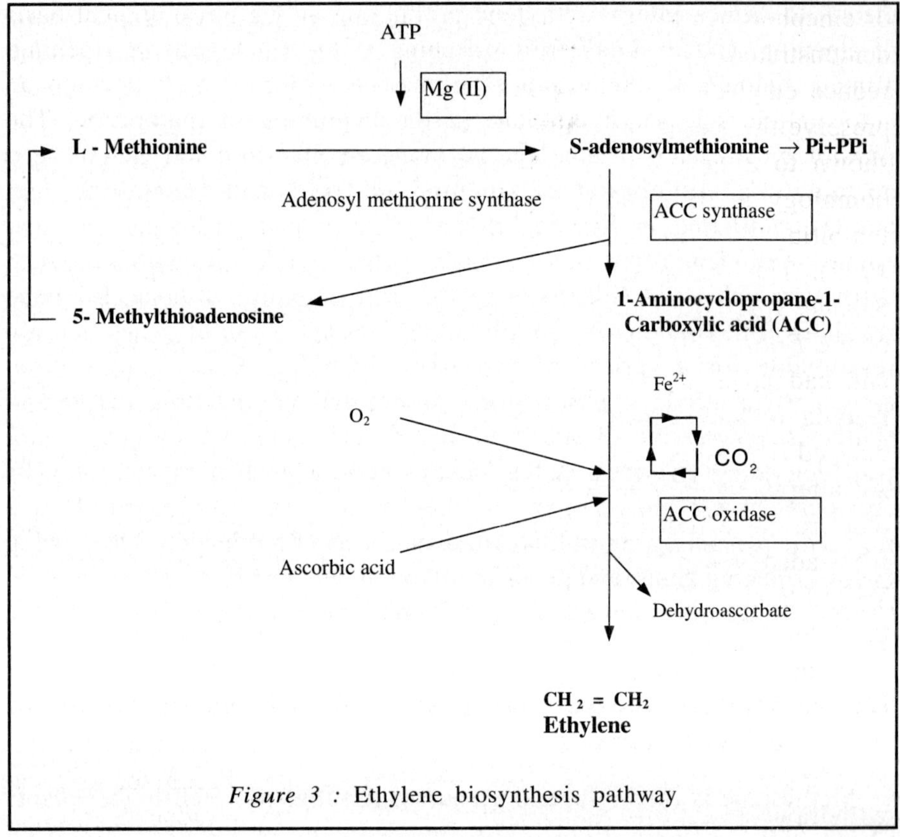

Figure 3 : Ethylene biosynthesis pathway

(Hamilton *et al.*, 1990) or by deamination of ACC (Klee *et al.*, 1991) represses tomato fruit ripening. Expression of ACC deaminase (the enzyme that degrades ACC to α-ketobutyric acid) via *Agrobacterium tumefaciens* mediated transformation led to significant decrease in ethylene production. In transgenic tomato plants expression of this gene under the control of CaMV 35S promoter and a pea *rbc* S3' region, ethylene synthesis is reduced up to 97% in leaves and 90% in fruits which delayed fruit ripening and the transgenic fruit remained firm for an indefinite period (Klee *et al.*, 1991). During tomato fruit ripening, two ACC synthase genes *le-acc* 2 and *le-acc* 4, are expressed (Rottman *et al.*, 1991). Expression of antisense RNA derived from the cDNA of *le-acc2* gene resulted in almost complete inhibition of m-RNA accumulation of both ripening induced ACC synthase genes (Oeller *et al.*, 1991). The antisense fruits never ripen unless exogenous ethylene is provided. It is possible to effectively stimulate ripening in this antisense phenotype by treating them with ethylene gas or supplying compounds such as ACC or ethephon, which leads to ethylene production in plants. These results demonstrated that antisense technology could be successfully used to reduce ethylene production and thereby to extend the shelf life and to preserve the quality of fruits and vegetables. Another gene, *e8* has been shown to effect ethylene synthesis indirectly. The gene *e8* has some homology to the family of dioxygenases, to which aco belongs and inhibiting *e8* expression increases ethylene synthesis (Penarrubia *et al.*, 1992). The transgenic tomato cultivar *Fresh World Farms® Endless Summer* (produced by DNAP), which is a genetically engineered version of the *Fresh World Farms®* tomato, was put on the test market in 1995 and had been engineered by cosupression of the ACC synthase gene leading to an extension of postharvest shelf life to 30-40 days. The delayed ripening fruit ripens as usual when exogenous ethylene is applied. An alternative approach was successfully tried by introducing enzymes that are not originally present in tomato. In one such approach a gene for S-adenosylmethionine hydrolase was isolated from the bacteriophage T3 and transferred to tomato (Ferro *et al.*, 1995). This enzyme hydrolyzes the intermediate AdoMet by which homoserine and MTA are released. MTA in turn is a potent inhibitor of ACC synthase. In another case, the enzyme gene for ACC deaminase from *Pseudomonas* was introduced in tomato, which catalyzes the opening of ACC cyclopropane ring to give α-ketobutyrate. In this case, the ethylene precursor ACC is withdrawn from the pathway, again resulting in a reduced ethylene synthesis. The genes targeted for manipulation of ethylene biosynthesis comprise AdoMet hydrolase, ACC synthase, N-ACC malonyltransferase, ACC oxidase and ACC deaminase (Table 4).

Transgenic tomatoes expressing S-adenosylmethionine decarboxylase, thus increasing the production of polyamines, were studied during the senescence phase (Mehta *et al.*, 1998). Ripening fruits from transgenic lines accumulate polyamines, whereas the red fruit from the wild type line carried through tissue culture had little, if any, spermidine and spermine. Fruits from some of these transgenic lines accumulated several fold higher lycopene, appeared to store better and showed delayed ripening and senescence compared to wild type plants. A new transgenic tomato cultivar Bioscein was bred with antisense efe cDNA (Ye *et al.*, 1999). It was derived from the cross of transgenic line D2 carrying an antisense DNA of ethylene forming enzyme and inbred line A53 of a common tomato cultivar. The hybrid had delayed fruit ripening, high yield, good adaptability, and high quality over control.

4.3. Genes involved in cell wall metabolism and their manipulation for texture improvement

Another aspect of ripening relates to the softening processes in the tomato fruits. Delay of softening has been achieved by gene suppression of enzymes that degrade the structural integrity of the cell tissue. Cellulose, hemicelluloses and pectins are cell wall components that contribute to textural characteristics. Polygalacturonase (PG) is the cell wall modifying enzyme that catalyzes the hydrolysis of polygalacturonic acid chains in unmethylated regions of pectin. It hydrolyzes the α-1, 4 linkages in the polygalacturonic acid component of cell walls. PG is synthesized specifically during ripening and is secreted into the intercellular space of pericarp cells. The function of PG during natural ripening process is to break down the pectin of the middle lamella between the fruit cell walls, thereby causing fruit softening. The first ripening cDNA to be cloned and sequenced originally called *tom6*, encoded polygalacturonase (Grierson *et al.*, 1986b). There appears to be one gene (Bird *et al.*, 1998) for the endopolygalacturonase that is synthesized *de novo* during tomato ripening. The *pg* gene is transcriptionally activated at the onset of ripening and the mRNA and protein are extremely abundant in ripening fruit. The best-known case of inhibition of Pectin degradation by means of genetic engineering is the antisense suppression of the *pg* gene, which is the effective principle in the transgenic cultivar 'Flavr-Savr' by Calgene. As a different approach, co-supression of the same gene was applied by Zeneca/Petoseed for the production of tomato giving a higher viscosity pulp, which was introduced into the British market as tomato paste and puree in February 1996. Pectin esterase (PE) is the second cell wall modifying enzyme that removes the methyl groups from methylated polyglacturonic acid in pectin, its action may be a pre-requisite for polygalacturonase attack. There are several isoforms of PE and these are now known to be encoded by two distinct multigene families. The first tomato *pe* cDNA sequence was reported by Ray *et al.*, (1998), but there are several related genes (Hall *et al.*, 1994). Antisense inhibition experiments revealed a separate gene family predicted to have a distinct DNA sequence (Hall *et al.*, 1993). One group of *pe* isoforms is expressed in ripe and unripe fruit and the second group is expressed in roots, stems, leaves and unripe fruit (Hall *et al.*, 1993). In addition to PE and PG, other cell wall modifying enzymes encoded by multigene families also occur in tomatoes, including cellulase (Lashbrook *et al.*, 1994) and galactanase (Carey *et al.*, 1995), but the precise role of each during ripening remains to be established. Tomato plants were transformed with an antisense endo-1, 4-ß glucanase (Cellulase), *cel2* transgene under the control of the constitutive cauliflower mosaic caulimovirus 35S promoter in order to suppress mRNA accumulation of *cel2*. In two independent transgenic lines, *cel2* mRNA abundance was reduced by > 95% in ripe fruit pericarp and by 80% in fruit abscission zones relative to non-transgenic controls. In both transgenic lines, the softening of *cel2* fruit pericarp measured using stress-relaxation analysis was indistinguishable from control fruit. No difference in ethylene evolution was observed between fruit of control and antisense *cel2* genotypes. However, in fruit abscission zones the suppression of *cel2* mRNA accumulation caused a significant increase in the force required to cause breakage of the abscission zone an increase of 27% in one transgenic line and 46% in the other transgenic line. The *Cel2* gene product contributes to cell wall disassembly occurring in cell separation during fruit abscission, but its role if any, in softening or textural changes occurring in fruit pericarp during ripening was not revealed by suppression of *Cel2* gene expression. Suppression of pectin methyl esterase (PE), involved in pectin modification, and of endo-1, 4-β glucanase, involved in fruit softening by degrading the major hemicellulosic polymer, xyloglucan, was also patented as approaches to inhibit fruit softening.

4.4. Genetic modification for taste, flavor and processing quality in tomato

Flavor results from the interaction of a large number of compounds with taste receptors, including simple carbohydrates, organic acids, salts, organic volatiles *viz.* aldehydes and ketones etc. In tomato, primarily simple sugars, organic acids and certain volatile compounds determine flavor. In other vegetable crops, very unusual and specific compounds have an influence on flavor, for example, terpenes in carrots, capsaicin in peppers, glucosinolates in crucifers and allylsulfides in onion and garlic.

The flavor of fresh market tomatoes is of continuous concern for consumers. Initially, the development of genetically engineered long shelf life tomatoes was claimed to improve the taste of the fruit because the tomatoes would ripen on the vine for a longer time and would, thus, accumulate more flavor than traditional cultivars, which are picked at the mature green stage (Vanderpan, 1994). However, from the beginning this was a matter of controversy because tomatoes properly harvested at the mature green stage will ripen into a product indiscreminable from vine ripened fruit. In addition, flavor in any fruit is a multiple factor trait dependent on the genetic background of a cultivar and may, thus, not be expected to undergo substantial improvement by modifying a single gene. Tomato plants were transformed with gene constructs containing a tomato alcohol dehydrogenase (ADH) cDNA coupled in a sense orientation with either the constitutive CaMV 35S promoter or the fruit specific tomato polugalacturonase promoter (Speirs *et al.*, 1998). Ripening fruits from plants transformed with the constitutively expressed transgene(s) had a range of ADH activities, some plants had no detectable activity, whereas others had significantly higher ADH activity, up to twice then that of control. Transformed plants with fruit specific expression of the transgene(s) also displayed a range of enhanced ADH activities in the ripening fruits, but no suppression was observed. Modified ADH levels in the ripening fruit influenced the balance between some of the aldehydes and the corresponding alcohols associated with flavor production. Hexanol and Z-3-hexenol levels were increased in fruits with increased ADH activity and reduced in fruit with low ADH activity. Concentrations of the respective aldehydes were generally unaltered. The phenotypes of modified fruit ADH activity and volatile abundance were transmitted to second-generation plants in accordance with the patterns of inheritance. In a preliminary taste trial, fruit with elevated ADH activity and higher levels of alcohols were identified as having a more intense ripe fruit flavor.

Solids content, which is important for processing of tomatoes is another quality trait that has been genetically modified by different approaches. Pectin esterase gene expression has been applied to increase soluble solids content by about 15%. The effect of genetic down regulation of polygalacturonase and pectin esterase activity on rheology and composition of tomato juice was studied by Errington *et al.*, (1998). Juice prepared from transgenic tomato plants containing antisense genes encoding polygalacturonase [PG(as)] and pectin esterase isoform 2 [PE(as)] were compared to those made of normal fruits. Juice prepared from PG (as) fruit was thicker than the normal fruit. This increased viscosity was shown to develop in a time dependent manner and may involve the continued action of PE in the post homogenized juice. Poretta *et al.*, (1998) reported improved viscosity, color and sensory attributes in transgenic fruits of tomato with reduced PG activity due to the expression of a PG antisense gene. Invertase gene suppression approach (Ohyama *et al.*, 1995; Klann *et al.*, 1996) was used by PetoSeeds to increase soluble solids that mainly consists of polysaccharides. Since the balance of sugars and acid has a major impact on the flavor of a fruit, attempts to modify the soluble solids are also of interest for the fresh market. The application of genetic modification yielded products with improved viscosity (40-60%), color (30-40%) and many sensory attributes in comparison with their conventional counterparts.

4.5. Carotenoid biosynthesis genes and their manipulation : designed food for health

Carotenoids derive their name from the main representative of their group, β-carotene, the orange pigment first isolated from carrots (*Daucus carota*) by Wackenroder, in the year 1831. Carotenoids provide the color pigments in a variety of fruits and vegetables. In tomato the red fruit color is provided by lycopene. Of the more than 600 naturally occurring carotenoids, 40 carotenoids have already been identified in vegetables (Table 6) out of which only α-, β- and ε-carotene possess vitamin A activity. These carotenes, alongwith γ-carotene and the lycopene and lutein, which do not convert to vitamin A, have been reported to offer protection against colorectal, breast, uterine and prostrate cancers. Carotenes have also been related to immune response and protection of the skin against ultraviolet radiation. Additionally, they give antioxidative protection to the glutathione Phase II detoxification enzymes in the liver and thus support elimination of carcinogenic chemicals, pollutants and other toxic chemicals from the body.

The second carotenoid subclass, the xanthophylls also comprise compounds of positive biological effects such as canthaxanthin (UV protection), cryptoxanthin, zeaxanthin or astaxanthin. These compounds seem to exhibit antioxidative protection of vitamin A, vitamin E and other carotenoids. By genetically manipulating the carotenoid pathway a range of different quality parameters such as contents of solids, vitamin A, antioxidative compounds, or fruit color may be altered. This has been attempted in tomato by gene-suppression or over-expression with HMG-CoA reductase or with phytoene synthase as targets respectively. Supression of the latter enzyme, which catalyzes the formation of the first carotene named phytoene, may lead to tomatoes that are not red but display a shade of yellow like certain peppers. On the other hand , over-expression of phytoene synthase may result in deeper red color and a higher antioxidative activity of the produce (Bird *et al.*, 1994).

The first step in the biosynthesis of carotenoids (Fig. 4) is catalyzed by phytoene synthase located in the chromoplasts. This enzyme catalyses the conversion of geranyl-geranyl diphosphate to phytoene which is subsequently converted to β-carotene and lycopene by phytoene

Table 6. Qualitative and quantitative carotenoid distribution in vegetables

Vegetables	Total carotenoids (µg/g fresh wt.)	β-carotene (µg/g fresh wt)	Major carotenoids
Asparagus (*Asparagus officinalis* L.)	8.5	4.3-7.0	β-Carotene, Lutein, Violaxanthin, Neoxanthin
Bitter gourd (*Momordica charantia* L.)	5.3	2.3	α-Carotene, β-Carotene, Zeinoxanthin, Lutein
French bean (*Phaseolus vulgaris* L.)	17.1	2-4	α-Carotene, β-Carotene, Lutein, Lutein 5,6-epoxide, Neoxanthin, Violaxanthin
Broccoli (*Brassica oleracea* var. *italica* Plenck.)	42.4	4.8	β-Carotene, Lutein, Isolutein, Luteoxanthin, Violaxanthin, Neoxanthin, Chrysanthemaxanthin
Cabbage (*Brassica oleracea* var. *captita* L.)	8.9	0.8	β-Carotene, Lutein, β-Carotene 5,6-epoxide, Neoxanthin, Violaxanthin, Chrysanthemaxanthin
Cauliflower (*Brassica oleracea* var. *botrytis* L.)	0.44	0.11	β-Carotene, Lutein, Violaxanthin, Neoxanthin
Carrots (*Daucus carota* L.)	54 - 124	76.0	α-Carotene, β-Carotene, J-carotene, β-Zeacarotene, r-carotene, Neurosporene
Cucumber (*Cucumis sativus* L.)	17.2	2.20	α-Carotene, β-Carotene, and Cryptoxanthin

Table 6. Continued

Vegetables	Total carotenoids (µg/g fresh wt.)	β-carotene (µg/g fresh wt)	Major carotenoids
Pepper (Green) (*Capsicum annuum* L.)	10.0	6.8	Capsanthin, Capsorubin, Cryptocapsin, β-Carotene
Pepper (Red)	127 - 284	1.27 - 2.84	β-Cryptoxanthin, Violaxanthin, Neoxanthin
Lettuce (*Lactuca sativa* L.)	68.0	10.8 - 24.5	β-Carotene, Lutein, Violaxanthin, Neoxanthin
Spinach (*Spinacia oleracea* L.)	69.0	40.0	β-Carotene, Lutein epoxide, Violaxanthin, Lutein, Antheraxanthin, Neoxanthin
Tomato (*Lycopersicon esculentum* L.)	70 - 190	7.8	Lycopene, β-Carotene, Phytoene, Phytofluene, γ-Carotene, Neurosporene

Source : Gross, J. (1991). Pigments in vegetables: Chlorophylls and Carotenoids, Van Nostrand Reinhold, New York.

desaturase and lycopene cyclase. Tomato phytoene synthase (*psy*) was the first gene for a carotenoid biosynthetic enzyme to be isolated from plants (Bartley *et al.*, 1992). The original cDNA was named *tom5* and the gene expressed in ripening fruit is called *psy1*. There is atleast one other related gene (*psy2*) expressed in green parts and flowers (Fray and Grierson, 1993), which is responsible for chloroplast carotenoid production prior to ripening. The concentration of *psy1* mRNA increases during ripening as the chloroplasts are transformed into chromoplasts (Grierson *et al.*, 1986a). Transgenic tomato fruits with low *psy1* were yellow rather than red due to lack of β-carotene and lycopene. In other experiments, *psy1* has been over-expressed in wild type tomatoes under control of 35 S promoter. In some transgenic plants sense suppression has been observed and in others a small increase in carotenoids has been found. In general, however, it has not yet been possible to bring about a major enhancement of carotenoids in tomatoes by over expressing *psy1*. In transgenic plants where the *psy1* transgene is strongly expressed, the plants were frequently reduced in stature and in severe cases the plants were dwarf. This dwarfism is probably due to utilisation of geranyl

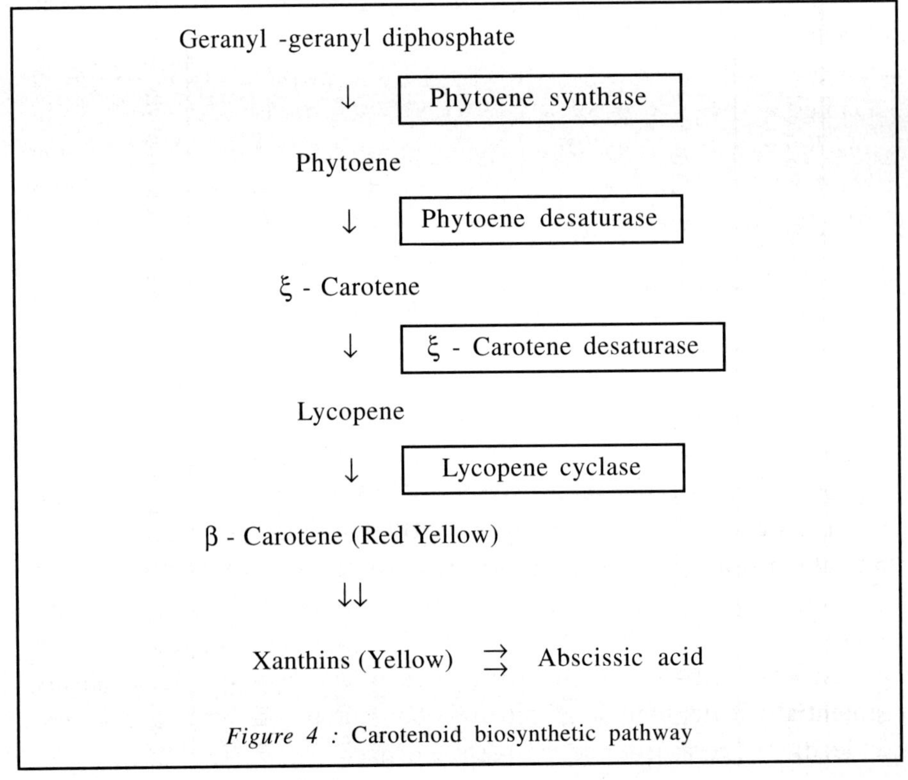

Figure 4 : Carotenoid biosynthetic pathway

geranyl diphosphate by the phytoene synthase, which prevents the normal production of gibberellins and phytol and other important metabolites (Fray *et al.*, 1995). Reduction in the level of gibberellins, caused dwarfism. The cytokinin levels of tomatoes have also been increased by over expressing an isopentenyl transferase gene under the control of *2a11* promoter. In fruits of some lines, high levels of the cytokinins were produced. This was correlated with the production of blotchy fruits with green and red patches (Martineau *et al.*, 1994). It was noticed that there was no correlation between the level of cytokinin and the stage of ripening, but the levels of total solids and soluble solids increased significantly in some of the transgenic lines (Martineau *et al.*, 1995).

4.6. Genetic modification for nitrogen assimilation in leafy vegetables

There is now a growing concern about the effect of N-fertilizer, especially nitrate, both on the environment and on human health. Nitrate and nitrite nitrogenous compounds occur as toxicants in green leafy vegetables with excessive nitrogen fertilization. The accumulation of nitrate, particularly in leafy vegetable crops consumed as green organs (Spinach, Amaranths, Lettuce and Cruciferous vegetables) could be detrimental to our health. The maximum admissible level of nitrite in vegetables is 1 mg Kg^{-1}. Nitrate content of leafy vegetables ranged from 30 to 270 mg kg^{-1} and in roots and tubers 31 to 2043 mg kg^{-1}. In China different vegetable samples contained 140.6 to 2762.5 mg kg^{-1} nitrate and 0.2 to 2.85 mg kg^{-1} nitrite.

4.6.1. Nitrates and human health

Nitrates and nitrites in vegetables may cause methemoglobinemia in babies, where due to the oxidation of ferrous iron in haemoglobin to ferric state, the oxygen carrying capacity of the red blood corpuscles is impaired. Other health problems associated with nitrate toxicity includes oral cancer, cancer of the colon, rectum or other gastrointestinal cancers, Alzhemier's disease, absorptive and secretive functional disorders of the intestinal mucosa, differentiation and apoptosis in intestinal crypts, reduced casein digestion, multiple sclerosis, neural tube defects and hypertrophy of thyroid. It has been shown that nitrate consumption leads to a decrease in the ascorbate/nitrate ratio in gastric juice, which regulates the synthesis of potentially carcinogenic N-nitroso compounds and decrease in the ratio leads to increased risk of gastric cancer.

The fact that plants over expressing NR have lower levels of foliar nitrate (Quillere *et al*., 1994) opens the possibility of improving the nutritional quality of edible plants like lettuce or spinach, which tend to accumulate this ion in their leaves. Thus, one can hope that the use of transgenic plants expressing deregulated forms of the enzymes involved in the nitrate assimilation pathway might lead to a higher nitrate uptake and to the reduction of the amounts of nitrate stored in the leaves. Genes coding for the NR apo-enzyme or NR cDNA clones have now been isolated from fifteen different plant species. The isolation of genomic or cDNA clones of NR has enabled the molecular analysis of the regulation of this major metabolic pathway.

4.6.2. Nitrate regulation

NR activity is induced by nitrate in plants. Nitrate is a major signal leading to the induction of the *nia* gene but also of the *nir* apoenzyme gene (*nii*) mRNA. Transgenic plants have been of great help in showing that the regulation by nitrate is transcriptional. It has also been shown that NR is regulated at the post-translational level by phosphorylation in the dark. A reversible inactivation of NR affecting its V_{max} but not its substrate affinities has been demonstrated when plants are transferred to dark (Huber *et al*., 1992). Phosphorylation of a serine residue is the first step involved in NR inactivation, but phospho-NR by itself is still active and becomes inactivated upon binding of an "inactivator protein"(Bachmann *et al*., 1995)

4.6.3. Nitrite reductase

Several genes or cDNAs coding for the NiR apoenzyme (Nii) have been cloned and sequenced from higher plants such as spinach (Back *et al*., 1988). Some plants contain only a single *nii* gene per haploid genome while other plant species contain two copies per haploid genome. Tobacco contains four genes, two from each ancestor (Kronenberger *et al*., 1993). The two genes coming from each ancestor are expressed differentially in leaves and roots.

Genetic transformation has allowed modulating the activity of the enzymes of the nitrate assimilation pathway. The analysis of plants expressing a deregulated form of NR has allowed the transcriptional regulation of the *nia* gene by environmental factors such as nitrate or light and by internal factors such as glutamine or sugars. The nitrate assimilation pathway appears to be of a surprising complexity. Indeed its first enzymatic step, nitrate reduction, is regulated both at the transcriptional and post-translational levels by light. *nir* itself is regulated at both, the transcriptional and post- translational levels by the nitrogen source (Crete *et al.*,1997). The use of CaMV 35S promoter allowed a modification of the tissue specificity of the expression of the enzymes involved in the nitrate pathway and could lead to a better understanding of the sink to source relationships with respect to Nitrogen metabolism.

4.7. Genetic modification of nutritional quality of seed proteins in leguminous vegetables

Seeds are the most important plant organs harvested by the humankind, in terms of their total yield and their use for food, feed and industrial raw material. They are the major source of dietary protein including the amino acids (lysine, threonine, cysteine and methionine), which are essential for human beings. Proportion of the total proteins in the dry seeds varies considerably, with higher levels in cultivated legumes (upto 40% dry weight in soybean) and lower levels in cereals (about 8-15 % dry weight). Osborne (1924), classified plant proteins on the basis of their solubility in water (albumins), dilute saline (globulins), alcohol water mixtures (prolamins) and dilute acid or alkali (gluteins). Albumin and globulin storage proteins are further classified on the basis of their sedimentation coefficients, with 2S albumins, 7-8S globulins and 11-12S globulins being widely distributed in legumes, crucifers, cucurbits and cereals.

Seed proteins are amongst the earliest proteins to be characterized and were one of the targets selected for molecular modifications with the aim to improve their composition and properties by genetic engineering. Saalbach *et al.* (1990) attempted to introduce additional methionine residues into *Vicia faba* legumins, but with disappointing results. They added four methionine residues by introducing a frame shift mutation at the c-terminal end of a gene and also constructed a methionine rich *Vicia faba* Legumin/Soybean glycinin fusion protein. Neither protein could be detected, although m-RNA was present. However, Utsumi (1993) and Takaiwa *et al.* (1995) had greater success based on initial testing of constructs in *E. coli*. Some of the mutants have enhanced contents of methionine and could contribute to increased nutritional quality.

4.8. Genetic modification of carbohydrate metabolism

All carbohydrates found in plants are ultimately derived from atmospheric CO_2, fixed in the chloroplasts during photosynthesis. The Calvin cycle produces trios-phosphate, which can be used for starch synthesis, and temporarily stored in the leaf chloroplast or translocated to the cytoplasm. There it can enter glycolysis or can be converted into sucrose, the major transport carbohydrate in most plants, and translocated to the sink tissues. These sink tissues are either (a) "Utilization sinks" - metabolically active rapid growing tissues like meristems and immature leaves and (b) "Storage sinks" such as seeds, tubers and roots, which deposit the carbohydrates in the form of storage compounds like starch, sucrose, fructans etc. (Sonnewald and Willmitzer, 1992). Potato plants transformed with an antisense construct of the sucrose transporter cDNA under the control of CaMV 35S promoter, showed a significant reduction in the expression of sucrose transporter (*sut1*) RNA, although in no case complete inhibition was observed. Plants with reduced *sut1* RNA levels show significant reduction in the transport of carbohydrates from the source leaves as compared to wild type plants. Due to reduced sucrose translocation, these plants are reduced in growth, root development and tuber yields. (Riesmeier *et al.*, 1994)

Starch biosynthesis occurs in the plastids of leaves (chloroplasts) and in the amyloplasts of storage organs such as potato tubers. Upon entry of sucrose in the sink cells, sucrose is cleaved predominantly by sucrose synthase and converted to glucose-1-phosphate, which is used for the synthesis of ADP-glucose (ADP-G), the major substrate for starch biosynthesis (Zrenner *et al.*, 1995). This reaction is catalyzed by ADP-glucose phosphorylase (ADP-G-PP) and is thought to be the rate-limiting step in starch biosynthesis. A potato tuber in which the expression of ADP-G-PP enzyme is inhibited via antisense inhibition, showed a dramatic reduction in starch content and accumulation of high level of soluble sugars (Muller-Rober *et al.* 1992). Further, studies have been performed to determine whether the starch content could be increased by introducing a heterologous *adp-gpp* (Stark *et al.*, 1992). Introduction of *E. coli adp-gpp* gene in potato and tomato, however, did not result in a significant increase in starch content. In later studies with mutant *E. coli adp-gpp*, transgenic potato plants contained on an average 35% more starch than control tubers, whereas some transgenic lines produced tubers with nearly 60% more starch (Stark *et al.*, 1992). This shows that it is possible to manipulate crop yield and the amount of key metabolites by genetically modifying regulatory steps in the biosynthetic process

4.9. Genetic modification of other vegetable crops

Besides tomato and potato, a variety of other transgenic vegetables have appeared, since 1995. These include broccoli, carrot, eggplant, pea, pepper and watermelon. After genetic modifications of quality parameters had been successfully tested in tomato, the respective approaches were extended to a variety of vegetables. In the year 1997, field release were conducted for melons and pepper with altered product quality. Ripening genes have been cloned from several other climacteric fruits, including *aco* (Balague *et al.*, 1993) and *psy* from melon (Karvouni *et al.*, 1995). A novel protein called fibrillin has been cloned from pepper, which appears to be associated with the carotenoids deposited in the chromoplasts of ripening pepper fruits (Deruere *et al.*, 1994).

Transgenic lettuce (*Lactuca sativa*) plants accumulating the iron storage protein ferritin were transformed with *Agrobacterium tumefaciens* carrying soybean ferritin cDNA (Goto *et al.*, 2000). The integration of the ferritin gene and expression levels in leaves were examined by Southern and Western-blot analysis, respectively. It was shown that transgenic lettuce plants contained iron levels ranging from 1.2 to 1.7 times that of the control plants, however the manganese content in transgenic lettuce plants was similar to that in the control. Enhanced growth of transgenic lettuce was observed at the early development stages, resulting in weights 27-42% greater than those of control plants and photosynthesis rates superior to those of the controls. They grew larger and faster during the period of three months from germination. These results demonstrate the possibility of producing lettuce plants with high yield, high iron content with rapid growth rate.

Successful expression of the *ada* (adenosine deaminase) gene was detected by Northern blot and enzyme activity, suggesting a potential application of ADA as a selectable marker in genetic transformation. The *bar* (herbicide resistance) gene was also introduced into hot pepper, and progenies of the transgenic plants were obtained. Stable inheritance (3:1 ratio) of *bar* was confirmed by PCR analysis. This highly efficient transformation system can be further used for the improvement of hot pepper quality using other useful genes.

5. CONCLUSIONS AND FUTURE PROSPECTS

Besides herbicide tolerance, disease and insect resistances quality traits are also amenable to genetic engineering. In the European union, transgenic potato with modified starch characteristics and male sterility (for use in hybrid seed production) respectively, have been approved for commercial use. Another promising prospect is genetic engineering of parthenocarpic fruit, which recently was exemplified for eggplant (Rotino *et al.*, 1997). In this case, a gene from the phytopathogenic bacteria *Pseudomonas syringae* pv. *Savastanoi* encoding synthesis of the auxin precursor indoleacetamide was put under the control of an ovule specific promoter and transferred into eggplants. The transgenic plants acquired the capacity to form seedless, marketable fruits in the absence of pollination under normal growth conditions. In addition, the transgenic plants developed marketable fruits without application of exogenous plant hormones under low temperature/light conditions, which are otherwise prohibitive for fruit set in untransformed eggplants. It can be expected that similar gene constructs will have impacts on parthenocarpic fruit production in other crops such as cucumber. After the initial standardization of the genetic engineering work in major vegetable crops, the future will certainly bring about the broadening of the range of crops that are subject to transgenic approaches. Softening inhibition, *e.g.*, with polygalacturonase or pectin methylesterase as targets will probably be conducted in more vegetables like carrots. Similarly ethylene biosynthesis will be altered by genetically modifying the expression of ACC synthase, deaminase, oxidase or AdoMet hydrolase in additional climacteric vegetables like melons and broccoli to increase the shelf life of the produce. Besides ethylene manipulation alteration of fruit and tissue color and vitamin A content by modifying carotenoid biosynthesis pathway will be more widely applied as well as reduction of bruising susceptibility in vegetables by suppression of polyphenol oxidase gene expression, *e.g.,* in potato. A promising improvement of gene transfer could come from new vectors, like Yeast Artificial Chromosome (YAC) and Binary Bacterial Artificial Chromosomes (BIBAC). The BIBAC vectors allow for the transfer of about 150 kb of foreign DNA (Hamilton *et al.*, 1996). This opens new horizons for the transfer of large gene clusters as well as quantitative trait loci. Thus, engineering of complex traits will be made feasible.

In future, genetically engineered crops will have a large potential of gaining a premium quality labeling on the market if quality traits of major public interest such as flavour, avoidance of common allergens or increased contents of antioxidants and secondary metabolites (recently identified as anticarcinogens) are consequently addressed as breeding goals.

REFERENCES

Bachmann M, McMichael RW, Huber JL, Kaiser WM and Huber SC (1995). Partial purification and characterization of a calcium-dependant protein kinase and an inhibitor protein required for inactivation of spinach leaf nitrate reductase. *Plant Physiol.*, **108** : 1083-1091.

Back E, Burkhardt W, Moyer M, Privalle L and Rothstein S (1988). Isolation of cDNA clones coding for spinach nitrite reductase : complete sequence and nitrate induction. *Mol. Gen. Genet.,* **212** : 20-26.

Balague C, Watson CF, Turner AJ, Rouge P, Picton S, Pech JC and Grierson D (1993). Isolation of a ripening and wound induced cDNA from *Cucumis melo* L. encoding a protein with homology to the ethylene forming enzyme. *Eur. J. Biochem.,* **212** : 27-34.

Barry CS, Blume B, Bouzayen M, Cooper W, Hamilton AJ and Grierson D (1996). Differential expression of the 1-aminocyclopropane-1-carboxylate oxidase gene family of tomato. *Plant J.,* **9** : 525-535.

Bartley GE and Scolnik PA (1995). Plant carotenoid pigments for photo protection, visual attraction and human health. *Plant Cell,* **7** : 1027-1038

Bartley GE, Vitanen PV, Bacot KO and Scolnik PA (1992). A tomato gene expressed during fruit ripening encodes an enzyme of the carotenoid biosynthesis pathway. *J. Biol. Chem.,* **267** : 5036 - 5039.

Bird CR, Grierson D and Schuch WW (1994). Modification of carotenoid production in tomatoes using pTOM5. U.S. Patent No. 5, 304, 478.

Bird CR, Smith CJS, Ray JA, Moureau P, Bevan MV, Bird AS, Hughes S, Morris PC, Grierson D and Schuch W (1998). The tomato polygalcturonase gene and ripening specific expression in transgenic plants. *Plant Mol. Biol.,* **11** : 651 - 662.

Carey AT, Holt K, Picard S, Wilde R, Tucker GA, Bird CR, Schuch W and Seymour GB (1995). Tomato exo-(1®4)-β-D-galactanase. *Plant Physiol.,* **108** : 1099-1107.

Cocking EC (1960). A method for the isolation of plants protoplasts and vacuoles. *Nature,* **187** : 927-929.

Courtney Gutterson N (1994). In genetic engineering of plant secondary metabolism (Eds BE Ellis, GW Kuroki and HA Stafford) pp 93-124, Plenum press, New York.

Crete P, Caboche M and Meyer C (1997). Nitrite reductase is regulated at the fast-transcriptional level by the nitrogen source in *Nicotiana plumbaginifolia* and *Arabidopsis thaliana. Plant J.,* **11** : 625-634.

Deruere J, Romer S, Dharlingue A, Backhaus RA, Kuntz M and Kamara B (1994). Fibril assembly and carotenoid over accumulation in chromoplasts: A model for supramolecular lipo protein structures. *Plant Cell,* **6** : 119-133.

Errington N, Tucker GA and Mitchell JR (1998). The effect of genetic down regulation of polygalacturonase and pectin esterase activity on rheology and composition of tomato juice. *J. Sci. Food Agri.,* **76** : 515-519.

Evans DA (1989). Somaclonal variation: Genetic basis and breeding application. *Genetics,* **5** : 46-50

Ferro AJ, Bestwick RK and Brown LR (1995). Genetic control of ethylene biosynthesis in plants using S-adenosylmethionine hydrolase. U.S. Patent No. 5, 416, 250.

Fray RG and Grierson D (1993). Identification and genetic analysis of normal and mutant phytoene synthase genes of tomato by sequencing, complementation and co-supression. *Plant Mol. Biol.,* **27** : 1143-1151.

Fray RG, Wallace A, Fraser PD, Valero D, Heddon P, Bramley P and Grierson D (1995). Constitutive expression of a fruit phytoene synthase gene in transgenic tomatoes causes dwarfism by redirecting matobolites from the gibberellin pathway. *Plant J.,* **8** : 693-701.

Goto F, Yoshiihara T and Saiki H (2000). Iron accumulation and enhanced growth in transgenic lettuce plants expressing the iron binding protein ferritin. *Theor. Appl. Genet.,* **100** : 5, 658-664.

Grandillo S, Ku HM and Tanskley SD (1996) Characterization of *fs8.1,* a major QTL influencing fruit shape in tomato. *Mol. Breed.,* **2** : 251-260.

Gray J, Picton S, Shabbeer J, Schuch W and Grierson D (1992). Molecular biology of fruit ripening and its manipulation with antisense genes. *Plant Mol. Biol.,* **19** : 69-87.

Grierson D, Maunders MJ, Slater A, Ray J, Bird CR, Schuch W, Holdsworth MJ, Tueker GA and Knapp JE (1986 a). Gene expression during tomato ripening. *Philosophical Transactions of the Royal Society of London.,* **314** : 399- 410.

Grierson D, Tucker GA, Keen J, Ray J, Bird CR and Schuch W (1986 b). Sequencing and identification of a cDNA clone for tomato polygalacturonase. *Nucl. Acids Res.,* **14** : 8595-8603.

Gross J (1991) Pigments in vegetables: Chyrophylls and Carotenoids. Van Nostrand Reinhold, New York.

Hall LN, Bird CR, Picton S, Tucker GA, Seymour GB and Grierson D (1994). Molecular characterization of cDNA clones representing pectin esterase isozymes from tomato. *Plant Mol. Biol.,* **25** : 313-318.

Hall LN, Tucker GA, Smith CJS, Watson CF, Seymour GB, Bundick Y, Boniwell JM, Fletcher JD, Ray JA, Schuch W, Bird CR and Grierson D (1993). Antisense inhibition of pectin esterase gene expression in transgenic tomatoes. *Plant J.,* **3** : 121 - 129.

Hamilton A, Lycett GW and Grierson D (1990). Antisense gene that inhibits synthesis of the hormone ethylene in transgenic plants. *Nature,* **346** : 284-287.

Hamilton CM, Fray A, Lewis C and Tanksley SD (1996). Stable transfer of intact high molecular weight DNA into plant chromosomes. Proc. Natl. Acad. Sci., USA, **93** : 9975-9979.

Heath DW and Earle ED (1995). Synthesis of high Eurecic acid rapeseed (*Brassica napus* L.). *Theor. Appl. Genet.,* **91** : 1129-1136

Heath DW and Earle ED (1997). Synthesis of low linolenic acid rapeseed (*Brassica napus* L.) through protoplast fusion. *Euphytica,* **93** : 339-343

Hovenkem JHM, Jacobson E, Ponstein AS, Visser RGF, Bijmult EW, De Veries JN, Withholt B and Feenstra WJ (1987). Isolation of an amylose free starch mutant of the potato (*Solanum tuberosum* L.), *Theor. Appl. Genet.,* **75** : 217-221.

Huber JL, Huber SC, Campbel WH and Redinbaugh MG (1992). Reversible light/ dark modulation of spinach leaf nitrate reductase activity involves protein phosphorylation. *Arch. Biochem. Biophy.,* **296** : 58-65.

James C (1997). Global status of transgenic crops in (1997). ISAAA Briefs No. 5. ISAAA: Ithaca, NY. pp 31.

Karp A (1991). On the current understanding of Somaclonal variation. Oxford Surv. *Plant Mol. Cell Biol.,* **7** : 1-58.

Karvouni Z, John I, Taylor JE, Watson CF, Turner AJ and Grierson D (1995). Isolation and characterization of a melon cDNA encoding phytoene synthase. *Plant Mol. Biol.,* **27** : 1153-1162.

Kende H (1993). Ethylene biosynthesis. *Ann. Rev. Plant Physiol. Plant Mol. Biol.,* **44** : 283-307.

Khanna KR (1991). Crop improvement in the perspective of agricultural advancement and the necessity for investigation at molecular level. In : *Biochemical Aspects of Crop Improvement,* (Ed. Khanna KR) CRC Press pp 3-36

Klann EM, Hall B and Bennett AB (1996). Antisense acid invertase (*TIVI*) gene alters soluble sugar composition and size in transgenic tomato fruit. *Plant Physiol.,* **112** : 1321-1330.

Klee HJ, Hayford MB, Kretzner KA, Barry GF and Kishore GM (1991). Control of ethylene synthesis by expression of a bacterial enzyme in transgenic tomato plants. *Plant Cell.,* **3** : 1187-1193.

Kronenberger J, Lepingle A, Caboche M and Vaucheret H (1993). Cloning and expression of distinct nitrite reductase, in tobacco leaves and roots. *Mol. Gen. Genet.,* **236** : 203-208.

Larkin PJ and Scowcraft WR (1981). Somaclonal variation-a novel source of variability from cell cultures form plant improvement. *Theor. Appl. Genet.,* **60** : 197-214.

Lashbrook CC, Gorizalez-Bosch C and Bennett AB (1994). Two divergent endo-β-1, 4 glucanase genes exhibit overlapping expression in ripening fruit and abscising flavours. *Plant Cell,* **6** : 1485-1493.

Martineau B, Houck CM, Sheehy RE and Hiatt WR (1994). Fruit specific expression of the *A. tumefaciens* isopentenyl transferase gene in tomato: effects on fruit ripening and defense related gene expression in leaves. *Plant J.,* **5** : 11-19.

Mattheij WM and Puite KJ (1992). Tetraploid potato hybrids through protoplast fusion and analysis of their performance in the field. *Theor. Appl. Genet.,* **83** : 887-894.

Mehta RA, Zhou D, Tucker M, Handa A, Solomos T, Mattoo AK, Kanellis AK, Chang C, Klee H, Bleecker AB and Pech JC (1998). Ethylene in higher plants: biosynthetic interactions with polyamines and high temperature mediated differential induction of NR versus TAE1 ethylene receptor. Biology and biotechnology of the plant hormone ethylene II. *Proceedings of the EU-TMR-Euroconference Symposium, Thira (Santorini),* Greece, 5-8 September.

Muller-Roher B, Sonnewald U and Willmitzer L (1992). Inhibition of the ADP-glucose phosphorylase in transgenic potatoes leads to sugar storing tubers and influences tuber formation and expression of tuber storage protein genes. *EMBO J.,* **11** : 1229-1230.

Oeller PW, Wong LM, Taylor LP, Pike DA and Theologis A (1991). Reversible inhibition of tomato fruit senescence by antisense 1-aminocyclopropane - 1 carboxylate synthase. *Science,* **254** : 427-439.

Ohyama A, Ito H, Sato T, Nishimura S, Imai S and Hirai M (1995). Supression of acid invertase activity by antisense RNA modifies the sugar composition of tomato fruit. *Plant Cell Physiol.,* **36** : 369-376.

Osborne TB (1924). The Vegetable Proteins. 2nd edn. London: Longmans, Green and Co.

Penarrubia L, Aguilar M, Margossian L and Fischer RL (1992). An antisense gene stimulated ethylene hormone production during tomato fruit ripening. *Plant Cell.,* **4** : 681-687.

Porretta S, Poli G and Minuti E (1998). Tomato pulp quality from transgenic fruits with reduced Polygalacturonase (PG). *Food Chem.,* **62** : 283 - 290.

Quillere IC, Duffose Y, Roux C, Foyer Caboche M and Morot Gaundry JF (1994). The effects of deregulation of NR gene expression on growth and nitrogen metabolism of *Nicotiana plumbaginifolia* plant. *J. Exp. Bot.,* **45** : 1205-1211.

Ray J, Knapp J, Grierson D, Bird C and Schuch W (1998). Identification and sequence determination of a cDNA clone for tomato pectin esterase. *Eur. J. Biochem.,* **174** : 119-124.

Reismeier JW, Willmitzer L and Frommer WB (1994). Evidence for essential role of the sucrose transporters in phloem loading and assimilate partitioning. *EMBO J.,* **13** : 1-7.

Rotino GL, Perri E, Zottini M, Sommer H and Spena A (1997). Genetic engineering of parthenocarpic plants. *Nature Biotech.,* **15** : 1398-1401.

Rottman WH, Peter GF, Oeller PW, Keller JA, Shen NF, Nagy BP, Taylor LP, Campbell AD and Theologis A (1991). 1-aminocyclopropane-1-carboxylate synthase in tomato is encoded by a multigene family whose transcription is induced during fruit and floral senescence. *J. Mol. Biol.,* **222** : 937-961.

Saalbach G, Jung R, Kunze G, Manteuffesl R, Saalbach I and Muntz K (1990). Expression of modified legume storage protein genes in different systems and studies on intracellular targeting of *Vicia faba* legumin yeast. In: *Genetic Engineering of crop plants,* (Eds Lycett GW and Grierson D), Butterworth pp 151-158. .

Scolink PA and Bartley GE (1993) Phytoene desaturase from *Arabidopsis. Plant Physiol.,* **103** : 1475.

Sheehy RE, Kramer M and Hiatt WR (1988). Reduction of polygalacturonase activity in tomato fruit by antisense RNA. *Proc. Natl. Acad. Sci. USA,* **85** : 8805- 8809.

Singh BD, Prasad BK, Srivastava K and Singh RP (1998). Gene transfer through somatic hybridisation. In : *Proceeding of National Conference on Plant Biotechnology, University of Mumbai,* pp. 46-62

Sonnewald U and Willmitzer L (1992). Molecular approaches to sink source interactions. *Plant Physiol.,* **99** : 1267-1270

Speirs J, Lee E, Holt K, Kim YD, Scott NS, Loveys B, Schuch W and Kim YD (1998). Genetic manipulation of alcohol dehydrogenase levels in ripening fruits affects the balance of some flavour aldehydes and alcohols. *Plant Physiol.,* **117** : 1047-1058.

Stark DM, Timmerman KP, Barry GF, Preiss J and Kishore GM (1992). Regulation of the amount of the starch in plant tissues by ADP-Glucose pyrophosphorylase. *Science,* **258** : 287-292

Takaiwa F, Katsube T, Kitagawa S, Hisaga T, Kito M and Utsumi S (1995). High level accumulation of soybean glycinin in vacuole derived protein bodies in the endosperm tissue of transgenic tobacco seed. *Plant Sci.,* **111** : 39-49.

Utsumi S, Kitagawa S, Katsube T, Kang IJ, Gidamis AB and Takaiwa F (1993). Studies, processing and accumulation of modified glycinins of soybean in the seeds, leaves and stems of transgenic tobacco. *Plant Sci.,* **92** : 191 - 202.

Vanderpan S (1994). A look inside a new tomato. *Sci. Food Agric. Environment,* **7** : 2-5.

Yang SF and Hoffman NE (1984). Ethylene biosynthesis and its regulation in higher plants. *Ann. Rev. Plant Physiol.,* **35** : 155-189.

Yang X and Fu H (1989). Hua 03: A high protein Indica rice. *Intl. Rice Res. Inst. News Lett.,* **14** : 14-15.

Ye ZB, Li HX and Liu XJ (1999). A new transgenic tomato cultivar-Bioscein bred with antisense *efe* cDNA. *China Vegetables,* **1** : 6-10.

Zrenner R, Salonoubat M, Willmitzer L and Sonnewald U (1995). Evidence of the crucial role of sucrose synthase for sink strength using transgenic potato plants (*Solanum tuberosum* L.) *Plant J.,* **7** : 97-107.

AUTHOR INDEX

SUBJECT INDEX